AF805514

Global Biorefinery Revolution

This book focuses on state-of-the-art approaches for biomass conversion into various products in a biorefinery. It examines biomass resources with a view to identifying, classifying, and elaborating on their global availability, supply chains, and logistics. This book also covers technologies for biomass conversion in biorefineries, the energy products of biomass processing, their properties, and prospective applications. It further explores the production of marketable industrial products from renewable biomass resources and discusses the policy and regulatory frameworks governing biorefinery businesses.

- Provides insights into the use of different biomass-based sources as feedstock and the conversion technologies used to achieve desired products.
- Incorporates regional case studies, policy frameworks, and a global perspective on biomass utilization.
- Discusses products derived from the thermochemical processing of biomass resources.
- Highlights cutting-edge technologies such as AI integration, biohydrogen production, and digital optimization.
- Emphasizes circular economy principles, carbon footprint reduction, and the social impact of biorefineries.

This book is aimed at graduate students and researchers in biotechnology, bioprocessing, and chemical engineering.

Global Biorefinery Revolution

Converting Biomass into Sustainable Energy and Value-Added Products

Edited by
Reckson Kamusoko, Raphael M. Jingura and Patrick Mukumba

CRC Press
Taylor & Francis Group
Boca Raton London New York

CRC Press is an imprint of the
Taylor & Francis Group, an **informa** business

Designed cover image: Getty Images

First edition published 2027
by CRC Press
2385 NW Executive Center Drive, Suite 320, Boca Raton FL 33431

and by CRC Press
4 Park Square, Milton Park, Abingdon, Oxon, OX14 4RN

CRC Press is an imprint of Taylor & Francis Group, LLC

© 2027 selection and editorial matter, Reckson Kamusoko, Raphael M. Jingura and Patrick Mukumba; individual chapters, the contributors

Reasonable efforts have been made to publish reliable data and information, but the author and publisher cannot assume responsibility for the validity of all materials or the consequences of their use. The authors and publishers have attempted to trace the copyright holders of all material reproduced in this publication and apologize to copyright holders if permission to publish in this form has not been obtained. If any copyright material has not been acknowledged please write and let us know so we may rectify in any future reprint.

Except as permitted under U.S. Copyright Law, no part of this book may be reprinted, reproduced, transmitted, or utilized in any form by any electronic, mechanical, or other means, now known or hereafter invented, including photocopying, microfilming, and recording, or in any information storage or retrieval system, without written permission from the publishers.

For permission to photocopy or use material electronically from this work, access www.copyright.com or contact the Copyright Clearance Center, Inc. (CCC), 222 Rosewood Drive, Danvers, MA 01923, 978-750-8400. For works that are not available on CCC please contact mpkbookspermissions@tandf.co.uk

For Product Safety Concerns and Information please contact our EU representative GPSR@taylorandfrancis.com. Taylor & Francis Verlag GmbH, Kaufingerstraße 24, 80331 München, Germany.

Trademark notice: Product or corporate names may be trademarks or registered trademarks and are used only for identification and explanation without intent to infringe.

ISBN: 978-1-041-09101-1 (hbk)
ISBN: 978-1-041-09566-8 (pbk)
ISBN: 978-1-003-65075-1 (ebk)

DOI: 10.1201/9781003650751

Typeset in Times
by SPi Technologies India Pvt Ltd (Straive)

Contents

About the Editors

Reckson Kamusoko is a research fellow at the University of Fort Hare and a renowned academic writer, author of several peer-reviewed journal articles and book chapters, with research interests in bioenergy, bioprocess engineering, and sustainable waste-to-energy technologies. He has over 15 years of experience in teaching, research, and academic service. He holds a PhD in Environmental Biotechnology. His work focuses on the development of innovative and low-cost solutions for converting biomass and organic waste into renewable energy and value-added products. He is particularly interested in biogas production, biorefinery systems, and technologies that promote sustainable resource utilization and environmental protection. His research aims to contribute to the advancement of clean energy solutions and support sustainable development, particularly in rural and resource-limited communities.

Raphael M. Jingura is a professor at Chinhoyi University of Technology and an accomplished academic writer, with numerous publications in refereed journals and book chapters. He has over 25 years of experience in teaching, research, and higher education management. His research focuses on the nexus of agricultural and energy systems for sustainable development. He has conducted extensive work on renewable energy and has made significant contributions to this field through his publications. His current research interests include the production of biomass as a feedstock for biofuels and biogas production.

Patrick Mukumba is a professor at the University of Fort Hare and a famous academic writer, who has made significant contributions to the field through his scholarly work, publishing several peer-reviewed journal articles and book chapters. He has served for more than 20 years in teaching, research, and university service. Professor Mukumba is a recognized expert in renewable energy technologies, including biomass and biogas systems, wind energy, solar photovoltaic, and hydropower. His research experience includes the design and construction of various biogas digesters. He is an active member of several professional organizations, including the South African Institute of Physics (SAIP), the South African Biogas Industry Association (SABIA), and the World Biogas Association (WBA).

Contributors

Adekunle A. Adeleke
Department of Mechanical Engineering
Nile University of Nigeria
Abuja, Nigeria

Fitri Agustina
Department of Industrial Engineering
Universitas Trunojoyo Madura
Bangkalan, Indonesia

Ifreke Akpan
Chemical Engineering Department, Faculty of Engineering
University of Port Harcourt
Rivers State, Nigeria

Cristina Barrera
Instituto de Ingeniería de Alimentos – FoodUPV
Universitat Politècnica de València
Valencia, Spain

Claudia Bas-Bellver
Instituto de Ingeniería de Alimentos – FoodUPV
Universitat Politècnica de València
Valencia, Spain

Indra Cahyadi
Industrial Engineering Department
University of Trunojoyo Madura
Bangkalan, Indonesia

Cristina Calderón-Tapia
Instituto de Ingeniería de Alimentos – FoodUPV
Universitat Politècnica de València
Valencia, Spain

Ion Lucian Ceapraz
Research team InTerACT UP 2018. C102
Polytechnic Institute Unilasalle
Beauvais, France

Claude Chammaa
Research team InTerACT UP 2018. C102
Polytechnic Institute Unilasalle
Beauvais, France

Chido H. Chihobo
Department of Fuels and Energy Engineering, School of Engineering Sciences and Technology
Chinhoyi University of Technology
Chinhoyi, Zimbabwe

Annie Chimphango
Department of Chemical Engineering
Stellenbosch University
Stellenbosch, South Africa

Kaitano Dzinavatonga
Faculty of Science
Zimbabwe Open University
Harare, Zimbabwe

Ifeanyichukwu Edeh
Chemical Engineering Department, Faculty of Engineering
University of Port Harcourt
Rivers State, Nigeria

Lanrewaju I. Fajimi
Chemical Engineering Department
University of Johannesburg
Johannesburg, South Africa

Seun Jesuloluwa
Department of Mechanical Engineering
Nile University of Nigeria
Abuja, Nigeria

Raphael M. Jingura
Department of Animal Production and Technology, School of Agricultural Sciences and Technology
Chinhoyi University of Technology
Chinhoyi, Zimbabwe

Reckson Kamusoko
Department of Computational Science, Physics Discipline, Faculty of Science and Agriculture
University of Fort Hare
Alice, South Africa

Beata Kilonzo
Institute for Rural Development
University of Venda
Thohoyandou, South Africa

Denzel C. Makepa
Department of Chemical Engineering
Stellenbosch University
Stellenbosch, South Africa

Tawanda P. Makopa
Department of Biological Sciences and Biotechnology
Botswana International University of Science and Technology
Palapye, Botswana

Ngonidzashe Mangoma
Department of Applied Biology and Biochemistry
National University of Science and Technology
Bulawayo, Zimbabwe

Stanley C. Msiska
Department of Chemical Engineering
Stellenbosch University
Stellenbosch, South Africa

Patrick Mukumba
Department of Computational Science, Physics Discipline, Faculty of Science and Agriculture
University of Fort Hare
Alice, South Africa

Sophie T. Mulaudzi
Department of Physics, Faculty of Sciences, Engineering and Agriculture
University of Venda
Thohoyandou, South Africa

Thembekile Ncube
Department of Applied Biology and Biochemistry
National University of Science and Technology
Bulawayo, Zimbabwe

Chizoma N. Adewumi
Department of Pure and Applied Chemistry, Faculty of Natural and Applied Sciences
Veritas University
Abuja, Nigeria

Petrus Nzerem
Department of Petroleum and Gas Engineering
Nile University of Nigeria
Abuja, Nigeria

KeChrist Obileke
Department of Computational Science, Physics Discipline, Faculty of Science and Agriculture
University of Fort Hare
Alice, South Africa

Matthew A. Onu
Chemical Engineering Department
University of Johannesburg
Johannesburg, South Africa

Oluwadamilola B. Oyinloye
Department of Industrial Chemistry
Nile University of Nigeria
Abuja, Nigeria

Richard K. Padi
Department of Food Process Engineering
University of Ghana
Accra, Ghana

Thilivhali Rasimphi
Department of Physics, Faculty of Science, Engineering and Agriculture
University of Venda
Thohoyandou, South Africa

Rishen Roopchund
Chemical and Materials Engineering Department
University of South Africa
Johannesburg, South Africa

Naadhira Seedat
Department of Chemical Engineering
University of Pretoria
Pretoria, South Africa

Lucía Seguí
Instituto de Ingeniería de Alimentos – FoodUPV
Universitat Politècnica de València
Valencia, Spain

Farooq Sher
Department of Engineering
Nottingham Trent University
Nottingham, United Kingdom

David Tinarwo
Department of Physics, Faculty of Sciences, Engineering and Agriculture
University of Venda
Thohoyandou, South Africa

Nerve Zhou
Department of Biological Sciences and Biotechnology
Botswana International University of Science and Technology
Palapye, Botswana

Paul Zvawanda
Thermodynamics Research Unit, School of Engineering
University of KwaZulu-Natal
Durban, South Africa

Preface

Large quantities of biomass are produced from anthropogenic practices, such as agricultural production, municipal waste management, and industrial activities. Biomass materials from agricultural, municipal, and industrial activities include forest residues, livestock waste, crop residues, energy crops, municipal solid waste, sewage sludge, and aquatic biomass. When poorly managed, some of these materials can be deleterious to the environment. Notably, biomass materials are a potent feedstock to produce multiple products. A biorefinery approach presents viable pathways for the conversion of biomass into various products.

It is well-established that biomass is a feedstock for valorization into multiple energy carriers, food, and several biochemicals in an integrated biorefinery. A wide range of products, which encapsulate biogas, bioethanol, biodiesel, biobutanol, pyrolytic products, single cell protein, biofertilizers, biochemicals, and bioactive compounds, among others, can be produced from biomass resources. These products have important uses in the energy, food, medical, and industrial value chains. Production of these products from biomass, using a biorefinery approach, is consistent with principles and practices of the circular economy and sustainable production.

The proposed edited book, *Global Biorefinery Revolution: Converting Biomass into Sustainable Energy and Value-Added Products*, focuses on the state-of-the-art approaches for biomass conversion into various products in a biorefinery. This book covers several topics that are subdivided into three sections. The first section deals with biomass resources with the view to identifying, classifying, and elaborating on their global availability, supply chains, and logistics. An overview of the biorefinery approach and the biobased business models is provided in this section. A life cycle assessment of bio-based refineries is also included in this section. The second section disseminates in-depth information on technologies for biomass conversion into various energy products in a biorefinery. Properties and prospective applications of the energy products are explored. Strategies for improved conversion processes and productivity are also covered in this section. The last section deals with the production of other marketable industrial products from renewable biomass resources. The value of the digestate waste as a by-product from biomass processing is highlighted in this section. Policy and regulatory frameworks that govern how biorefinery businesses operate are presented in this section.

1 Overview of Biomass as a Feedstock

Raphael M. Jingura and Reckson Kamusoko

1.1 INTRODUCTION

In a global economy largely reliant on fossil fuels, the shift to renewable energy gained momentum in the new millennium. This transition is driven by the need to prevent climate change and its harmful impacts. Countries worldwide are motivated to adopt renewable fuels. A persistent search for alternative energy sources characterizes the transition. The ultimate aim is to diversify the energy mix by including effective and sustainable sources. In these global efforts, biomass is a resource that is extensively exploited as a sustainable energy source. Biomass serves as a feedstock for biofuels and other products in a biorefinery. Biofuels help to reduce greenhouse gas emissions and serve as an alternative to fossil fuels (Openshaw 2000; Demirbas 2010; Lo et al. 2021; Duca & Toscano 2022; Cavelius et al. 2023). Furthermore, biofuels contribute to the achievement of Sustainable Development Goals (SDGs), particularly SDG 7 (affordable and clean energy) and SDG 13 (climate action) of the United Nations (Nazari et al. 2020).

Biomass is among the most valuable and versatile resources on Earth. Extensive evidence exists in the literature on initiatives promoting biomass as a renewable energy source (Bridgwater 2006; Demirbas 2009b; Ahorsu et al. 2018; Cotana et al. 2022; Olayokun et al. 2024; Panchani et al. 2024). According to the World Bioenergy Association (WBA) annual report for 2024, bioenergy constituted about 9% of the global energy supply in 2022 (WBA 2024a). In the same year, bioenergy accounted for 39 EJ out of the 422 EJ of global energy consumption (WBA 2024a). Furthermore, several international initiatives are promoting the utilization of biomass as a feedstock for energy production. The contribution of different biomass sources to the global bioenergy component is shown in Figure 1.1.

The term 'biomass' has varied definitions. For example, Bracmort (2013) and Dayton and Thomas (2020) identified 14 different definitions of the term in the US legislation between 2004 and 2010. Ecologically, biomass refers to the amount of living material in a specific habitat, measured either by the weight of organisms per unit area or the volume of organisms per unit volume of habitat (Dayton & Thomas 2020). From an energy perspective, biomass is organic matter that can be converted into fuel (Bracmort 2013; Dayton & Thomas 2020). According to the European Renewable Energy Directive (EU 2024 p. 7),

DOI: 10.1201/9781003650751-1

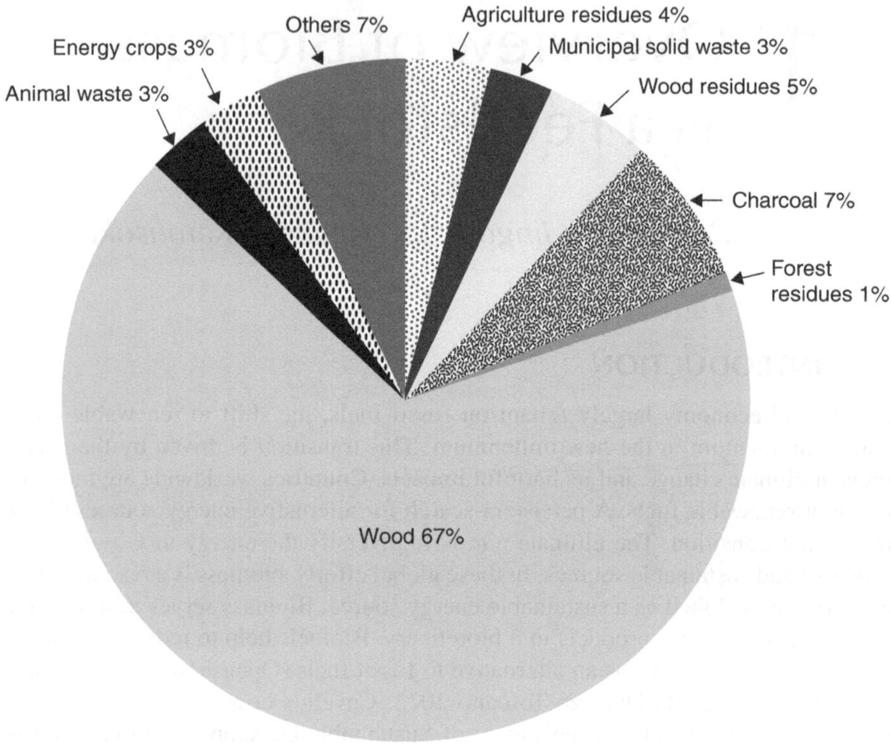

FIGURE 1.1 Percentage of biomass sources in the bioenergy combination.

(Adapted from FAO 2022 and IPCC 2022.)

> biomass means the biodegradable fraction of products, waste and residues of biological origin from agriculture, including vegetal and animal substances, from forestry and related industries, including fisheries and aquaculture, as well as the biodegradable fraction of waste, including industrial and municipal waste of biological origin.

A commonality among these definitions is that biomass is organic matter from plant and animal materials. The elemental composition of biomass includes carbon, hydrogen, oxygen, and nitrogen.

The term 'feedstock' refers to raw materials or inputs used in a process to produce goods. Biofuel feedstocks are any living, dead, or decomposed plant or animal materials suitable for processing and conversion to biofuels by means of different processes (Ziolkowska 2020). Biomass is a raw material or input for processes that convert it into biofuels, bioproducts, and bioenergy (Dayton & Thomas 2020; Cavelius et al. 2023). It is through these conversion processes that biomass is transformed into energy. Therefore, biomass is a resource whose characterization is essential to understand its physical and chemical properties and optimize its utilization efficiency. Biomass is a heterogeneous feedstock, requiring various conversion technologies to produce diverse end-products. The main methods for converting biomass into energy products are chemical, mechanical, thermal, and biological. The choice of feedstock and the type of end-product determine the conversion method to be used. The conversion method influences the type of marketable end-product, which may include electricity, heat, liquid and solid biofuels, and chemicals.

This chapter offers an overview of biomass as a feedstock for energy production. The different types of biomass are described to provide a comprehensive overview of biomass typology. Additionally, the characterization of biomass is presented to illustrate its physical and chemical properties. The advantages and challenges linked to the use of biomass as a feedstock are also discussed.

1.2 TYPES OF BIOMASS

Biomass feedstock for energy production can be sourced from various origins that produce different types of materials. These materials may come from plants grown specifically for energy or from plant parts, residues, processing waste, and material generated by animal and human activities. By its nature, biomass is a heterogeneous material. The types of biomass vary across regions, depending on the agricultural, forestry, and human practices in that area.

Biomass can be classified in several taxonomic ways. The simplest classification divides biomass into aquatic or terrestrial categories. Aquatic biomass includes microalgae and macroalgae, while terrestrial biomass encompasses all land-based plant- and animal-related materials. One classification system divides biomass into four generations, each with its distinctive characteristics. The four generations of biomass are shown in Figure 1.2.

1.3 SOURCES OF BIOMASS

There are several sources of biomass feedstock for energy production. Anthropogenic activities generate large amounts of biomass material. These sources are mainly plant and animal materials. Currently, commercially produced biofuels are the first- and second-generation types.

1.3.1 First-Generation Biomass

First-generation biomass is food or edible crops such as wheat, barley, maize, and soybean. They constitute approximately 4% of the total bioenergy production (Inam & Deniz 2023). These crops generally contain fat, sugar, and starch. Cereal grains like maize and wheat are rich in starch, while oilseeds such as soybean are rich in oils. They are mainly converted into biofuels such as biodiesel and bioethanol through processes that convert sugars, starches, and oils into liquid fuels.

1.3.2 Second-Generation Biomass

Second-generation biomass is non-food materials, which are generally lignocellulosic. These materials include agricultural and forestry residues, energy crops, municipal and food wastes, as well as other waste streams. They constitute about 96% of total biofuel production (Inam & Deniz 2023). Valorization of these wastes

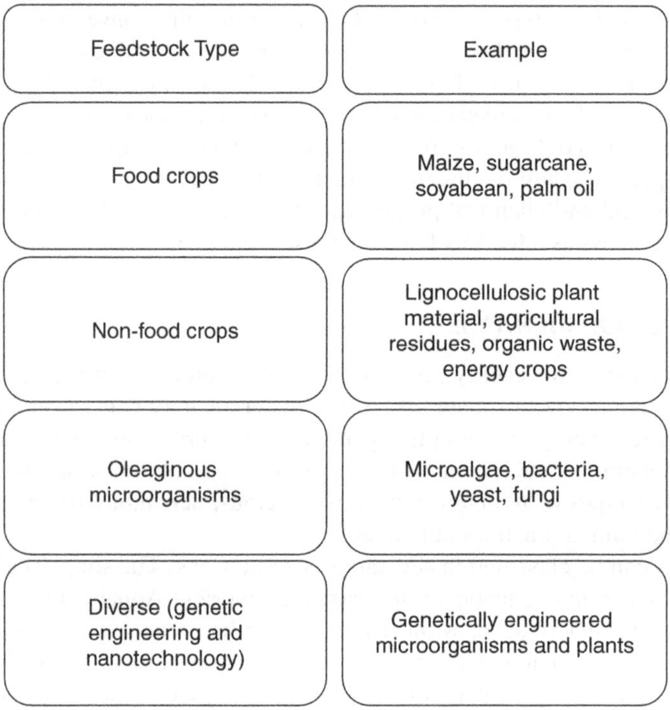

FIGURE 1.2 Types of biomass feedstock.

and residues through biofuel production has gained momentum in circular economies. The various types of second-generation biomass are shown in Table 1.1.

1.3.2.1 Biowastes

Biowaste refers to any waste originating from biodegradable organic materials generated through various domestic and commercial activities (Jiménez-Rosado et al. 2023). Biowastes show wide variability and include a broad spectrum of materials (Table 1.1). In 2020, animal, crop, and municipal solid waste (MSW) contributed 3%, 4%, and 3%, respectively, to biomass sources used in bioenergy (Figure 1.1). In 2022, industrial and municipal wastes contributed 2% and 3%, respectively, to the global domestic biomass supply (WBA 2024b).

TABLE 1.1
Examples of Second-Generation Biomass

Biowastes	Forest Products	Energy Crops
- Animal waste	- Forest residues	- Short-rotation woody crops
- Agricultural residues	- Wood	- Herbaceous woody crops
- Municipal solid waste	- Wood residues	- Forage crops
- Sewage sludge		
- Food waste		

The livestock industry is extensive and diverse. Livestock species include chickens, pigs, cattle, sheep, and goats, which are the most prevalent. Livestock generate approximately 13 billion tons of waste annually. This waste represents a significant resource available in many countries. Animal waste, also known as livestock waste, comprises urine and manure along with all other organic materials produced during farming, processing, transportation, and marketing of livestock. Livestock produce an estimated 3.12 billion tons of manure each year (Ródenas 2022). Other animal waste streams include unused feed or residues, effluents, hatchery refuse, wastewater from milk processing, and slaughterhouse waste (Girotto & Cossu 2017). Slaughterhouse waste is produced during the industrial processes of slaughtering animals and processing meat. Wastewater from the dairy industry is biodegradable and contains high concentrations of organic compounds (Cvetkovića et al. 2014).

Crop production is a vital economic sector in many countries. Vast amounts of food and fiber crops are produced throughout the world. Farming and processing of these crops produce substantial quantities of residues. Crop yields are of particular importance because they affect the amount of residues generated (Perlack et al. 2005). The ratio of crop residues to grain is a key variable that affects the quantity of residues produced. A wide variation in residue-to-grain ratios exists in the literature (Perlack et al. 2005). Agricultural residues are materials left after harvesting and processing crops. These include leaves, stems, shells, stalks, husks, straws, and other parts associated with agricultural products (Bareschino et al. 2021). The characteristics of agricultural residues depend on the post-harvest processing methods and the age or stage of plant growth at harvest. Several billion metric tons of agricultural residues are produced annually. Some estimates suggest that the figure is around 5 billion metric tons per year (Shinde et al. 2022). The benefits of utilizing agricultural residues as a biomass source include cost-effectiveness, reduced greenhouse gas emissions, minimizing environmental pollution, and employment creation (Zhang et al. 2021).

MSW consists of waste generated in municipalities. This waste is produced by households, schools, shops, industry, offices, and various other urban activities. It is a heterogeneous material comprising organic and inorganic matter. This waste includes items such as cardboard, fruits, yard trimmings, wood, vegetables, furniture, leftover food, paper, cloth, and non-renewable materials like plastics, glass, and metal (Kaur et al. 2023). The organic portion is typically separated for bioenergy production. According to the World Bank, over two billion tons of MSW are produced globally each year, a figure projected to increase by 73% to nearly four billion tons by 2050 (Zhang et al. 2025). Per capita MSW generation is estimated at 0.5 kg/day, though it varies across different cities (Jingura & Matengaifa 2009). This results in vast quantities of MSW worldwide and represents a significant biomass resource for energy generation. However, around 23% of MSW remains uncollected, while 33% is openly dumped (Zhang et al. 2025).

Sewage sludge is a by-product of wastewater treatment. It is mainly produced during primary (physical or chemical), secondary (biological), and tertiary treatments (Mahmoudi et al. 2024). Globally, the average daily production of sewage sludge per person is estimated at 35–85 g of dry solids per capita per day (Werle & Sobek 2019). Sewage sludge contains water (97%–98%), organic matter, inorganic substances, and toxic contaminants such as metallic trace elements and heavy metals. It also contains microorganisms (Yilmaz et al. 2018). As a rich source of organic matter, it has been widely utilized for energy production.

Another source of biomass feedstock for energy production is food waste. This waste is generated during food production, processing, and consumption. Its composition varies considerably depending on the source (Chen et al. 2023). Food waste includes scraps of protein, lipid, and carbohydrate foods. Common sources are households, restaurants, and food processing plants. According to the United Nations Environmental Program (UNEP), about 1 billion tons of food waste were generated globally in 2022 (UNEP 2024). About 60% of the food waste came from households, 28% from food services, and 12% from retail. This amounts to 19% of food available to consumers being wasted (UNEP 2024). Therefore, food waste is a resource available for bioenergy value chains.

1.3.2.2 Forest Products

One dominant source of biomass feedstock is woody plants, which are highly lignocellulosic. Wood has been the primary source of energy for mankind for many years. To this day, wood remains one of the most widely utilized renewable energy sources in the world (Mu et al. 2023). Data in Figure 1.1 attest to this fact. Woody biomass, especially fuelwood and charcoal from forests, provides basic energy services for cooking and space heating. About 2.3 billion people (29% of the world's population) relied on woody biomass for energy purposes in 2021 (FAO 2024). Most (82%) fuelwood is produced and used in Africa, Asia, and South America (FAO 2024).

Forest products can be categorized into three groups: primary, secondary, and tertiary products (Perlack et al. 2005). Primary products encompass logging residues from harvesting operations, residues from forest management such as thinning and land clearing operations, residues from the removal of excess biomass from timberlands and other forestlands, and fuelwood extracted from forestlands (Perlack et al. 2005). Secondary products include wood processing mill residues and pulping liquors (black liquor) from pulp and paper production. Processing mill residues include sawdust, chips, and bark. Tertiary products include urban wood residues from construction and demolition debris, tree trimmings, packaging wastes, and consumer durables (Perlack et al. 2005).

World roundwood removals are about 4 billion m^3 annually (FAO 2024). Around half of which has been used directly as fuelwood or in the production of charcoal and pellets (FAO 2024). In 2022, global wood pellet production was 48 million tons (WBA 2024b). In 2023, global pellet production reached 47 million tons, showing some stagnation in production (FAO 2024). Of all the wood used as fuel worldwide, about 17% is converted to charcoal (FAO 2017). In 2018, over 53 million tons of wood charcoal were produced worldwide (Nabukalu & Glere 2020). Wood charcoal is predominantly sourced from the tropical forests of Brazil, Nigeria, Tanzania, and the Democratic Republic of Congo.

1.3.2.3 Energy Crops

The global demand for biomass feedstocks has created a need for alternative sources of biomass. In this effort, energy crops are among the potential options. Energy crops are plants cultivated specifically to produce biomass feedstock for bioenergy production. Advantages of energy crops include high productivity, high potential for carbon sequestration and soil improvement, and high energy density. The assessment of the global technical potential for energy crop production indicates that the energy potential is expected to grow from 160 EJ/year in 2000 to 180–200 EJ/year by 2050 and to 220–270 EJ/year by 2100 (Hayashi et al. 2015). Energy crops possess traits such as high biomass yield, fast growth rates, and efficient conversion of sunlight into chemical energy (Dida 2024). Energy crops are classified in various ways. Energy crops can also be categorized based on their chemical composition and the specific pathways for biofuel generation. Carbohydrate-rich crops such as sugarcane and maize are largely used for bioethanol production. Oilseed crops such as jatropha are used for biodiesel production.

Lignocellulosic crops consist of woody and herbaceous plants. Woody plants are highly lignocellulosic. Lignocellulose is the most abundant form of biomass on Earth (Chen & Peng 2013), accounting for approximately 57% of the planet's biogenic carbon (Bar-On et al. 2018). The most common woody plants include shrubs and trees such as poplar (*Populus* spp.), eucalyptus (*Eucalyptus* spp.), and willow (*Salix* spp.). Willow has gained recognition as a highly promising biomass crop due to its easy propagation, rapid growth in short-rotation coppice cycles, and low fertilizer requirements (Dida 2024). The crop yields of woody plants vary depending on species, site conditions, and management practices. Short-rotation coppice systems for poplar and willow annually produce yields between 10 and 15 tons/ha in the Northern Hemisphere (Schwerz et al. 2020). Generally, woody plant yields can range from five to over 20 tons per hectare annually, depending on species and silvicultural practices (Schwerz et al. 2020). For example, eucalyptus species such as *Eucalyptus globulus* and *Eucalyptus occidentalis* have demonstrated yields of up to 22.2 tons/ha/year under optimal conditions (Sochacki et al. 2007). This represents a significant amount of biomass produced by woody plants.

Herbaceous plants are vascular plants characterized by stems that are not woody, but soft and succulent. They are usually harvested annually after taking two to three years to reach maturity. Common examples include switchgrass (*Panicum virgatum*), Miscanthus (*Miscanthus* spp.), reed canary grass (*Phalaris arundinacea*), sorghum (*Sorghum bicolor*), energy cane (*Saccharum* spp.), and elephant grass (*Pennisetum purpureum*) (Agostini et al. 2015). Energy cane is bred for high biomass and fiber content, making it suitable for biofuel production. Despite their smaller physical stature compared to trees, herbaceous crops store roughly twice the amount of CO_2 underground when compared to woody crops (Agostini et al. 2015). Typical yields in tons per hectare include approximately 23 for *Miscanthus* (Lee et al. 2014), 40 for sorghum (Mullet et al. 2014), 20–80 for elephant grass (Heuzé et al. 2020), 150–200 for energy cane (Salassi et al. 2014), and 9–13 for switchgrass (Chaganti et al. 2021).

Oilseed crops are cultivated for their oil-rich seeds. The oil content in these seeds varies significantly. The extracted oil can be used directly as raw oil or converted into biodiesel or other biofuels. Edible oilseed crops include rapeseed (*Brassica napus*), which has an oil content of 38%–45%, soybean (*Glycine max*) with 16%–24%, and sunflower (*Helianthus annuus*) with 20%–30%. Jatropha (*Jatropha curcas*) produces non-edible oil specifically grown for energy production. Since the 2000s, jatropha has become a popular energy crop for biofuel production in many arid and semi-arid areas. Its dry seeds contain between 32% and 40% oil (Jongh & van der Putten 2010). Other non-edible oilseeds include castor bean (*Ricinus communis*), hemp (*Cannabis sativa*), neem (*Azadirachta indica*), and pongame (*Pongamia pinnata*). The oil content of seeds from these plants is approximately 50%, 35%, 30%, and 30%–40%, respectively (Ahmad et al. 2011). Some biofuel project promoters have shifted to other oleaginous crops such as *Balanites aegyptiaca*. *Balanites aegyptiaca* seeds have an oil content between 46% and 55%, making them a good source for energy production.

1.3.3 Third-Generation Biomass

Third-generation biomass is derived from oleaginous microorganisms, which include macroalgae, microalgae, bacteria, yeast, and fungi (Leong et al. 2018). Algae have 2–4 times higher photosynthesis rates than land plants, leading to quicker biomass accumulation (Li et al. 2008).

Microalgae are microscopic, photosynthetic organisms found in aquatic environments. They are the fastest-growing photosynthesizing organisms. Compared with terrestrial biomass, microalgae have higher photosynthetic efficiency, resulting in faster growth rates and potentially better CO_2 sequestration (Brennan & Owende 2009). The oil yield from algae exceeds that of the best-performing plant oils by over 200 times (Demirbas 2009a). Algae can be divided into two functional groups. Phototrophic algae utilize CO_2 and light for photosynthesis, while heterotrophic algae require organic carbon sources and oxygen for growth via aerobic respiration (Biller 2018). Approximately 46 tons of oil per hectare per year can be produced from diatom algae (Demirbas 2009a). Algal oil is an excellent feedstock for biodiesel production.

Macroalgae, also known as seaweed, are macroscopic, multicellular marine algae. They are larger than microalgae, reaching over 50 m in length, and exhibit greater diversity in the oceans (Pereira 2021). Macroalgae are perhaps the most promising non-consumable biofuel source, as they can grow rapidly in saline water, under adverse conditions, and in salty environments. Its growth rate is approximately 20–30 times faster than fodder crops (Godvin et al. 2021). Most macroalgae have a lipid content typically less than 5% of their dry weight, but there are species with more than 10% oil content (El Maghraby & Fakhry 2015). Macroalgae can be used to produce various solid, liquid, and gaseous biofuels. Other aquatic biomass, such as duckweed, waterweed, and water hyacinth, are often invasive species and are found in abundance. These can be cultivated specifically for biofuel production. They have been reported to double their mass in 24 hours, yield up to 20 tons of dry matter per hectare annually, and utilize wastewater nutrients (Biller 2018).

1.3.4 Fourth-Generation Biomass

Fourth-generation biomass is produced using advanced tools, such as genetic engineering and nanotechnology, to enhance the desired traits of organisms and advanced materials used in biofuel production (Leong et al. 2018; Bhagat et al. 2025; Sulis et al. 2025). Candidate organisms mainly include yeast, bacteria, and algae. Currently, advanced genetic engineering tools are also being utilized to develop transgenic plants (Sulis et al. 2025). Examples include plants with changed energy-storing cell walls and reproductive organs. This represents functional reorganization of plants. Functional reorganization for directing carbon flow through specific pathways or for directly constructing the final bioenergy compounds is the focus of the new generation of bioenergy crops (Bhagat et al. 2025). There are also endeavors to explore the potential role of synthetic biology and artificial pathways for fourth-generation biofuel production (Pfleger & Takors 2023; Bracmort (2013)).

1.4 ADVANTAGES AND DISADVANTAGES OF BIOMASS

The preceding sections described various types of biomass materials that are feedstocks for bioenergy production. However, biomass, as a feedstock for bioenergy, has both advantages and disadvantages. As a heterogeneous group of feedstocks, biomass presents a wide range of operational and logistical challenges that span the four generations of biomass.

1.4.1 Advantages of Biomass

Biomass improves energy security by producing biofuels, which lessen reliance on fossil fuels. It also aids in job creation throughout the biofuel value chain and helps to decrease greenhouse gas emissions (López 2024). The main benefit of biomass is that it is a renewable energy source. This supports the sustainability goal of lowering carbon emissions by using carbon-negative technology. Presently, first-generation biomass accounts for 4%, while second-generation biomass makes up 96% of commercial biofuel production (Inam & Deniz 2023). First-generation biomass generally contains fat, sugar, and starch, making it simple to process without needing costly or advanced techniques (Inam & Deniz 2023). This straightforward processing provides economic advantages in designing and operating processing facilities. Additionally, as food crops, first-generation crops have experienced significant productivity improvements, leading to high yields per hectare. As a result, less land is required to produce the necessary biomass.

An advantage of second-generation biomass is that it offers reduced competition with food crops. This is important in a world that faces food crises in some regions. Furthermore, second-generation biomass enhances sustainability by utilizing waste materials. They also promote the concept of a biorefinery, where multiple products are generated from the same feedstock (López 2024). There are a few supply-side bottlenecks as second-generation biomass benefits from abundant availability due to the large volume of residues produced in crop, food, and fiber value chains. Lignocellulosic wastes, food wastes, industrial wastes, or municipal solid wastes are inexpensive because they are considered waste that needs disposal. Therefore, second-generation biomass is a low-cost material that supports sustainable waste management and promotes circular economies.

Third-generation biomass exhibits a high photosynthesis rate, thereby enhancing its production efficiency. This biomass can be produced on non-arable land, reducing competition with other crops (López 2024). Algae can also be produced in non-freshwater. Algae contain high levels of lipids, protein, and sugar, making them established substrates for bioenergy production. They produce more oil than terrestrial plants. The ease of propagation and high yields make algae attractive as a feedstock. Microalgae offer several advantages, including easier availability, greater diversity, lower lignin content, higher growth rates, simpler pre-treatment, less cultivation land requirement, and no conflicts concerning food versus fuel (Inam & Deniz 2023).

Fourth-generation biomass is an emerging field still largely in the research and development phase. Its primary advantage is genetic modification to enhance the yield of valuable components such as lipids, which enter the bioenergy value chain. This enhances biofuel production with substantial economies of scale. Genetic modification can improve the adaptability to stressful conditions, thereby enhancing resilience in biomass production.

1.4.2 Disadvantages of Biomass

The primary drawback of first-generation biomass is the food versus fuel dilemma (López 2024). As these are food crops, this creates issues related to diverting food to the energy sector, competing with the food supply chain (Ahorsu et al. 2018; Dar et al. 2021). The cultivation of first-generation biomass also involves the use of fertilizers, pesticides, herbicides, and other chemicals, which is a disadvantage. Its life cycle generates significant greenhouse gas emissions. Moreover, depending on arable land for crop production, it can contribute to deforestation and biodiversity loss (López 2024).

The processes for converting second-generation biomass are complex and expensive. Since second-generation biomass is mainly lignocellulosic, it requires a pre-treatment stage involving substantial technological progress and infrastructure development (López 2024). Consequently, directly converting second-generation biomass into bioenergy is challenging, and its viability is hindered by high production costs and the need for significant investments in conversion facilities.

Third-generation biomass has yet to achieve commercial production due to several factors, including low production volume, high costs, and energy-intensive processing (Inam & Deniz 2023). Algae are highly sensitive to environmental conditions such as light, temperature, and nutrients. This makes production challenging and results in high variability in yields (Lee & Lavoie 2013). Producing microalgae requires specialized facilities and is costly, with scaling up limited by these factors. Contamination of open ponds by extraneous species frequently destroys the algae and restricts yields (Lee & Lavoie 2013; Inam & Deniz 2023). Algal biomass is a complex mixture of lipids, proteins, and carbohydrates, necessitating multi-step processing (Lee & Lavoie 2013). This reduces conversion efficiency.

Large-scale production of fourth-generation biomass has not yet been achieved due to high costs, low yields, and environmental and health-related issues (Inam & Deniz 2023; López 2024). This is because fourth-generation biomass often involves genetically modified organisms. There are economic and technical barriers to its production. Most fourth-generation biomass remains in the experimental stage, with high research and development costs and limited commercial viability (López 2024). Furthermore, regulatory and operational challenges exist. Many countries lack robust frameworks for synthetic biology and carbon-negative technologies, hindering implementation. Since the primary goal is carbon-negative energy, the efficiency of carbon capture in fourth-generation biomass may not always fully offset emissions (Ahorsu et al. 2018).

1.5 BIOMASS CHARACTERIZATION

Biomass is made up of naturally occurring components that vary in their properties. Biomass characterization is the process of analyzing the physical, chemical, and biological properties of biomass feedstocks (Mekonnen et al. 2015; Biswas et al. 2022; Shadangi et al. 2023). These characteristics determine the suitability of different feedstocks for producing various biofuels through a range of conversion pathways. The actual analytical methods are outside the scope of this chapter. A biomass characterization scheme is shown in Figure 1.3.

1.5.1 Energy Density of Feedstocks

Energy density describes the amount of energy stored in a system, substance, or region of space per unit of volume or mass. For biomass feedstock, the main focus is on the energy stored within the material. There are different types of energy density. Gravimetric energy density refers to the energy stored per unit mass (e.g., MJ/kg). It

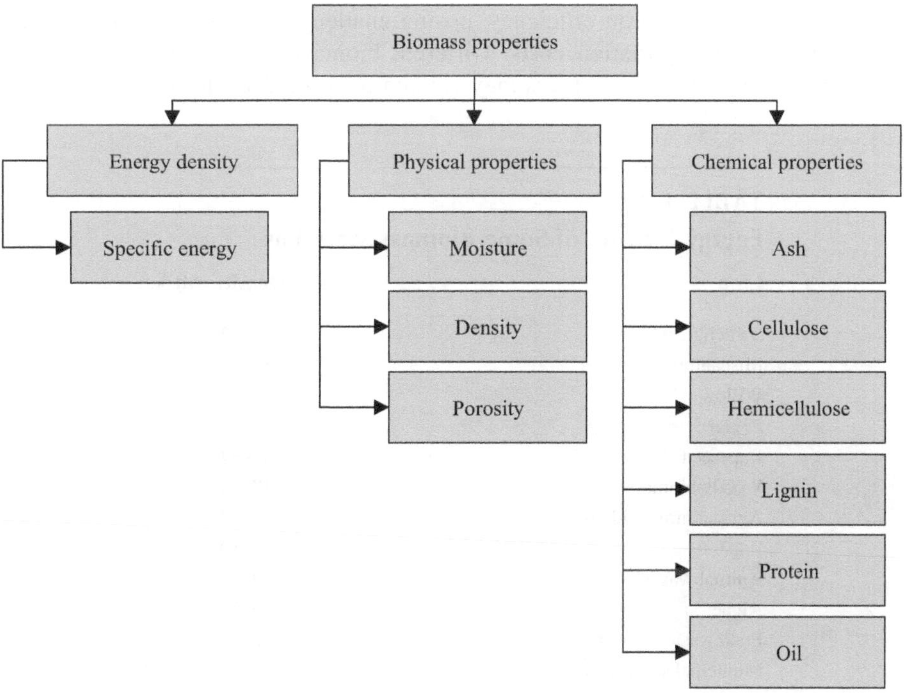

FIGURE 1.3 Biomass characterization scheme.

is also called specific energy. This measure helps indicate the total energy stored in biomass materials and is useful for assessing the biofuel potential of various materials. The gravimetric energy density of some biomass materials is shown in Table 1.2.

The energy density of biomass materials varies depending on their composition and the treatment method used. For example, the high moisture content of some food wastes dilutes their energy level. When torrefied, their energy density increases due to the removal of moisture and volatiles. As shown in Table 1.2, oilseeds, because of their high oil content, have very high specific energy compared to other materials. Lignocellulosic materials derive their energy value from lignin and cellulose, which have a gravimetric energy density of 25–27 and 17–18 MJ/kg, respectively.

1.5.2 Physical Properties

Physical properties are a key aspect of biomass feedstock. Biomass feedstocks have various physical properties that influence supply chain logistics, such as how they are handled, transported, and stored (Eisenbies et al. 2019). Physical properties also impact the processes for converting biomass into bioenergy. Key physical properties include moisture content, density, and porosity (Demirbas 2009b; Williams et al. 2017; Eisenbies et al. 2019). It is important to distinguish these from properties related to the fuel preparation process. This chapter focuses on the physical properties of feedstocks in their natural state before processing.

Biomass is a hygroscopic material and therefore contains free moisture. Moisture impacts the quality of feedstocks in various ways. These include reducing the calorific value and combustion efficiency, posing challenges in handling and storage, and increasing transportation costs. Different biomass feedstocks naturally have different moisture levels. Biomass may contain between 5% and 99% moisture by

TABLE 1.2
Energy Density of Some Biomass Materials

Crop	Energy Density (MJ/kg)
Switchgrass	18–22
Miscanthus	17–20
Willow	18–20
Poplar	18–22
Rapeseed	33–37
Woody biomass	18–19
Agricultural residues	15–17
Bagasse	15–19
Animal waste	15–20
Algae	20–25
Food waste	12–18
Municipal solid waste	7–15
Oilseed oil	37–42

weight. For example, sewage sludge is a semi-solid material with high water content, often ranging from 93% to 99%, while wood waste can have moisture content from 5% to 30%. Fresh macroalgae have moisture content ranging from 70% to 90% (Cesário et al. 2018).

Bulk density influences supply chain logistics, such as collection, transportation, and how material is stored and handled (Ebadian et al. 2018). Bulk density is the ratio between the weight of a material and the volume it occupies. This property affects how materials are packed. Materials with low bulk density require larger storage space. Agricultural residues have a bulk density ranging from 80 to 100 kg/m^3, while woody material is around 150–200 kg/m^3 (Ebadian et al. 2018). In practice, this means crop residues pose more logistical challenges in packing than woody biomass.

Biomass feedstocks vary in porosity due to differences in cellular structure and composition. Porosity indicates the proportion of void spaces, such as pores, vessels, and cell lumens, within the material. Higher porosity increases the surface area of materials. A larger surface area improves water uptake, enhances heat transfer, and can facilitate faster decomposition or reactions (Edeh et al. 2023). Woody biomass can have porosity ranging from 50% to 70%, depending on the type of wood, from hard to soft.

1.5.3 Chemical Composition

Biomass feedstocks vary significantly in their chemical composition, just as they vary in their types. Their chemical composition is a function of many variables, including species, genetics, production conditions, harvesting, collection, and storage methods. Due to the numerous and complex influences of these factors, biomass is largely a heterogeneous material. The chemical composition influences the processes used to convert biomass into energy. Understanding this composition helps to optimize and select appropriate conversion processes. The chemical composition of biomass is a complex combination of structural, non-structural, and inorganic components, as shown in Figure 1.4. The main structural components of plant biomass are the biopolymers like cellulose, hemicellulose, and lignin, which form the cell wall. The non-structural components are the extractives, which are not chemically bound to the cell wall polymers. Extractives include lipids, proteins, simple sugars, starches, pectin, resins, phenolic compounds, hydrocarbons, and other substances. These extractives vary widely depending on the type of biomass feedstock. Ash represents the mineral-based components of biomass feedstock.

The proportions of cell wall components, extractives, and ash vary depending on the type of feedstock. There is considerable variation in composition among different feedstock types, such as between woody and herbaceous materials. Significant variation also exists within a feedstock type. For instance, there is notable compositional variability between different anatomical parts of the same plant. Lignocellulosic biomass contains a high fraction of cellulose in the heartwood, while the bark contains a larger percentage of lignin, and the extractives are evenly distributed (Williams et al. 2017). In crop residues, maize stover contains most of the extractives in the leaves and internodes (Williams et al. 2017). This compositional variation is so

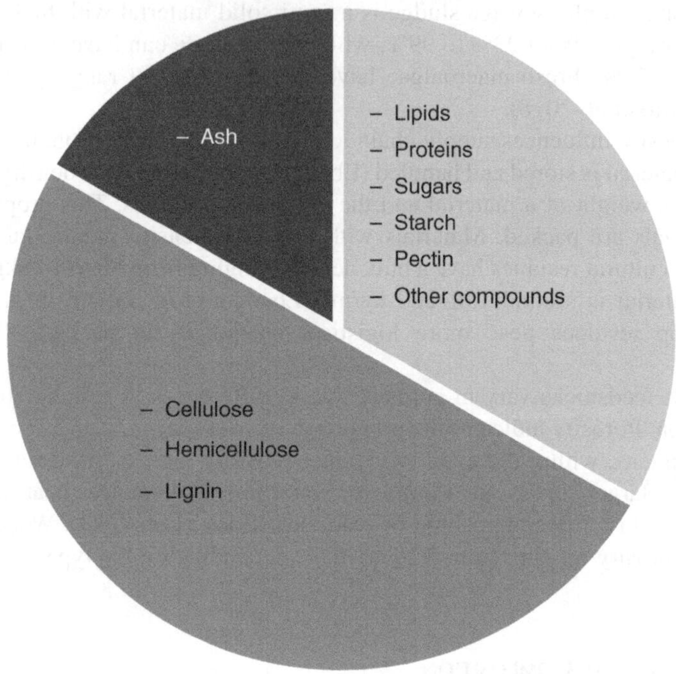

FIGURE 1.4 Chemical components of biomass.

substantial that the conversion equipment must be operated under specific conditions tailored to the type of feedstock. There is a link between biofuel technology and feedstock type.

Cell wall components are abundant in lignocellulosic materials such as wood and crop residues. By their nature, wood products, forest residues, crop residues, and energy crops are all lignocellulosic. Generally, in feedstocks such as woody biomass and crop residues, the proportion of the cell wall ranges from 70% to 90%, extractives from 5% to 30%, and ash from 0.2% to 10% (Kenney et al. 2013; Williams et al. 2017; Silva et al. 2024). In herbaceous plants, the cell wall proportion is around 60%, and in MSW, it is approximately 50% (Williams et al. 2017).

Ash content varies widely among different feedstocks. For example, wheat straw can have ash content exceeding 10%, whereas wood typically contains 1% or less. Sources of ash include structural ash associated with the plant cell walls, vascular ash associated with the plant cell extracts, and introduced ash resulting from soil contamination (Kenney et al. 2013). Structural and vascular ashes together comprise a fraction Kenney et al. (2013) call 'physiological ash', which is a property of the feedstock. Introduced ash arises from the feedstock handling processes such as harvesting, transportation, handling, and storage (Kenney et al. 2013).

The chemical composition of selected biomass feedstocks is shown in Table 1.3. These chemical composition figures fluctuate based on various factors. For example, in plants, cell wall components increase as maturity progresses; thus, younger plants are less lignocellulosic than mature ones. For heterogeneous feedstocks like MSW, the content of cellulose, hemicellulose, and lignin depends on the amount and type of lignocellulosic material present. This is shown by the wide ranges given in Table 1.3.

Cellulose is the most abundant organic polymer on Earth. It forms the main structural component of plant cell walls, including in algae. Its content ranges from 38% to 50% by weight in woody and herbaceous feedstocks (Biswas et al. 2022). In other feedstocks such as MSW and sewage sludge, the content varies considerably. Cellulose is a polysaccharide composed of repeating glucose units linked in a large straight-chain macromolecule. These glucose units are connected by β-(1,4)-glycosidic bonds. Cellulose is insoluble in most solvents and has limited accessibility to acid and enzymatic hydrolysis (Demirbas 2009a).

The second major component of biomass is hemicellulose. It is a heteropolymer made up of several polysaccharides such as xylans, galactoglucomannans, arabinogalactans, glucans, and galactans. Hemicellulose contains different monosaccharides, including xylose, mannose, glucose, galactose, rhamnose, arabinose, and uronic acids. The polymer chains of hemicellulose are short branches, typically 500–3,000 sugar units long. Unlike crystalline cellulose, hemicellulose is amorphous. Because of its amorphous structure, hemicellulose is more susceptible to hydrolysis than crystalline cellulose. Structurally, hemicellulose cross-links with cellulose microfibrils, enhancing the mechanical strength and flexibility of biomass materials. The hemicellulose content in various biomass feedstocks is shown in Table 1.3.

Lignin is one of the most abundant organic compounds on Earth, after cellulose and chitin. It is most commonly derived from wood and is a vital part of the cell walls of plants, especially in tracheids, xylem fibers, and sclereids (Demirbas 2009a). Lignin is a complex compound composed of aromatic alcohols. It is typically

TABLE 1.3
Chemical Composition of Selected Feedstocks (%)

Feedstock	Cellulose	Hemicellulose	Lignin	Ash
Maize stover	30–40	20–25	12–17	3–11
Wheat straw	30–45	20–30	5–24	4–23
Miscanthus straw	30–50	15–40	5–30	1–9
Switchgrass	30–45	21–30	12–28	2–4
Poplar wood	40–50	20–30	20–35	0.5–4
Willow wood	35–65	15–35	15–30	1–2
Municipal solid waste	20–40	16–30	12–20	15–25
Sewage sludge	10–30	10–30	10–16	20–50
Food waste	40–50	20–60	10–25	5–12
Animal manure	2–6	2–28	0–6	9–30

Sources: Kenney et al. (2013), Nayak & Mukherjee (2015), Mensah et al. (2021), Williams et al. (2017), Mushtaq et al. (2023).

TABLE 1.4
Composition of Some Vegetable Oils

	Fatty Acid Composition (%)				
	Palmitic	Stearic	Oleic	Linoleic	Linolenic
Plant Oil	16:0[a]	18:0	18:1	18:2	18:3
Rapeseed	5	1.5	57	20	8
Soybean	10	4	23	54	7
Jatropha	12	4	44	34	0.5
Sunflower	6	3	30	58	0.1
Sesame	9	6	40	43	0
Groundnut	9	3	59	20	0.5
Algae	16	2	27	22	16
Maize	11	2	28	56	0.8
Cottonseed	22	3	17	55	0.3

Sources: Haas & Wagner (2011), Fabiszewsk & Białecka-Florjańczyk (2015), Azeez et al. (2019), Fediol (2025).

[a] Number of carbon atoms: double bonds.

described as a three-dimensional polymer made of monolignols, which are phenyl-propane units. These units are derived from coniferyl alcohol, sinapyl alcohol, and p-coumaryl alcohol. Lignin intertwines with the cellulose and hemicellulose fractions of the cell wall and, functionally, binds these components together to provide strength, rigidity, and resistance to decomposition. The lignin content of various feedstocks is shown in Table 1.3.

Proteins are among the naturally occurring macromolecules found in biomass. They are copolymers of amino acids linked by peptide bonds. Their length can range from 2 to 50 amino acids to thousands of amino acids. Although biomass is often associated with energy production from structural carbohydrates such as cellulose and hemicellulose, the protein content is more relevant to creating protein-rich biocomposites, which have numerous industrial applications (Mekonnen et al. 2015). These include the production of animal feeds.

Plant oils are used to produce liquid fuels such as biodiesel. They consist of triglycerides, molecules formed by glycerol and three fatty acids. The fatty acids may be saturated or unsaturated. Common fatty acids found in plant oils include palmitic acid, stearic acid, oleic acid, linoleic acid, and linolenic acid. The proportions of these fatty acids vary considerably between different oils. These are shown in Table 1.4.

1.6 CONCLUSION

The basis for pursuing new, alternative, and sustainable fuels has been well-established in this millennium. This has led to the drive to exploit biomass as a renewable energy feedstock. Furthermore, biorefineries offer vast opportunities to valorize biomass. The fact that biomass is a sustainable alternative energy source is also well-established. Biomass is one of the most abundant resources on Earth. It is a heterogeneous material composed of organic matter of plant and animal origin. Its chemical composition depends on various factors, including species, genetics, production conditions, harvesting, collection, and storage methods. This heterogeneity provides several pathways through which biomass can be converted into bioenergy. Contributing around 9% to the global energy mix, there are extensive opportunities to increase this proportion by improving the supply side of biomass. This chapter has presented a concise review of different types of biomass and their characterization.

The overview highlights the four generations of biomass. The types of biomass vary considerably in composition. The main chemical components are cellulose, hemicellulose, and lignin. First-generation biomass includes food crops rich in sugars, starch, and lipids, mainly used to produce liquid biofuels such as biodiesel and bioethanol. The fact that they are food crops is the primary disadvantage of first-generation biomass. Second-generation biomass comprises crop and animal residues, wastes, and energy crops. These materials are generally lignocellulosic, and their conversion into energy represents the valorization of crop and animal wastes. Third-generation biomass mainly consists of aquatic organisms, with algae as the primary component. Algae are efficient at photosynthesis and rich in oil, making them a viable feedstock. Fourth-generation biomass involves genetically engineered microorganisms and plants. These materials are modified to improve production efficiency and offer opportunities to expand the biomass resource base.

The physical and chemical properties of biomass affect supply chain logistics, such as how they are handled, transported, and stored. Understanding the types and characteristics of biomass is also important for designing and operating conversion processes. This is important for the functioning of a biorefinery. The main challenge lies with highly lignocellulosic materials such as crop residues. Such materials require a pre-treatment stage, which can considerably increase costs. In this context, fourth-generation biomass offers a new class of materials capable of overcoming the limitations of other types of biomass.

REFERENCES

Agostini, F., A.S. Gregory, & G.M. Richter. 2015. Carbon sequestration by perennial energy crops: Is the jury still out? *Bioenergy Res.* 8: 1057–1080.

Ahmad, M., M.A. Khan, M. Zafar, & S. Sultana. 2011. Biodiesel from non-edible oil seeds: A renewable source of bioenergy. In *Economic Effects of Biofuel Production*, ed. M.A. dos Santos Bernardes, 259–279. London: Intech.

Ahorsu, R., F. Medina, & M. Constantí. 2018. Significance and challenges of biomass as a suitable feedstock for bioenergy and biochemical production: A review. *Energies* 11: 3366.

Azeez, A., A. Fasakin, & J. Orege. 2019. Production, characterization and fatty acid composition of *Jatropha curcas* biodiesel as a viable alternative to conventional diesel fuel in Nigeria. *Green Sustain. Chem.* 9: 1–10.

Bareschino, P., E. Marrasso, & C. Roselli. 2021. Tobacco stalks as a sustainable energy source in civil sector: Assessment of techno-economic and environmental potential. *Renew. Energy* 175: 373–390.

Bar-On, Y.M., R. Phillips, & R. Milo. 2018. The biomass distribution on Earth. *Proc. Natl. Acad. Sci. USA* 115: 6506–6511.

Bhagat, S.P., B.D. Wagh, S.M. Satpute, M.B. Gunjal, S.K. Kasar, & P.B. Roylawar. 2025. Fourth-generation biofuels: Genetic engineering for sustainable energy. In *Recent Trends in Lignocellulosic Biofuels and Bioenergy. Clean Energy Production Technologies*, ed. M.A. Dar, H.M. Zabed, & M. Shahnawaz, 419–438. Singapore: Springer.

Biller, P. 2018. Hydrothermal liquefaction of aquatic feedstocks. In *Direct Thermochemical Liquefaction for Energy Applications*, ed. L. Rosendahl, 101–125. Cambridge, MA: Woodhead Publishing.

Biswas, B., B.B. Krishna, M.K. Kumar, R.K. Sukumaran, & T. Bhaskar. 2022. Biomass characterization. In *Advanced Biofuel Technologies*, ed. D. Tuli, S. Kasture, & A. Kuila, 151–175. Cambridge, MA: Elsevier.

Bracmort, K. 2013. *Biomass: Comparison of Definitions in Legislation*. Washington, DC: Congressional Research Service. https://digital.library.unt.edu/ark:/67531/metadc271997/m1/1/high_res_d/R40529_2013Nov26.pdf

Brennan, L., & P. Owende. 2009. Biofuels from microalgae - A review of technologies for production, processing, and extractions of biofuels and co-products. *Renew. Sustain. Energy Rev.* 14: 557–577.

Bridgwater, T. 2006. Biomass for energy. *J. Sci. Food Agric.* 86: 1755–1768.

Cavelius, P., S. Engelhart-Straub, N. Mehlmer, L. Lercher, A. Awad, & T. Brück. 2023. The potential of biofuels from first to fourth generation. *PLoS Biol.* 21: e3002063.

Cesário, M.T., M.M.R. da Fonseca, M.M. Marques, & M.C. de Almeida. 2018. Marine algal carbohydrates as carbon sources for the production of biochemicals and biomaterials. *Biotechnol. Adv.* 36: 798–817.

Chaganti, V.N., G. Ganjegunte, M.N. Meki, J.R. Kiniry, & G. Niu. 2021. Switchgrass biomass yield and composition and soil quality as affected by treated wastewater irrigation in an arid environment. *Biomass Bioenergy* 151: 106160.

Chen, P., & L. Peng. 2013. The diversity of lignocellulosic biomass resources and their evaluation for use as biofuels and chemicals. In *Biological Conversion of Biomass for Fuels and Chemicals: Explorations from Natural Utilization Systems*, ed. J. Sun, S.-Y. Ding, & J.D. Peterson, 83–113. Cambridge: The Royal Society of Chemistry.

Chen, Y., L. Pinegar, J. Immonen, & K.M. Powell. 2023. Conversion of food waste to renewable energy: A techno-economic and environmental assessment. *J. Clean. Prod.* 385: 135741.

Cotana, F., V. Coccia, G. Cavalaglio, M. Barbanera, & A. Petrozzi. 2022. Biomass-based systems. In *Polygeneration Systems*, ed. F. Calise, M.D. D'Accadia, L. Vanoli, & M. Vicidomini, 137–192. Cambridge, MA: Academic Press.

Cvetkovića, S., T.K. Radoičić, B. Vukadinović, & M. Kijevčanin. 2014. Potentials and status of biogas as energy source in the Republic of Serbia. *Renew. Sustain. Energy Rev.* 31: 407–416.

Dar, R.A., M. Parmar, E.A. Dar, R.K. Sani, & U.G. Phutela. 2021. Biomethanation of agricultural residues: Potential, limitations and possible solutions. *Renew. Sustain. Energy Rev.* 135: 110217.

Dayton, D.C., & D.F. Thomas. 2020. *Analytical Methods for Biomass Characterization and Conversion*. Cambridge, MA: Elsevier.

Demirbas, A. 2009a. Biomass feedstock. In *Biotechnology – Volume VII: Fundamentals in Biotechnology*, ed. H.W. Doelle, J.S. Rokem, & M. Berovic, 315–325. Paris: UNESCO-EOLSS.

Demirbas, A. 2010. *Biorefineries: For Biomass Upgrading Facilities*. London: Springer.

Demirbas, M.T. 2009b. Biorefineries for biofuel upgrading: A critical review. *Appl. Energy* 86: 151–161.

Dida, G. 2024. Biotechnology towards energy crops. *CABI Agric. Biosci.* 5: 45.

Duca, D., & G. Toscano. 2022. Biomass energy resources: Feedstock quality and bioenergy sustainability. *Resources* 11: 57.

Ebadian, M., M.E. Shedden, E. Webb, S. Sokhansanj, M. Eisenbies, T. Volh, J. Heavey, & K. Hallen. 2018. Impact of parcel size, field shape, crop yield, storage location, and collection equipment on the performance of single-pass cut-and-chip harvest system in commercial shrub willow fields. *Bioenergy Res.* 11: 364–381.

Edeh, I.G., O. Masek, & F. Fusseis. 2023. 4D structural changes and pore network model of biomass during pyrolysis. *Sci. Rep.* 13: 22863.

Eisenbies, M.H., T.A. Volk, O. Therasme, & K. Hallen. 2019. Three bulk density measurement methods provide different results for commercial scale harvests of willow biomass chips. *Biomass Bioenergy* 124: 64–73.

El Maghraby, D.M., & E.M. Fakhry. 2015. Lipid content and fatty acid composition of Mediterranean macro-algae as dynamic factors for biodiesel production. *Oceanologia* 57: 86–92.

European Union. 2024. Directive (EU) 2018/2001 of the European Parliament and of the Council of 11 December 2018 on the promotion of the use of energy from renewable sources. http://data.europa.eu/eli/dir/2018/2001/2024-07-16 (accessed 15 August 2025).

Fabiszewsk, A.U., & E. Białecka-Florjańczyk. 2015. Factors influencing synthesis of extracellular lipases by *Yarrowia lipolytica* in medium containing vegetable oils. *J. Microbiol. Biotechnol. Food Sci.* 4: 231–237.

FAO. 2017. The charcoal transition. https://www.uncclearn.org/wp-content/uploads/library/charcoal.pdf (accessed 21 July 2025).

FAO. 2022. World food and agriculture: Statistical yearbook 2022. https://www.fao.org/3/cc2212en/cc2212en.pdf (accessed 10 August 2025).

FAO. 2024. The state of the world's forests 2024. https://openknowledge.fao.org/server/api/core/bitstreams/768ba59e-c692-47c3-9a13-3c3c10993396/content/src/html/wood-production-record-levels.html#gsc.tab=0 (accessed 21 July 2025).

Fediol. 2025. Table of fatty acid composition of selected oils and fats. https://www.fediol.eu/data/fatty%20acids.pdf (accessed 29 July 2025).

Girotto, F., & C.A. Cossu. 2017. Animal waste and waste animal by-products generated along the livestock breeding and meat food chain. *Waste Manag.* 70: 1–2.

Godvin, S.V., K.M. Dinesh, A. Pugazhendi, A.K. Bajhaiya, P. Gugulothu, & B.J. Rajesh. 2021. Biofuel production from Macroalgae: Present scenario and future scope. *Bioengineered* 12: 9216–9238.

Haas, M.J., & K.M. Wagner. 2011. Simplifying biodiesel production: The direct or in situ transesterification of algal biomass. *Eur. J. Lipid Sci. Technol.* 113: 1219–1229.

Hayashi, A., K. Akimoto, F. Sano, & T. Tomoda. 2015. Evaluation of global energy crop production potential up to 2100 under socioeconomic development and climate change scenarios. *J. Jpn Inst. Energy* 94: 548–554.

Heuzé, V., G. Tran, S. Giger-Reverdin, & F. Lebas. 2020. Elephant grass (*Pennisetum purpureum*). Feedipedia, a program by INRAE, CIRAD, AFZ and FAO. https://www.feedipedia.org/node/395 (accessed 15 July 2025).

Inam, A., & I. Deniz. 2023. Introduction to bioenergy: Current status, merits & demerits. In *Bioenergy Engineering: Fundamentals, Methods, Modelling, and Applications*, ed. K.P. Shadangi, P.K. Sarangi, K. Mohanty, I. Deniz, & A.R. Gollakota, 1–15. Cambridge, MA: Woodhead Publishing.

IPCC. 2022. *Climate Change 2022: Mitigation of Climate Change*. IPCC. https://www.ipcc.ch/report/ar6/wg3/downloads/report/IPCC_AR6_WGIII_FullReport.pdf (accessed 8 July 2025).

Jiménez-Rosado, M., V. Perez-Puyana, & A. Romero. 2023. The use of biowaste for the production of biodegradable superabsorbent materials. *Curr. Opin. Food Sci.* 49: 100975.

Jingura, R.M., & R. Matengaifa. 2009. Optimization of biogas production by anaerobic digestion for sustainable energy development in Zimbabwe. *Renew. Sustain. Energy Rev.* 13: 1116–1120.

Jongh, J.A., & E. van der Putten. 2010. *The Jatropha Handbook: From Cultivation to Application*. Eindhoven: FACT Foundation.

Kaur, A., R. Bharti, & R. Sharma. 2023. Municipal solid waste as a source of energy. *Mater. Today Proc.* 81: 904–915.

Kenney, K.L., W.A. Smith, G.L. Gresham, & T.L. Westover. 2013. Understanding biomass feedstock variability. *Biofuels* 4: 111–127.

Lee, D.K., A. Parrish, & T.B. Voigt. 2014. Switchgrass and giant Miscanthus agronomy. In *Engineering and Science of Biomass Feedstock Production and Provision*, ed. Y. Shastri, A. Hansen, L. Rodriguez, & K. Ting, 37–60. New York: Springer Science + Business Media.

Lee, R.A., & J.-M. Lavoie. 2013. From first- to third-generation biofuels: Challenges of producing a commodity from a biomass of increasing complexity. *Anim. Front.* 3: 6–11.

Leong, W.-H., J.-W. Lim, M.-K. Lam, Y. Uemura, & Y.-C. Ho. 2018. Third generation biofuels: A nutritional perspective in enhancing microbial lipid production. *Renew. Sustain. Energy Rev.* 91: 950–961.

Li, Y., M. Horsman, N. Wu, C.Q. Lan, & N. Dubois-Calero. 2008. Biofuels from microalgae. *Biotechnol. Prog.* 24: 815–820.

Lo, S.L.Y., B.S. How, W.D. Leong, S.Y. Teng, M.A. Rhamdhani, & J. Sunarso. 2021. Techno-economic analysis for biomass supply chain: A state-of-the-art review. *Renew. Sustain. Energy Rev.* 135: 110164.

López, R. 2024. Understanding first, second, third and fourth generation biofuels: Explore the different generations of biofuels, from first to fourth. https://www.jord.one/blogs/understaning-first-second-third-and-fourth-generation-biofuels (accessed 31 July 2025).

Mahmoudi, A., S.A. Mousavi, & P. Darvishi. 2024. Performance and recent development in sewage sludge-to-bioenergy using microbial fuel cells: A comprehensive review. *Int. J. Hydrog. Energy* 50: 1432–1455.

Mekonnen, T.H., M. Misra, & A.K. Mohanty. 2015. Processing, performance, and applications of plant and animal protein-based blends and their biocomposites. In *Biocomposites: Design and Mechanical Performance*, ed. M. Misra, J.K. Pandely, & A.K. Mohanty, 201–235. Cambridge, MA: Woodhead Publishing.

Mensah, M.B., H. Jumpah, N.O. Boadi, & J.A.M. Awudza. 2021. Assessment of quantities and composition of corn stover in Ghana and their conversion into bioethanol. *Sci. Afr.* 12: e00731.

Mu, L., R.Y. Wang, P.W. Xie, Y.L. Li, X.K. Huang, H.C. Yin, & M. Dong. 2023. Comparative investigation on the pyrolysis of crop, woody, and herbaceous biomass: Pyrolytic products, structural characteristics, and CO_2 gasification. *Fuel* 335: 126940.

Mullet, J., D. Morishige, R. McCormick, S. Truong, J. Hilley, B. McKinley, R. Anderson, S.N. Olson, & W. Rooney. 2014. Energy sorghum – A genetic model for the design of C4 grass bioenergy crops. *J. Exp. Bot.* 65: 3479–3489.

Mushtaq, S., F. Jamil, M. Hussain, A. Inayat, P. Akhter, K. Majeed, M.S. Khurram, M. Aslam, J. Lee, & Y.-K. Park. 2023. Unearthing the potential sustainability of cellulose and exploring its source, fate, and recovery. *Environ. Eng. Res.* 28: 230054.

Nabukalu, C., & R. Glere. 2020. *Global Charcoal Consumption and the Question of Energy Security*. Klein Centre for Energy Policy. https://kleinmanenergy.upenn.edu/commentary/blog/global-charcoal-consumption-and-the-question-of-energy-security/ (accessed 18 July 2025).

Nayak, S., & A.K. Mukherjee. 2015. Management of agricultural wastes using microbial agents. In *Waste Management: Challenges, Threats and Opportunities*, ed. R.P. Singh, & A. Sarkar, 65–91. Hauppauge, NY: Nova Science Publishers.

Nazari, M.T., J. Mazutti, L.G. Basso, L.M. Colla, & L. Brandli. 2020. Biofuels and their connections with the sustainable development goals: A bibliometric and systematic review. *Environ. Dev. Sustain.* 23: 11139–11156.

Olayokun, O. E., B.E. Anyaegbuna, A.O. Onokwai, D.I. Faola, I. Muslim, A.O. Eruobodo, & A.O. Adiatu. 2024. Review of biomass as a renewable energy for sustainable environment. Paper presented at the *2024 International Conference on Science, Engineering and Business for Driving Sustainable Development Goals (SEB4SDG)*, Omu-Aran, April 2–4, 2024.

Openshaw, K. 2000. A review of *Jatropha curcas*: An oil plant of unfulfilled. *Biomass Bioenergy* 19: 1–15.

Panchani, M., S. Narang, R. Jamwal, & M.P. Thakur. 2024. Biomass as source of energy for sustainable development. In *Futuristic Trends in Renewable & Sustainable Energy*, ed. P. Sudheer, M. R. Gurrala, C. L. Bhaskara, S. Ali, G. G. R. Sekhar, S.P. Umayal, P. Pogaku et al., 115–128. Karnataka: Iterative International Publishers.

Pereira, L. 2021. Macroalgae. *Encyclopedia* 1: 177–188.

Perlack, R.D., L.L. Wright, A.F. Turhollow, R.L. Graham, B.J. Stokes, & D.C. Erbach. 2005. *Biomass as Feedstock for a Bioenergy and Bioproducts Industry: The Technical Feasibility of a Billion-Ton Annual Supply*. US Department of Energy and US Oak Ridge National Laboratory. https://www.osti.gov/servlets/purl/1216415/ (accessed 16 July 2025).

Pfleger, B.F., & R. Takors. 2023. Recent progress in the synthesis of advanced biofuel and bioproducts. *Curr. Opin. Biotechnol.* 80: 102913.

Ródenas, M.M. 2022. How waste mismanagement drives biodiversity loss and accelerates climate risk. https://www.fairr.org/resources/reports/index-2021-how-animal-waste-mismanagement-drives-biodiversity-loss-and-accelerates-climate-risk (accessed 17 July 2022).

Salassi, M.E., K. Brown, B.M. Hilbun, K.A. Gravois, T.B. Mark, & L.L. Falconer. 2014. Farm-scale cost of producing perennial energy cane as a biofuel feedstock. *Bioenergy Res.* 7: 609–619.

Schwerz, F., D.D. Neto, B.O. Caron, C. Nardini, J. Sgarbossa, E. Eloy, A. Behling, E.F. Elli, & K. Reichardt. 2020. Biomass and potential energy yield of perennial woody energy crops under reduced planting spacing. *Renew. Energy* 153: 1238–1250.

Shadangi, K.P., P.K. Sarangi, & A.K. Behera. 2023. Characterization techniques of biomass: Physico-chemical, elemental, and biological. In *Bioenergy Engineering: Fundamentals, Methods, Modelling, and Applications*, ed. K.P. Shadangi, P.K. Sarangi, K. Mohanty, I. Deniz, & R.K. Gollakota, 51–66. Cambridge, MA: Woodhead Publishing.

Shinde, R., D.K. Shahi, P. Mahapatra, C.S. Singh, S.K. Naik, N. Thombare, & A.K. Singh. 2022. Management of crop residues with special reference to the on-farm utilization methods: A review. *Ind. Crop. Prod.* 181: 114772.

Silva, T.A.L., L.H.R. Varão, & D. Pasquini. 2024. Lignocellulosic biomass. In *Handbook of Biomass*, ed. S. Thomas, M. Hosur, D. Pasquini, & C.J. Chirayil, 105–142. Singapore: Springer.

Sochacki, S.J., R.J. Harper, & K.R.J. Smettem. 2007. Estimation of woody biomass production from a short-rotation bio-energy system in semi-arid Australia. *Biomass Bioenergy* 31: 608–616.

Sulis, D.B., N. Lavoine, H. Sederoff, X. Jiang, B.M. Marques, K. Lan, C. Cofre-Vega, R. Barrangou, & J.P. Wang. 2025. Advances in lignocellulosic feedstocks for bioenergy and bioproducts. *Nat. Commun.* 16: 1244.

UNEP. 2024. *Food Waste Index Report 2024*. UNEP. https://unstats.un.org/unsd/envstats/fdes/EGES11/9UNEP_Food%20Waste%20Index.pdf (accessed 13 July 2025).

WBA. 2024a. *Global Bioenergy Statistics Report 2024*. Stockholm: WBA. https://www.worldbioenergy.org/uploads/241023_GBS_Report.pdf

WBA. 2024b. *Global Bioenergy Statistics 2022*. Stockholm: WBA. https://www.worldbioenergy.org/uploads/221223%20WBA%20GBS%202022.pdf (accessed 3 July 2025)

Werle, S., & S. Sobek. 2019. Gasification of sewage sludge within a circular economy perspective: A Polish case study. *Environ. Sci. Pollut. Res.* 26: 35422–35432.

Williams, C. L., R.M. Emerson, & J.S. Tumuluru. 2017. Biomass compositional analysis for conversion to renewable fuels and chemicals. In *Biomass Volume Estimation and Valorization for Energy*, ed. J.S. Tumuluru, 251–v270. London: Intech.

Yilmaz, E., M. Wzorek, & S. Akcay. 2018. Co-pelletization of sewage sludge and agricultural wastes. *J. Environ. Manag.* 216: 169–175.

Zhang, B.Q., J.L. Xu, Z.X. Lin, T. Lin, & A.P.C. Faaij. 2021. Spatially explicit analyses of sustainable agricultural residue potential for bioenergy in China under various soil and land management scenarios. *Renew. Sustain. Energy Rev.* 137: 110614.

Zhang, M., K. Ionkova, & S. Boukerche. 2025. How the World Bank is tackling the growing global waste crisis. https://blogs.worldbank.org/en/sustainablecities/how-the-world-bank-is-tackling-the-growing-global-waste-crisis# (accessed 18 July 2025).

Ziolkowska, J.R. 2020. Biofuels technologies: An overview of feedstocks, processes, and technologies. In *Biofuels for a More Sustainable Future*, ed. J. Ren, A. Scipioni, A. Manzardo, & H. Liang, 1–19. Cambridge, MA: Elsevier.

2 Classification and Logistics of Biomass Supply Chain

Fitri Agustina and Indra Cahyadi

2.1 INTRODUCTION

In the current era, most human activities rely on energy availability. When energy is limited or disrupted, many daily activities and industrial operations cannot run effectively. Energy also has an important role in producing goods and services. Therefore, interruptions in energy supply may reduce national economic productivity. At present, the dominant sources of energy are fossil fuels, which were formed from ancient organisms and microorganisms that were buried and transformed over millions of years. To improve energy security, many countries try to diversify their energy sources and reduce dependence on a single type of fuel.

The continuous increase in global energy demand, together with the need to reduce climate change impacts, has encouraged stronger attention to renewable and sustainable energy sources. Biomass is considered a promising option because it is widely available and distributed across many regions. It can be converted into different forms of energy as well as value-added products. In contrast to fossil fuels, biomass is a renewable resource derived from biological materials and organic residues. For this reason, biomass is often discussed as a potential contributor to greenhouse gas reduction and the transition toward a circular economy. The advantage of biomass is commonly linked to its biogenic carbon cycle. During growth, plants absorb CO_2, and during biofuel conversion or combustion, CO_2 is released back to the atmosphere. Under this assumption, biomass-based fuels are frequently described as carbon-neutral (Saravanan et al. 2025).

Biomass utilization is important not only for producing bioenergy (e.g. bioethanol, biodiesel, biogas, and electricity) but also for producing biochemicals and bioproducts through integrated biorefinery systems. In recent decades, progress in conversion technologies, such as combustion, pyrolysis, gasification, and fermentation, has supported more efficient and cleaner energy pathways. This development also led to the classification of bioenergy into first, second, and third generations, where each generation aims to address limitations of earlier approaches, especially related to food security, feedstock variation, and environmental sustainability.

However, the long-term sustainability and large-scale implementation of biomass depend strongly on effective supply chain management. Biomass supply chains are

DOI: 10.1201/9781003650751-2

complex because they involve many connected stages, including feedstock cultivation, harvesting, pretreatment, storage, transportation, conversion, and distribution to end users. Compared with conventional industrial supply chains, biomass systems face specific challenges such as seasonal availability, geographically dispersed resources, low bulk density, relatively high transportation costs, and exposure to uncertainty in supply, demand, and environmental conditions. In large-scale bioenergy projects, the biomass supply chain is a critical element, and in many cases, it becomes a major barrier when seeking initial investment for developing energy-crop systems.

For these reasons, efficient logistics planning and supply chain models are needed to maintain stable feedstock availability, control total costs, and reduce environmental impacts. Decision-making is required at strategic, tactical, and operational levels to coordinate activities across the supply chain. These decisions should be supported by modelling tools that can represent system complexity and manage uncertainty. Addressing these supply chain issues is important for improving the competitiveness of bioenergy systems and strengthening their contribution to sustainable development and global energy security.

2.2 UNDERSTANDING BIOMASS, BIOENERGY, AND BIOREFINERY

Biomass has strong potential as a renewable resource for producing different forms of energy, such as electricity, heat, and fuels, through several conversion systems (Roudneshin & Sosa 2024). In general, biomass refers to organic material originating from living organisms. It includes both living and dead biological matter located above or below ground. The biomass resource base consists of many plant species from terrestrial and aquatic ecosystems, and it also includes byproducts and residues from agriculture, forestry, and industry, including animal waste. Examples of biomass resources are wood, Napier grass, rapeseed, water hyacinth, seaweed, chlorella, sawdust, straw, rice husks, kitchen waste, and animal byproducts.

Bioenergy is energy produced from biomass and mainly comes from biodegradable materials. At present, traditional biomass sources, such as firewood, charcoal, and animal waste, are still widely used for bioenergy. In many regions, especially in developing countries, traditional biomass remains a major energy source for domestic activities (Danso-Boateng & Achaw 2022). Although this approach is relatively simple, it often has low efficiency and can create negative impacts on human health and the environment. Therefore, newer bioenergy production methods are increasingly developed using cleaner and more efficient technologies, including pyrolysis, gasification, direct combustion, and anaerobic fermentation. Through these conversion routes, biomass can be transformed into solid fuels (e.g. firewood, wood chips, pellets, charcoal, and briquettes), liquid fuels (e.g. bioethanol, biodiesel, and bio-oil), gaseous fuels (e.g. biogas, syngas, and hydrogen), or directly into heat as part of the production process (Igwebuike et al. 2024). These technologies can be applied at the household level, small enterprises, and large industrial facilities. Bioenergy development can support energy diversification and improve national energy security. In addition, it is often considered an important strategy to reduce greenhouse gas emissions, support a circular economy, and promote more sustainable economic growth based on local resources (Danso-Boateng & Achaw 2022).

An early and widely cited definition from the U.S. National Renewable Energy Laboratory (NREL) dates to 1990 and describes a "biorefinery" as the process of using biomass to produce fuels and various bioproducts. In this concept, a biorefinery is an integrated system that links feedstock production, biomass processing, and end use, in a way that is comparable to petroleum refining. However, instead of using crude oil, it uses biomass to produce multiple high-value outputs, including fuels, basic chemicals, materials, and renewable energy (Ahmed & Makkawi 2025). In practice, biorefineries may combine several types of operations, such as mechanical processes (e.g. grinding and extraction), thermochemical conversion (e.g. combustion, gasification, and thermal decomposition), chemical processing (e.g. hydrolysis and transesterification for biodiesel), and biochemical pathways (e.g. fermentation). These operations can be integrated in one facility to produce final products and energy simultaneously.

2.3 ADVANCEMENT IN BIOENERGY PRODUCTION

The bioenergy production process is still developing until today. Existing pathways continue to change in response to technological progress and the needs of different stakeholders. The development of bioenergy generation shows a more complex approach over time, not only in terms of feedstock selection but also in processing technologies and sustainability targets. Table 2.1 compares bioenergy in the first, second, and third generations, adapted and elaborated from Igwebuike et al. (2024).

TABLE 2.1
Evolution of Bioenergy Production

	Bioenergy Form		
Criteria	**First Generation**	**Second Generation**	**Third Generation**
Type of biomass	- Food material	- Non-food materials, industrial by-products, crop and forest waste, and non-food energy crops	- Aquatic materials (microalgae, seaweed, cyanobacteria, etc.)
Feedstock properties	- Homogeneous	- Various	- Homogenous
Availability	- Available in large quantities in certain areas, such as corn in the United States	- Significant advantages in reducing greenhouse gas emissions	
Technology	- Well-established, cost-effective, user-friendly, and serving as an exemplary illustration of enzyme technology	- Relatively novel and not yet fully established technologies, offering excellent opportunities for cost reduction - Examples include bioprocessing methods such as pyrolysis, Fischer-Tropsch synthesis, and various catalytic processes	
Nature of technology	- Fixed processing	- Flexible processing	

(Continued)

TABLE 2.1 (Continued)
Evolution of Bioenergy Production

	Bioenergy Form		
Criteria	**First Generation**	**Second Generation**	**Third Generation**
Conversion rate	- Low conversion rate	- Cheaper	
Carbon emissions generated	- Emission presence from harvesting, machinery, fertilizer, and distribution	- Minor emissions	- Minor emissions
Products	- Biodiesel, bioethanol, and biogas	- Facilitates production of more value-added products such as heat, gas, electricity, and biofuels - Cellulosic ethanol and Fischer-Tropsch fuels - Biochemicals include gums, resins, rubber, waxes, terpenes, terpenoids, steroids, triglycerides, tannins, plant acids, and alkaloids	- Biodiesel
Process and manufacturing development costs	- Costly option for energy development	- Research is still ongoing, but not available at a commercial scale	
Conclusion	- Competition for food, water, and land use renders the system unsustainable - Raw materials cannot be stored for long periods	- There is potential for sustainable energy and chemical production - However, there are techno-economic issues, including the variation of raw materials, high costs for intermediate processes, and other operational challenges in each type of process	

2.3.1 First-Generation Bioenergy

First-generation bioenergy uses established technologies and mainly relies on food-based crops. Bioethanol is commonly produced by fermenting starch or sugar crops, such as cassava, sugarcane, sugar beet, and corn. Biodiesel is produced from vegetable oils and animal fats, for example, soybean oil, rapeseed, and sunflower oil.

In general, the process involves enzymatic digestion to convert starch into sugars, followed by fermentation. However, many researchers argue that first-generation bioenergy can compete with the food supply and may threaten food security. In addition, the large use of land and fertilizers can create serious environmental impacts. Although this technology is relatively affordable and widely applied, its long-term sustainability is still questioned because of social and ecological concerns (Awudu & Zhang 2012; Igwebuike et al. 2024). These issues encouraged the development of second-generation bioenergy based on lignocellulosic resources.

2.3.2 Second-Generation Bioenergy

Second-generation bioenergy aims to reduce the limitations of the first generation by using non-food biomass that contains high lignocellulose content. Typical feedstocks include straw, rice husks, empty fruit bunches from oil palm, sugarcane bagasse, agricultural residues, forestry residues, and energy grasses. The process usually starts with pretreatment to disrupt the lignin-cellulose structure. After that, biochemical processes, such as enzymatic hydrolysis and fermentation, can be applied to produce bioethanol. In addition, thermochemical routes like gasification can be used, and downstream synthesis pathways (e.g. Fischer-Tropsch) can convert syngas into liquid fuels. The main advantage of this generation is that it reduces direct competition with food crops and can utilize abundant waste materials. However, important challenges remain, including low process efficiency, variation in feedstock quality, and high pretreatment costs. Even though many pilot projects have been developed, large-scale commercialization is still limited due to weak economic feasibility. Nevertheless, this generation is promising for biomass-rich countries, including those in Southeast Asia (Igwebuike et al. 2024).

2.3.3 Third-Generation Bioenergy

Third-generation bioenergy mainly uses microalgae and marine algae as feedstocks. These organisms can grow quickly, do not require arable land, and may accumulate high lipid content, which is useful for biodiesel production. Some studies also discuss the potential of algae to support low-carbon pathways because algae can absorb CO_2 during growth, although the overall climate performance depends on cultivation and processing conditions. However, algae-based bioenergy is still in a developing stage, especially regarding efficient mass cultivation, harvesting, oil extraction, and cost reduction. Algae can offer a higher yield per ha compared with many land plants, and the byproducts of algae cultivation may also be used for fertilizers or chemical products. Although commercialization is still limited, third-generation bioenergy is considered important for future development because it can reduce land-use conflicts and may provide better ecological performance if technological barriers can be addressed (Kazmi et al. 2025).

While Table 2.1 explains the differences in bioenergy development, Habibi et al. (2023) emphasize that the three generations also share several common objectives. These similarities include:

1. Improving energy security,
2. Receiving support from agricultural and industrial sectors,

3. Reducing dependence on imported oil,
4. Responding to demand and public expectations,
5. Contributing to lower greenhouse gas emissions.

Based on the comparison of differences and similarities above, bioenergy development can provide both short-term and long-term benefits. Therefore, it is important to define a clear direction and roadmap, especially for advancing second- and third-generation bioenergy.

2.4 OVERVIEW OF THE BIOMASS AND BIOFUEL SUPPLY CHAIN

A supply chain refers to the flow of materials, information, and financial resources from the source to the final user. It typically includes suppliers, producers, distribution centers, and consumers. Supply chain management coordinates these actors so that products can be produced and delivered in the required quantity, at the right time and location, while maintaining quality and service standards and minimizing total costs. Supply chain performance depends on the coordination and integration among entities, supported by effective product and information flows, digital technologies, and cross-functional collaboration (Anwar et al. 2025).

The biomass supply chain specifically manages bioenergy production activities, starting from biomass collection and continuing until biomass reaches the conversion facilities, from harvesting activities to the final manufacturing stage (Achillas & Bochtis 2021). This supply chain structure includes biomass suppliers (from single or multiple sources), storage facilities (at one or more intermediate points), and biorefinery sites (one or more locations). It also involves preliminary processing stages and transportation using one or more transport modes (Roudneshin & Sosa 2024). In general, the biomass supply chain can be viewed as two related processes:

1. Production planning and control, which includes activities, such as planting, harvesting, and initial processing,
2. Distribution and logistics, which covers storage, transportation, and delivery of biomass.

Habibi et al. (2023) proposed a conceptual model describing key operations in the biomass supply chain. This model is illustrated in Figure 2.1. In this model, the biomass supply chain is divided into three main segments: upstream, midstream, and downstream. The upstream segment covers the full process from biomass production until delivery to the conversion facility. The conversion stage is complex because biomass is used as the input, while outputs include bioenergy and other co-products. Pretreatment is important to reduce moisture and prepare biomass through methods such as baling and pelletization, which can improve bulk density and handling. Because biomass production follows seasonal calendars, storage is often needed to match supply availability with conversion plans. Storage may occur in fields or forests (e.g. as piles), on farms, in centralized storage sites, or at conversion facilities. The midstream segment refers to the conversion process inside the biorefinery, while the downstream segment focuses on storing bioenergy products and distributing them to consumers.

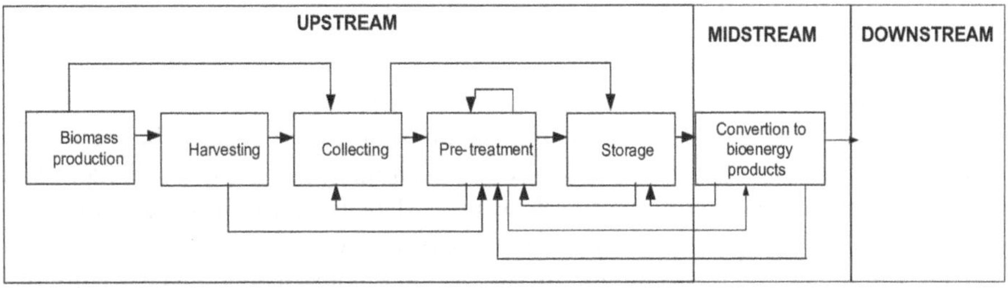

FIGURE 2.1 Biomass supply chain network.

(Adapted from De Meyer et al. 2015.)

In analyzing the streams within the supply chain, it is important to note that the biomass supply chain mainly focuses on transporting biomass from agricultural fields to conversion facilities (Vazifeh et al. 2023; Roudneshin & Sosa 2024). In this sense, it can be viewed as a specific stream within the broader biofuel supply chain, which continues downstream until the final products reach consumers and markets. Habibi et al. (2023) explain that the biofuel supply chain generally consists of five segments: (1) feedstock production, (2) biomass logistics, (3) biofuel production, (4) biofuel distribution, and (5) biofuel end users. To ensure the flow across these segments, several operational activities are required, including feedstock selection, harvesting and initial treatment, conversion processes, and transportation (Tumuluru 2024). Recent studies highlight that connecting upstream biomass activities with downstream stages, such as distribution and consumption, is essential for building sustainable renewable energy systems. Therefore, modern biofuel supply chains require efficient logistics and synchronized production-distribution strategies that are aligned with market demand and environmental sustainability, and should be managed as an integrated system (Talitha et al. 2023).

Agustina et al. (2024) proposed a palm oil-based biodiesel supply chain network. Using a network perspective helps to improve understanding of system interactions and supports identification of current operating conditions, including critical and vulnerable activities. In their model, the biodiesel supply chain is divided into three streams based on network topology: upstream, midstream, and downstream. The upstream network covers the procurement and delivery of fresh fruit bunches (FFB) from multiple oil palm plantations to palm oil mills. At the mills, FFB are processed into crude palm oil (CPO), which becomes the main feedstock for biodiesel production. The midstream network links CPO supply with biorefineries that produce biodiesel. In the biorefinery, CPO is converted into biodiesel (fatty acid methyl ester-FAME) through esterification and/or transesterification processes. The downstream network then distributes biodiesel to consumers through dispensing stations. Before final distribution, biodiesel is typically blended with petroleum diesel at a blending station. In the Indonesian context, a common policy blend is B30, which contains 30% biodiesel and 70% diesel fuel.

The operational activities in the biomass supply chain are as follows.

2.4.1 Selection of Feedstocks

Numerous factors affect the selection of feedstocks for bioenergy production. In general, an ideal biomass feedstock has several desirable characteristics. These include fast growth rate, low water requirement, low input of nutrients and other resources, resistance to diseases and pests, ability to grow on marginal or poor soils, and high biomass yield. In addition, higher bulk or mass density can improve logistics efficiency and support more stable conversion performance.

Lignocellulosic biomass mainly consists of three components, namely cellulose, hemicellulose, and lignin. Cellulose and hemicellulose can be converted into fermentable sugars, while lignin is commonly used as an energy source to generate heat and electricity (Roudneshin & Sosa 2024). Ng and Maravelias (2017) classify lignocellulosic biomass feedstocks into four categories:

a. Energy crops or perennial grasses like miscanthus and weeds,
b. Agricultural waste or crop residues, such as straw, bagasse, husks, corn cobs, and many more,
c. Wood waste, including short-rotation woody plants (e.g. *Populus, Salix*, and *Eucalyptus*), milling residues, and logging, and
d. Urban waste.

The selection of suitable feedstocks and their associated processes is crucial, as a significant portion of bioenergy costs depends on the price of the raw materials. Therefore, the selection of raw materials is strongly influenced by local conditions and availability, potentially involving a combination of different raw materials that possess similar qualities. As a result, a diverse range of suppliers may offer biomass, which will ultimately enhance the system's complexity.

Besides classification based on chemical composition and physical form, biomass can be grouped by its origin, such as forest and non-forest biomass. Duca and Toscano (2022) highlight that this classification is important for designing supply chains and supporting sustainable biomass management strategies. Non-forest biomass typically comes from energy crops and agricultural residues, whereas forest biomass originates from roundwood, logging residues, wood chips, and sawmilling byproducts, such as sawdust and shavings. This classification helps in selecting suitable conversion technologies, assessing regional resource potential, and identifying effective logistics strategies. Table 2.2 presents the classification of biomass based on its origin.

In several European countries, such as Germany and Denmark, research shows that miscanthus cultivation can be effective on marginal land. Different miscanthus varieties were reported to produce around 17–30 tons of dry matter per ha per year on less fertile land in Europe (Awty-Carroll et al. 2023). In the United States, switchgrass (*Panicum virgatum*) is widely studied as a promising biomass source because it can adapt to different crop rotations and can provide relatively stable yields. In Southeast Asia, biofuel production from palm oil and castor oil has increased in recent years. However, evidence from Myanmar suggests that the expansion of castor oil plants as an energy-crop feedstock created negative impacts.

TABLE 2.2
Types of Biomass Based on Sources

Forest	Non-Forest	
Bark	Corn stover	*Juniperus virginiana*
Saw dust	Switchgrass	Cotton residue
Woody	Sorghum straw	Oil palm
Roundwood pulpwood	*Arundo donax*	Pineapple
Wood chips	Bagasse and cane residue	Rapeseed
Commercial wood	Miscanthus	Mesotrophic grassland
Shavings	Poplar	Willow
Roadside logging debris	Rice straw	Tall herb
Logging residuals	Agave residue	Corn silage
Thinning residues	Barley straw	Thistle oil and residues
Thinnings	Jatropha	
Prunings	Wheat straw	
Grasses	Pecan nut residue	
Chips/shavings	Safflower	

In particular, the use of fertile land for energy crops instead of food crops was associated with higher food insecurity in some rural areas. As a consequence, castor-based biofuel feedstock is considered less efficient in this context and has been reduced or discontinued.

Within Southeast Asia, biofuel production is strongly dependent on palm oil, and Indonesia is the leading producer of this commodity. In the period 2022–2023, Indonesia's CPO production reached 45.5 million tons, with an average oil yield of about 5,000 L/ha/year, which is substantially higher than many alternative oil crops (Nugrahapsari et al. 2024). Palm oil is an important raw material for both cooking oil and biodiesel, and its demand continues to increase. Afriyanti et al. (2016) assessed scenarios for Indonesia's palm oil production up to 2050 and suggested that Indonesia could supply a large share of global and Asian demand (39%–60%) using currently available land, without requiring deforestation.

Palm oil biomass also has a relatively high calorific value and shows good potential for thermochemical conversion, such as pyrolysis and gasification. These processes can produce biochar, bio-oil, and synthesis gas, although the output composition depends on the specific biomass type and operating conditions (Parthasarathy et al. 2024). In oil palm cultivation and processing, only about 10% of the total biomass is converted into palm oil, while the remaining 90% becomes biomass residues. Typically, around five tons of FFB produce one ton of crude oil (Awoh et al. 2023). Major palm oil biomass residues include oil palm trunks, oil palm fronds, empty fruit bunches, mesocarp fiber, and palm kernel shells. Nabizadeh et al. (2025) summarized possible products from palm oil biomass conversion, and these outputs are presented in Table 2.3.

TABLE 2.3
Value-Added Products from Palm Oil Biorefinery

Products and Conversion Outputs
Bio-ethanol, biodiesel, biogas, electricity, briquets, pellets, dried long fiber, short fiber, gasification, compost, steam, refined palm, charcoal, stearin/olein/fatty acid distillate, bioplastics, biochar, bioadsorbent, biocomposite, biovanillin, bio-butanol, glycerol, methanol, hydrogen, higher alcohols, glycerin, and aromatics.

Source: Garcia-Nunez et al. (2016).

2.4.2 Harvesting and Preliminary Processing of Biomass

Harvesting activities can only be carried out within a limited time period because biomass availability is strongly influenced by seasonal patterns and cropping calendars. In most cases, freshly harvested biomass cannot be directly converted into bioenergy efficiently without additional processing. Therefore, pre-treatment is required to improve biomass properties and enhance conversion performance. Common objectives of biomass pre-treatment include reducing moisture content and particle size, increasing bulk or energy density, improving accessibility of carbohydrates by disrupting lignin structures, and facilitating the conversion of hemicellulose and cellulose into fermentable sugars. Roudneshin and Sosa (2024) classify biomass pretreatment into three main categories:

a. Physical treatment, such as mechanical size reduction, drying, and densification,
b. Physicochemical treatment, such as ammonia fiber explosion, liquid hot water treatment, and steam explosion (including SO_2-assisted steam explosion),
c. Chemical treatment, such as acid treatment, alkaline treatment, and organosolv processes.

Harvesting is a major contributor to the total cost of delivered biomass. The harvesting cost depends on the specific method and equipment used, as well as local field conditions. Helal et al. (2023) reported that harvesting and pretreatment expenses can represent around 30%–45% of total biomass supply chain costs. This relatively high proportion is largely explained by the influence of biomass energy density and transport distance on collection, drying, and transportation requirements. When biomass has high moisture content and low density, transportation becomes less efficient and more costly. For this reason, to improve cost efficiency, pretreatment facilities are often located close to biomass sources. This strategy reduces the need to transport biomass in its high-moisture, low-density form over long distances and can improve overall supply chain performance.

2.4.3 Biomass Processing

After initial processing, biomass can be prepared for conversion into bioenergy, including heat, electricity, or biofuels. Various conversion pathways are available, and the selected processing approach is usually adapted to the characteristics of the specific biomass type, in order to improve conversion efficiency. The choice of a suitable processing method depends strongly on biomass chemical composition and key properties, such as heating value, moisture content, ash content, alkali metal concentration, carbon-to-nitrogen ratio, the polysaccharide-to-lignin ratio, oil content, and volatile matter (Beringer et al. 2011). For example, biomass with relatively high lignin and low ash content is generally more suitable for thermochemical routes, such as pyrolysis or gasification. In contrast, biomass that contains high proportions of polysaccharides is often more appropriate for biochemical pathways, including enzymatic hydrolysis and fermentation (Pahnila et al. 2023). Hydrothermal liquefaction is commonly discussed as a promising option for wet biomass because it can process high-moisture feedstocks directly, thereby reducing the energy needed for drying and potentially improving overall efficiency without reducing product quality. Furthermore, reactor design and operating conditions should consider the thermal decomposition behavior of major lignocellulosic components. Hemicellulose typically decomposes at around 150°C–200°C, cellulose at approximately 240°C–350°C, and lignin over a wider range of about 300°C–500°C. These temperature ranges are important for optimizing yields and energy content of products, such as biochar, bio-oil, and syngas in thermochemical systems.

The concept of a lignocellulosic biorefinery emphasizes using each biomass component to produce bioenergy in multiple forms as well as other value-added products. This approach encourages the integration of different biomass processing technologies, in order to maximize total biomass utilization and minimize waste, often by applying substrate-specific processing methods (Roudneshin & Sosa 2024). In general, biomass processing techniques can be grouped into three broad categories: thermochemical, chemical, and biochemical methods. Thermochemical processes convert biomass into solid, liquid, and gaseous fuels. Biochemical conversion uses biological mechanisms, such as enzymatic reactions and fermentation, while chemical conversion produces fuels through different chemical routes, including processes for upgrading biomass-derived intermediates (Shahzad et al. 2024).

2.4.4 End-Users

At the downstream stage, the main activities involve storing and distributing biofuels to end users. Current biofuel supply chain management gives strong attention to inventory optimization at depots, appropriate storage conditions (e.g. temperature, pressure, and humidity), and protection of fuel quality to prevent degradation during storage and handling. Distribution to consumers is generally conducted through coordinated systems, such as fuel terminals, distribution networks, and in some cases pipelines. When these downstream operations are integrated, the supply chain can reduce logistics costs and potentially lower emissions associated with transportation. In addition, downstream planning is closely related to supply chain resilience, because robust storage and distribution arrangements help maintain a stable biofuel supply under changing operational conditions and disruptions (Habibi et al. 2023).

2.5 BIOMASS SUPPLY CHAIN COMPARED TO INDUSTRIAL SUPPLY CHAIN

When assessing bioenergy production, it is important to apply a systems perspective that clearly explains the relationships among biomass resources, supply systems, conversion technologies, and the energy services provided. In practice, integrating these components can create different levels of system complexity. The design of a bioenergy production system therefore requires optimization of the supply chain or network structure according to the specific conditions of the production context, such as climate, topography, available raw materials, and technology. Effective biomass harvesting also depends on efficient storage and transportation, supported by supply chain and operations management as well as the use of appropriate technologies. For this reason, supply chain and logistics issues are strongly linked to the feasibility and performance of bioenergy production systems. In general, supply chain management aims to identify the most suitable supply chain configuration by planning and coordinating key activities, including sourcing, procurement, conversion processes, and logistics.

A comparison between biomass supply chains and conventional industrial supply chains indicates that biomass systems require a more comprehensive approach. The reason being that they are influenced by spatial dispersion, seasonal variability, and feedstock characteristics. Biomass supply chains typically include cultivation, harvesting, pretreatment, transportation, handling, and storage. These activities usually end at the biorefinery or conversion plant gate, although in some cases they can extend further to downstream distribution for end-users (Achillas & Bochtis 2021). In many settings, the biomass supply area is large and geographically dispersed, and planning is carried out over long time horizons. In addition, differences between input and output can occur due to natural degradation during storage and the limited time window for harvesting, which makes synchronization difficult. This situation differs from industrial supply chains, which are often more synchronized and structured across long-, medium-, and short-term planning levels and typically generate higher added value along the chain (Achillas & Bochtis 2021; Balanay & Halog 2024). These characteristics emphasize the need for bioenergy supply chain design to be adapted to local conditions, including climate, feedstock availability, and distributed technologies, in order to improve efficiency and sustainability (Achillas & Bochtis 2021; Balanay & Halog 2024). Table 2.4 presents key differences between biomass supply chains and industrial supply chains in terms of activities, spatial coverage, planning, and operational characteristics.

Despite several fundamental differences, both industrial and biomass supply chains can be represented as networks (graphs), where nodes indicate locations and arcs describe the flow of products between these locations. In both systems, supply chain performance depends on having a stable, cost-efficient, and reliable supply of raw materials with acceptable quality each year. Logistics is important in enabling the movement of biomass from production areas to processing facilities. Although different transportation modes can be used, in practice there are often constraints related to the availability of fleets and trip frequency. These constraints are influenced by vehicle range, travel distance, and regulations on driving time. For sites with limited access, such as forest locations, land transport is often the only feasible option, and trucks are commonly preferred because they provide higher flexibility.

TABLE 2.4
Biomass and Industrial Supply Chains

Criteria	Biomass Supply Chain	Industrial Supply Chain
Activity	- The typical stages encompass planting, harvesting, preprocessing, transportation, handling, and storage - These steps generally conclude at the beginning of a conversion unit, often a biorefinery that produces biofuels and chemical products or a bioenergy facility that generates electricity, heat, and cooling - Several papers also refer to a distribution stage that delivers products from the conversion plant to the end-user	- Procurement of materials and services, manufacturing or production, and distribution to customers
Region	- Widespread	- It depends on the type of product
Planning	- Long term - The input (biomass production) and output (conversion process) are not aligned, requiring timely harvesting of crops once they reach readiness due to degradation	- Long, medium, and short-term
Characteristics	- Availability varies by season - It has a limited time frame influenced by harvest seasons, weather conditions, and the intention to replant - Low-density raw materials	- The formation of supply chain networks, processes for enhancing value-added services, and management components are all critical aspects

Source: Rentizelas et al. (2009).

2.6 DECISION-MAKING IN THE BIOMASS SUPPLY CHAIN

In biomass supply chain design and planning, decision-making is typically complex and hierarchical. Biomass supply chains face uncertainty in both supply and demand. This uncertainty is mainly caused by the characteristics of biomass itself, such as its vulnerability to degradation, large volume and low density, and strong seasonal availability.

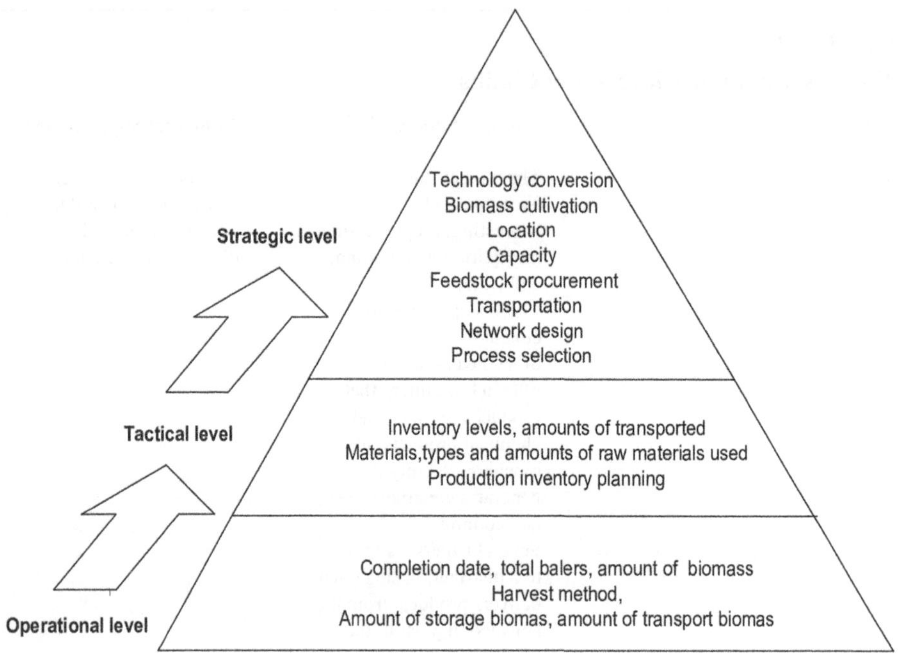

FIGURE 2.2 Hierarchy of decision-making in the biomass supply chain.

(Adapted from Agustina et al. 2018.)

Therefore, biomass supply chain planning should include all major activities, starting from sourcing and cultivation and continuing to harvesting, pretreatment, storage, transportation, and conversion into energy or bioproducts. This planning is commonly supported by an integrated supply chain network model. Such a model can connect strategic decisions (e.g. facility location and capacity), tactical decisions (e.g. harvesting rights, delivery planning, and contract arrangements), and operational decisions (e.g. daily transport scheduling and inventory control). In addition, seasonal supply patterns and demand fluctuations should be considered, in order to improve sustainability and cost efficiency (Helal et al. 2023).

More broadly, biomass supply chain planning covers all activities related to biomass supply and conversion processes (Shahzad et al. 2024). In supply chain management, these decisions are usually grouped into three levels: strategic decisions (long-term, typically five years or more), tactical decisions (medium-term, commonly six to twelve months), and operational decisions (short-term, weekly, or daily). Figure 2.2 presents the decision-making hierarchy applied in the design and planning of biomass supply chains.

2.6.1 Strategic Decision-Making

Strategic-level decisions have a major influence on long-term supply chain performance (typically five years or more) and are often difficult to change once implemented. These decisions usually involve multiple stakeholders, especially actors responsible for financing, regulating, or developing bioenergy production systems. At this level, large capital investments are commonly needed to build infrastructure and establish supply chain networks (Habibi et al. 2023). Strategic planning is also important for designing an efficient supply chain that aligns with organizational objectives and strengthens competitiveness through appropriate biomass supply chain configurations (Roudneshin & Sosa 2024). Typical strategic decisions in biomass supply chains include selecting conversion technologies, determining planting locations and supply capacities, designing the supply chain network, selecting major processing pathways, and planning the overall transportation structure.

2.6.2 Tactical Decision-Making

At the tactical level, decision-making focuses on translating strategic plans into medium-term actions and evaluating activities through cost-benefit considerations. This level aims to ensure that biomass availability is sufficient to meet production plans and scheduling requirements. Tactical decisions may include aggregate production planning, inventory policies, fleet allocation, and the selection of collection, storage, pretreatment, and transportation strategies.

2.6.3 Operational Decision-Making

At the operational level, decisions are made on a short-term basis, often daily. These decisions relate to immediate execution activities, such as determining the required biomass quantity, scheduling harvesting operations and methods, managing biomass storage, and planning vehicle routing and scheduling to ensure timely delivery of biomass to conversion facilities.

2.7 KEY ISSUES IN BIOMASS LOGISTICS AND SUPPLY CHAIN

2.7.1 Uncertainty in Biomass Supply Chains

2.7.1.1 Understanding Biomass Supply Chain Uncertainty

The characteristics of biomass create several sources of uncertainty that must be managed in the biomass supply chain. Wang et al. (2014) argue that uncertainty is a complex and interrelated phenomenon and can appear at any stage of the supply chain. In a similar view, van der Vorst and Beulens (2002) describe supply chain uncertainty as a condition, where decision makers face difficulties because goals are not clear, information about the supply chain or its environment is incomplete, data processing capability is limited, the impacts of decisions are hard to predict, or effective control mechanisms are weak.

Childerhouse and Towill (2004) introduced the uncertainty circle, which represents major uncertainty sources as segments of a circle, namely, process, supply, demand, and control. This concept emphasizes the importance of understanding both the causes and consequences of uncertainty, including internal and external relationships and flows across upstream and downstream parts of the supply chain. Figure 2.3

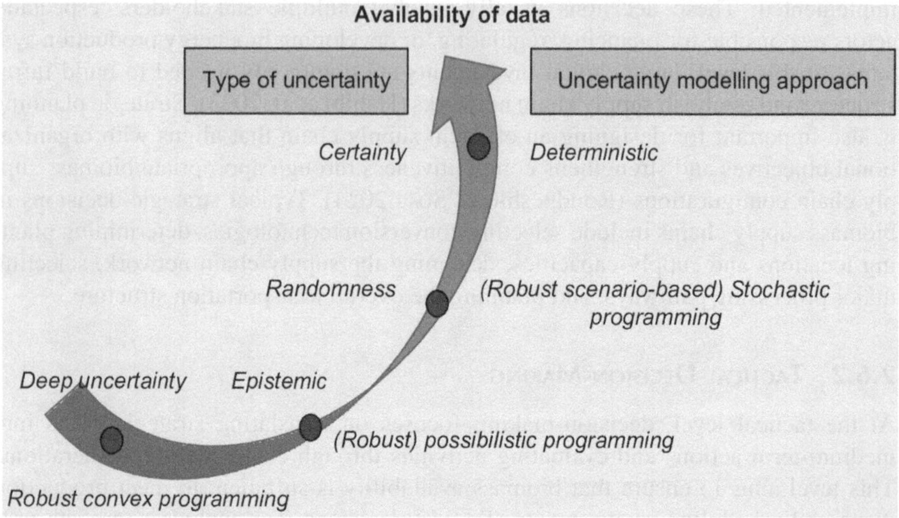

FIGURE 2.3 Taxonomy of uncertainty and modelling approaches.

(Adapted from Bairamzadeh et al. 2018.)

summarizes common types of uncertainty and the modelling approaches that are typically used to address them.

The element of random uncertainty requires sufficient and reliable historical data to determine its probability distribution. In modelling, the uncertain inputs are treated as stochastic parameters, which may be continuous or discrete, depending on the case. Much of the existing literature focuses on this type of uncertainty. In contrast, epistemic uncertainty reflects limited knowledge about input data and is often represented using qualitative descriptions or expert judgement. This type of uncertainty is commonly modelled using fuzzy sets or interval-based representations. For example, Mohagheghi et al. (2024) developed a supply chain network model that considers epistemic uncertainty by using fuzzy parameters for biomass harvest levels and bioethanol demand. Finally, deep uncertainty refers to conditions, where probability distributions cannot be determined because key information is missing or highly contested. In such cases, modelling approaches often combine different methods (e.g. robust approaches alongside stochastic representations) to support tactical decision-making under uncertainty, including uncertainty related to biomass quality and availability.

2.7.1.2 Sources of Uncertainty

Awudu and Zhang (2012) emphasize that optimal performance in biomass supply chain management depends on incorporating uncertainty into decision-making. For this reason, identifying the main sources of uncertainty in biomass supply chains is an essential step. In general, uncertainty in biomass systems can be grouped into five categories: (1) supply, (2) transportation and logistics, (3) production and operations, (4) demand and pricing, and (5) other uncertainties.

a. Biomass supply uncertainty: A large share of uncertainty in biomass supply chains originates from variability in biomass productivity, quality, harvest rate, land rental, and overall availability. This uncertainty may arise from (1) changes in feedstock quality and location, (2) uncertain weather conditions, crop pests and diseases, as well as farmers' production decisions, (3) limited land resources, (4) differences in crop management practices, (5) supply disruptions caused by unforeseen events, and (6) regional variation in key resources, such as water availability and CO_2 exposure. To reduce supply uncertainty, common strategies include maintaining safety stock and developing supply relationships with multiple suppliers.
b. Transportation and logistics: Transportation and logistics uncertainty refer to difficulties in delivering biomass efficiently and at a reasonable cost. This uncertainty is influenced by factors, such as transportation distance, capacity limitations of transport modes, and variability in travel time. As a result, logistics uncertainty can lead to higher shipping costs, inconsistent feedstock availability, and increased inventory and warehousing costs.
c. Production and operations: Production and operational uncertainty create challenges in meeting target output levels and production schedules. This uncertainty is often linked to equipment reliability, process performance, and conversion efficiency, especially, when new bioenergy technologies are adopted. Such conditions may cause production delays, higher downtime, and reduced operational stability.
d. Demand and price: Demand uncertainty refers to unpredictable changes in the timing and quantity of market demand. Price uncertainty refers to the risk that product prices will change due to market fluctuations. Key drivers include: (1) macroeconomic instability, (2) changes in policy and regulation, (3) the relationship between bioenergy prices and fossil fuel prices (gasoline and diesel), (4) the influence of government incentives, and (5) competing demand from other sectors (e.g. livestock feed or industrial uses).
e. Other uncertainties: In addition to the categories above, biomass supply chains may experience uncertainty due to unexpected disruptions and environmental issues. These may include disruptions from extreme events, limitations in environmental impact assessment methods (e.g. uncertainty in greenhouse gas emission estimates), regulatory and policy changes, and climate-related variability.

2.7.1.3 Modelling Uncertainty in Biomass Supply Chain

Biomass supply chain involves uncertainty across all segments, including upstream, midstream, and downstream stages. To reduce the negative impacts of this uncertainty, it is necessary to incorporate uncertainty explicitly into decision-making. Biomass supply chain modelling studies commonly aim to (1) predict system performance, (2) identify the most efficient supply chain configuration, (3) optimize component sizes to reduce total costs, or (4) optimize the scheduling of supply chain activities (Stellingwerf et al. 2022).

Chen et al. (2023) classify uncertainty modelling approaches into four main categories: (1) analytical methods, (2) artificial intelligence models, (3) simulation models, and (4) hybrid methods. Among these approaches, analytical methods are most frequently applied to represent uncertainty in supply chain management, mainly because they provide structured optimization frameworks and transparent decision rules.

a. Analytical methods: Analytical methods address uncertainty-related problems using formal mathematical and optimization techniques. Common approaches include stochastic mixed-integer linear programming (SMILP), integer stochastic programming (ISP), stochastic mixed-integer nonlinear programming (SMINLP), Markov decision processes (MDP), and linear programming with scenario generation (LP-SG) (Awudu & Zhang 2012). Stochastic programming, especially the two-stage stochastic model is discussed by Mohagheghi (2024). In this approach, the first stage represents decisions that must be made before uncertainty is revealed, while the second stage includes recourse decisions made after additional information becomes available. Robust optimization is applied by Roudneshin and Sosa (2024) to support decision-making under uncertain parameters without requiring full probability distributions. In addition, fuzzy programming is used to represent epistemic uncertainty, when input values are imprecise or based on expert judgement (Stellingwerf et al. 2022). Quantitative performance metrics commonly used in analytical models are summarized in Table 2.5.

TABLE 2.5
Performance Measures for Quantitative Analysis in Biomass Supply Chain Models

Minimizing	Maximizing
- Expected total cost/annual cost	- Expected total profit/annual profit
- The expected value of risk	- Expected annual income
- Greenhouse gas emissions/CO_2/ environmental impact	- Expected net present value
- Annual transportation cost	- Number of generated job opportunities/social activities
- The shortest distance	- Total social welfare
- Inter-annual variations in harvest	- Return on investment
- Greenhouse gas emission savings	- Carbon fixation
- The land use change (social) of the fuel sector	- The service level
- Investment cost	
- Annual purchasing cost	

TABLE 2.6
Advantages and Disadvantages of Simulation Methods for Biomass Supply Chains

Advantages	Disadvantages
a. Simulation provides high modeling flexibility (flows, events, priorities, waiting times, resource conflicts, etc.)	a. It necessitates a thorough explanation of the chain, including its process, flow, and events
b. Random generation manages stochastic events	b. Takes a considerable amount of time for large supply chains
c. Large and complex supply chains can be modeled along with their dynamics	c. Strategic decisions like chain structure and facility location are the only ones for which simulation is functional
d. Simulation is particularly suitable for operational and tactical levels	d. The results obtained are not optimal
e. Models are relatively straightforward to understand and can be adjusted by skilled users	e. Model errors are challenging to identify (users always receive indicators at the end)

Source: Ba et al. (2016).

b. Simulation: Simulation methods are well-suited for systems with complex and dynamic structures, especially when performance indicators and operational events are random. For this reason, simulation is widely used to analyze supply chain problems that involve uncertainty. Simulation approaches can be classified in several ways, for example, static or dynamic, discrete or continuous, and deterministic or stochastic. The selection of a simulation type depends on the characteristics of the system and the decision context being examined.

In biomass supply chain studies, simulation models commonly combine process representation with economic calculations to estimate total costs, energy use, and greenhouse gas emissions. A key advantage of simulation is that it can reproduce detailed activities along the supply chain, including harvesting, transportation, and storage. Therefore, simulation can provide practical support for tactical and operational decision-making, such as evaluating alternative logistics policies, storage strategies, and scheduling options under uncertainty. Several studies, including Stellingwerf et al. (2022), applied Monte Carlo simulation to represent uncertainty and used dynamic simulation to capture changes over time. Table 2.6 summarizes the main advantages and disadvantages of simulation approaches in biomass supply chain modelling.

2.7.2 Storage: When It Is Characterized by Seasonal Availability

Storage is a critical element in biomass logistics because it connects the seasonal availability of biomass feedstocks with the continuous operating requirements of processing facilities, such as biorefineries. For instance, corn stalk that is harvested annually may require storage for up to nine months. In forestry-based industries, including pulp, pellet, and bioenergy production, it is common to maintain on-site biomass inventories for days or weeks to ensure a stable supply and uninterrupted operations (Helal et al. 2023; Roni et al. 2023).

Large-scale storage of granular biomass can create serious fire risks, mainly due to self-heating. Self-heating may occur because of biological activity and internal oxidation reactions, which can raise internal temperatures and, in extreme cases, lead to spontaneous ignition. To reduce this risk, practical control measures, such as ventilation and material agitation, are often recommended, particularly when temperature monitoring indicates that internal heat is approaching unsafe levels (Liu et al. 2017).

Fire hazards during storage are also influenced by environmental conditions, such as humidity, and by the specific biomass type. Some reactive biomass materials can ignite within a short period, even within a few days, if storage conditions are unfavorable. Risk reduction strategies include selecting appropriate storage sites, managing storage capacity, improving stack or pile design to support heat dissipation, and limiting storage duration, especially for moist or highly reactive biomass through effective inventory management practices (Font & Font 2025).

In general, the prevention of biomass self-heating and self-ignition should address both physical and chemical mechanisms. This includes controlling moisture exchange, reducing reactivity, improving heat transfer and dissipation, and using safer biomass forms, such as densified pellets. Moisture control is important because excessive moisture can accelerate biological activity and increase heating. As a result, predictive mathematical models can support planning and operational decisions for safer storage design and management (Helal et al. 2023).

Beyond safety risks, quality variability of biomass feedstocks can create major challenges for storage and logistics. High moisture content, low calorific value, and rapid degradation can reduce fuel quality and increase handling difficulty. Addressing these issues may require more advanced and controlled storage systems, such as dedicated terminals as well as careful coordination of inventory and logistics to minimize quality deterioration during storage (Piqueiro et al. 2023).

Storage challenges also occur for bio-oil produced from pyrolysis. A key limitation is its relatively high oxygen content, often reported around 35%–50%, compared with crude oil. This characteristic contributes to chemical instability and high acidity, which can increase corrosion risks. Therefore, the storage and transport of bio-oil commonly require specialized, corrosion-resistant materials to maintain safety and preserve product quality (Habibi et al. 2023). One important upgrading approach that can improve storage and handling is torrefaction, which thermally treats biomass (approximately 180°C–350°C) and causes partial devolatilization. The treated biomass is commonly densified into pellets, resulting in improved water resistance, lower biological degradation risk, and higher energy density. These improvements can reduce storage losses and support more efficient transportation and handling (Wild & Calderón 2021).

2.7.3 Combining Multiple Biomass Supply Chains

Biomass resources are geographically dispersed, seasonal, heterogeneous in quality, and vulnerable to degradation during collection and storage. These inherent characteristics create major challenges, when different biomass streams, such as agricultural residues, wood residues, and agro-industrial wastes, are integrated into a single supply system. Empirical and modelling studies show that early-stage decisions (e.g. drying or densification) can significantly affect both costs and material quality in later stages, including transportation and conversion. Stellingwerf et al. (2022) note that integrating multiple biomass supply chains introduces a set of operational and planning challenges, which can be summarized as follows:

a. Uncertain and seasonal suppliers: A key challenge is the uncertain availability of biomass feedstocks. Supply is often constrained by harvest seasons and originates from multiple sources, such as farms, plantations, and processing facilities. Although combining supply chains may improve year-round availability, coordination becomes more complex because different feedstocks have different harvesting calendars, collection points, and quality standards. Consequently, many studies recommend multi-level network models and optimization approaches that balance cost efficiency with supply reliability. Recent work suggests that establishing collection centers or biohubs serving multiple feedstock sources can lower the cost per unit of delivered energy. However, such hubs and their supporting information systems typically require substantial investment (Valipour et al. 2024).
b. Material quality and storage losses: After collection, usually biomass must be stored before it can be transported or processed. During storage, biomass can experience dry-matter losses, which increase costs and may raise greenhouse gas emissions. Research indicates that poorly managed storage can increase emissions by approximately 1%–53%, depending on the storage conditions and feedstock type. Moisture is a critical driver because higher moisture content accelerates biological degradation. Therefore, many studies recommend controlling moisture, often targeting levels below around 20%, depending on the material, to preserve quality and reduce losses. Control strategies include covered or closed storage systems, moisture management, and wet ensiling techniques (Wzorek et al. 2024).
c. Preprocessing and densification as strategies to reduce transport costs: Because biomass has relatively low volumetric energy density, transport costs per unit of energy can become high, particularly over long distances. Preprocessing methods such as drying, pelletizing, briquetting, and torrefaction can increase energy density and improve stability during transport. Studies report that torrefaction and densification may reduce transportation cost per unit of energy in specific cases, and pelletization can lower shipping costs due to improved volumetric efficiency compared with loose or long-form biomass. However, these savings must be evaluated against capital investment, operating costs, and energy requirements of preprocessing (Valipour et al. 2024).
d. Network structure: The role of bio-hubs, local terminals, and clustering: Integrating multiple supply chains often leads to alternative network topologies, such as decentralized systems with many collection points and depots, or more centralized systems with fewer but larger bio-hubs. Bio-hubs can aggregate feedstocks, perform preprocessing (e.g. drying, grinding, and densification), and coordinate deliveries to conversion plants. The literature

highlights that hubs may improve economies of scale and supply chain resilience, but hub location decisions must account for average collection distance, road capacity, and land or infrastructure costs. Clustering algorithms and routing heuristics are often applied to reduce transport distance and cost, while limiting emissions (Valipour et al. 2024).

e. Logistics and transportation infrastructure: Transportation mode selection (truck, rail, and barge) strongly affects supply costs and the economically feasible transport distance. Rail and barge are usually more efficient for large volumes and longer distances, but they depend on transshipment facilities and infrastructure access. Trucks offer flexibility for local collection and last-mile delivery. Therefore, modal selection should consider feedstock properties and network design. Studies commonly find that the economic distance threshold for truck transport ranges from tens to hundreds of kilometers, depending on factors, such as feedstock density and fuel prices. Integrating nearby biomass sources and using hubs can shift the cost structure and improve overall logistics performance (Valipour et al. 2024).
f. Resilience, regulation, and sustainability: Integrated biomass supply chains must also address socio-environmental impacts, including land-use change and effects on local communities. Disruptions, such as extreme weather events, pandemics, and price volatility, demonstrate the need for resilient supply chain design. Common mitigation measures include supply diversification, strategic inventory, and long-term contracts. In addition, sustainability and traceability standards are increasingly important to meet regulatory requirements and market expectations (Chidozie et al. 2023).
g. Digitalization and chain coordination (ICT and data): Integration across multiple biomass supply chains requires timely data exchange on variables, such as stock availability, moisture content, storage temperature, vehicle status, scheduling, and contract performance. The use of digital platforms, IoT sensors (e.g. moisture and temperature monitoring in biomass stacks), and real-time scheduling systems can reduce losses and improve operational efficiency. Recent modelling studies also incorporate demand and supply uncertainty using empirical data to reduce operational risks and support more reliable planning (Ogunrewo & Nwulu 2024).

Overall, integrating diverse biomass supply chains offers clear opportunities to stabilize supply and reduce renewable energy costs. Consolidation through bio-hubs and the adoption of densification technologies can improve efficiency and lower the cost per unit of delivered energy. However, successful implementation requires careful management of feedstock quality, adequate infrastructure investment, and strong coordination of transportation and information flows across the network.

2.7.4 Sustainable Biomass Supply Chain

Future research should place stronger emphasis on comprehensive sustainability analysis. Managing a sustainable biomass supply chain requires effective control of all stages, including the flow of materials, information, and financial resources, in order to improve profitability, reduce environmental impacts, and increase social benefits. Sustainability-oriented initiatives may strengthen biomass supply chains by improving their competitiveness and supporting more inclusive participation of stakeholders. Although sustainability is a broad concept, many studies apply the three-pillar framework, which evaluates sustainability based on economic, environmental, and social dimensions. Because of this framework, a large portion of the literature focuses on developing sustainable supply chain models that integrate sustainability indicators across these dimensions.

In addition, incentive programs can encourage more sustainable biomass utilization, particularly for feedstocks that do not compete with food production, such as second-generation biomass. Government policies and regulatory frameworks are important to ensure the sustainability of biomass supply chains, especially when economic trade-offs must be addressed in bioenergy development. For example, policy interventions may determine subsidy levels for raw materials and shape incentive schemes or profit-sharing mechanisms among stakeholders in the supply chain.

Green supply chain management provides a relevant perspective by showing how green technologies and practices may reduce costs and improve operational efficiency. Although green supply chain concepts can be defined in different ways, they generally emphasize reducing environmental impacts throughout the chain. Typical components of green supply chain management include green design, green production, green supply, green logistics, and reverse logistics (Islam et al. 2017). In this context, the biomass supply chain is often considered part of a green supply chain because it aims to use renewable raw materials and adopt technologies that can lower emissions and improve cost-effectiveness. Green logistics focuses on assessing and minimizing the environmental impacts of logistics activities, for example, through optimized transportation networks and better scheduling. Furthermore, reverse logistics is important for strengthening economic, environmental, and social performance by supporting reuse, recycling, remanufacturing, and repair activities. Integrating these circular practices into biomass supply chains is necessary to reduce waste and improve overall sustainability.

REFERENCES

Achillas, C., & D. Bochtis. 2021. Supply chain management for bioenergy and bioresources: Bridging the gap between theory and practice. *Energies* 14: 60977.

Afriyanti, D., C. Kroeze, & A. Saad. 2016. Indonesia palm oil production without deforestation and peat conversion by 2050. *Sci. Total Environ.* 557: 562–570.

Agustina, F., I. Vanany, & N. Siswanto. 2018. Biomass supply chain design, planning and management: A review of literature. *Proceeding of 2018 IEEE International Conference on Industrial Engineering and Engineering Management (IEEM)*: 884–888.

Agustina, F., I. Vanany, & N. Siswanto. 2024. Resilience assessment model for biodiesel supply chain: An Indonesian case study. *Biofuels* 5: 1117–1130.

Ahmed, A., & Y. Makkawi. 2025. The biorefinery paradigm: Technologies, feedstocks, and retrofitting for future sustainable energy. *Energies* 18: 5919.

Anwar, U.A.A., A. Rahayu, L.A. Wibowo, M.A. Sultan, & T. Aspiranti. 2025. Supply chain integration as the implementation of strategic management in improving business performance. *Discov. Sustain.* 6: 101.

Awoh, E.T., J. Kiplagat, S.K. Kimutai, & A.C. Mecha. 2023. Current trends in palm oil waste management: A comparative review of Cameroon and Malaysia. *Heliyon* 9: e21410.

Awty-Carroll, D., E. Magenau, M. Al Hassan, E. Martani, M. Kontek, P. van der Pluijm, C. Ashman, E. de Maupeou, J. McCalmont, G.-J. Petrie, C. Davey, K. van der Cruijsen, V. Jurišić, S. Amaducci, I. Lamy, A. Shepherd, J. Kam, A. Hoogendam, M. Croci, O. Dolstra, A. Ferrarini, I. Lewandowski, L.M. Trindade, A. Kiesel, & J. Clifton-Brown. 2023. Yield performance of 14 novel inter- and intra- species *Miscanthus* hybrids across Europe. *GCB Bioenergy* 154: 399–423.

Awudu, I., & J. Zhang. 2012. Uncertainties and sustainability concepts in biofuel supply chain management: A review. *Renew. Sust. Energ. Rev.* 16: 1359–1368.

Ba, B.H., C. Prins, & C. Prodhon. 2016. Models for optimization and performance evaluation of biomass supply chains: An operations research perspective. *Renew. Energy* 87: 977–989.

Bairamzadeh, S., M. Saidi-Mehrabad, & M.S. Pishvaee. 2018. Modelling different types of uncertainty in biofuel supply network design and planning: A robust optimization approach. *Renew. Energy* 116: 500–517.

Balanay, R., & A. Halog. 2024. Bioenergy updates and prospects for decarbonization in the ASEAN Region: A review on logistical concerns and potential solutions. *WIREs Energy Environ.* 13: e499.

Beringer, T., W. Lucht, & S. Schaphoff. 2011. Bioenergy production potential of global biomass plantations under environmental and agricultural constraints. *GCB Bioenergy* 3: 299–312.

Chen, L., T. Dong, & J. Peng. 2023. Uncertainty analysis and optimization modeling with application to supply chain management: A systematic review. *Mathematics* 11: 2530.

Chidozie, B.C., A.L. Ramos, J.V. Ferreira, & L.P. Ferreira. 2023. Residual agroforestry biomass supply chain simulation insights and directions: A systematic literature review. *Sustainability* 15: 9992.

Childerhouse, P., & D. R. Towill. 2004. Reducing uncertainty in European supply chains. *J. Manuf. Technol. Manag.* 15: 585–598.

Danso-Boateng, E., & O.W. Achaw. 2022. Bioenergy and biofuel production from biomass using thermochemical conversions technologies – A review. *AIMS Energy* 10: 585–647.

De Meyer, A., D. Cattrysse, & J.V. Orshoven. 2015. A generic mathematical model to optimize strategic and tactical decisions in biomass-based supply chains (OPTIMASS). *Eur. J. Oper. Res.* 245: 247–264.

Duca, D., & G. Toscano. 2022. Biomass energy resources: Feedstock quality and bioenergy sustainability. *Resources* 11: 57.

Font, R., & A. Font. 2025. Self-ignition of biomass: Parameters and scaling. *Energy Fuel* 39: 7455–7466.

Garcia-Nunez, J.A., N.E. Ramirez-Contreras, D.T. Rodriguez, E. Silva-Lora, C.S. Frear, C. Stockle, & M. Garcia-Perez. 2016. Evolution of palm oil mills into bio-refineries: Literature review on current and potential uses of residual biomass and effluents. *Resour. Conserv. Recycl.* 110: 99–114.

Habibi, E., R.K. Chakrabortty, & A. Abbasi. 2023. Towards facing uncertainties in biofuel supply chain networks: A systematic literature review. *Environ. Sci. Pollut. Res.* 30: 100360–100390.

Helal, M.A., N. Anderson, Y. Wei, & M. Thompson. 2023. A review of biomass-to-bioenergy supply chain research using bibliometric analysis and visualization. *Energies* 16: 1187.

Igwebuike, C.M., S. Awad, & Y. André. 2024. Renewable energy potential for second-generation biomass feedstock bioethanol production: A review. *Molecules* 29: 1619.

Islam, S., N. Karia, F.B.A. Fauzi, & M. Soliman. 2017. A review on green supply chain aspects and practices. *Manag. Mark.* 12: 12–36.

Kazmi, A., T. Sultana, A. Ali, A. Nijabat, G. Li, & H. Hou. 2025. Innovations in bioethanol production: A comprehensive review of feedstock generations and technology advances. *Energ. Strat. Rev.* 57: 101634.

Liu, Z., S. Wang, & Y. Ouyang. 2017. Reliable biomass supply chain design under feedstock seasonality and probabilistic facility disruptions. *Energies* 10: 1895.

Mohagheghi, O., R. Shiva, & E. Hassannayebi. 2024. Stochastic optimization approaches to biomass-to- bioenergy supply chain network design under demand uncertainty. *Oper. Res. Forum* 6: 137.

Nabizadeh, R., I.L. García, Z. Bahmani, A.N. Baghani, S. Sadjadi, K. Yaghmaeian, B. Kamarehie, A. Mahvi, & M. Yunesian. 2025. Energy and cost analysis of bottom waste cooking oil (B-WCO) based solid alcohol biofuel and supernatant waste cooking oil (S-WCO) based biodiesel. *Heliyon* 11: e42180.

Ng, R.T.L., & C.T. Maravelias. 2017. Design of biofuel supply chains with variable regional depot and biorefinery locations. *Renew. Energy* 100: 90–102.

Nugrahapsari, R.A., Harianto, R. Nurmalina, & A. Fariyanti. 2024. The position of Indonesia's palm oil in the vegetable oil international market the position of Indonesia's palm oil in the vegetable oil international market. *IOP Conf. Ser. Earth Environ. Sci.* 1379: 012014.

Ogunrewo, O.F., & N.I. Nwulu. 2024. Optimizing biomass supply chains: A probabilistic approach to managing uncertainties in Southwest Nigeria. *Clean. Eng. Technol.* 22: 100785.

Pahnila, M., A. Koskela, P. Sulasalmi, & T. Fabritius. 2023. A review of pyrolysis technologies and the effect of process parameters on biocarbon properties. *Energies* 16: 6936.

Parthasarathy, P., M. Alherbawi, M. Shahbaz, H.R. Mackey, & G. McKay. 2024. Conversion of oil palm waste into value-added products through pyrolysis: A sensitivity and techno-economic investigation. *Biomass Convers. Biorefinery* 14: 9667–9687.

Piqueiro, H., R. Gomes, & J.P. De Sousa. 2023. Managing disruptions in a biomass supply chain: A decision support system based on simulation/optimization. *Sustainability* 15: 7650.

Rentizelas, A.A., A.J. Tolis, & I.P. Tatsiopoulos. 2009. Logistics issues of biomass: The storage problem and the multi-biomass supply chain. *Renew. Sust. Energ. Rev.*13: 887–894.

Roni, M.S., Y. Lin, D.S. Hartley, D.N. Thompson, A.N. Hoover, & R.M. Emerson. 2023. Importance of incorporating spatial and temporal variability of biomass yield and quality in bioenergy supply chain. *Sci. Rep.* 13: 6813.

Roudneshin, M., & A. Sosa. 2024. Optimizing agricultural waste supply chains for sustainable bioenergy production: A comprehensive literature review. *Energies* 17: 2542.

Saravanan, A., Y.P. Ragini, S. Karishma, R.V. Hemavathy, & M. Jyotsna. 2025. A review on advancing sustainable energy: The role of biomass and bioenergy in a circular economy. *Sustain. Futures* 10: 100835.

Shahzad H.M.A., Z. Asim, S.J. Khan, F. Almomani, K.A. Mahmoud, M.R. Ul Mustafa, & K. Rasool. 2024. Thermochemical and biochemical conversion of agricultural waste for bioenergy production: An updated review. *Discov. Environ.* 2: 134.

Stellingwerf, H.M., X. Guo, & E. Annevelink. 2022. Logistics and supply chain modelling for the biobased economy: A systematic literature review and research agenda. *Front. Chem. Eng.* 4: 778315.

Talitha, T., W.A. Jauhari, & M. Hisjam. 2023. Review of sustainable biofuel supply chain model design. *E3S Web Conf.* 465: 02022.

Tumuluru J.S. 2024. An overview of preprocessing and pretreatment technologies for biomass. In *Densification Impact on Raw, Chemically and Thermally Pretreated Biomass: Physical Properties and Biofuels Production*, vol 6, ed. J.S. Tumuluru, 1–14. Singapore: World Scientific.

Valipour, M., F. Mafakheri, B. Gagnon, R. Prinz, D. Bergstr, M. Brown, & C. Wang. 2024. Integrating bio-hubs in biomass supply chains: Insights from a systematic literature review. *J. Clean. Prod.* 46: 142930.

van der Vorst J.G., & A.J. Beulens. 2002. Identifying sources of uncertainty to generate supply chain redesign strategies. *Int. J. Phys. Distrib. Logist. Manag.* 32: 409–430.

Vazifeh, Z., F. Mafakheri, C. An, & F. Bensebaa. 2023. A game theoretic approach to contract-based enviro-economic coordination of wood pellet supply chains for bioenergy production. *Sustain. Energy Res.* 10: 17.
Wang, M., F. Jie, & A. Abareshi. 2014. The measurement model of supply chain uncertainty and risk in the Australian courier industry. *Oper. Supply Chain Manag. Int. J.* 7: 89–96.
Wild, M., & C. Calderón. 2021. Torrefied biomass and where is the sector currently standing in terms of research, technology development, and implementation. *Front. Energy Res.* 9: 678492.
Wzorek, M., A. Król, R. Junga, J. Małecka, E. Yilmaz, & A. K. Więcek. 2024. Effect of storage conditions on lignocellulose biofuels properties. *Sci. Rep.* 14: 15192.

3 Fundamentals of Biorefinery

Reckson Kamusoko and Raphael M. Jingura

3.1 INTRODUCTION

Over the past few years, a significant paradigm shift towards a circular economy has been observed (de Jong & Jungmeier 2015). This has been necessitated by the decline in fossil reserves, the rise in fuel prices, global population growth and the destructive impact of climate change (Amoah et al. 2019; Khoaele et al. 2024). The goal is to manage waste and create products that can compete in the global market with conventional petroleum products. Global greenhouse gas (GHG) emissions are on the rise, with 2023 reaching a new record high of 53 billion metric tons of carbon dioxide (CO_2) equivalent ($GtCO_2e$) (Tiseo 2025). Nowadays, human emissions of GHGs are the major drivers of global climate change. Petroleum fuels are the primary source of energy for transportation, heating and power generation. These fuels account for around 80% of the total GHG emissions (Antar et al. 2021). This has forced researchers to seek for low-cost, sustainable, and renewable alternatives to meet the demands of conventional fuels and chemicals. In 2015, the United Nations adopted the Sustainable Development Goals (SDGs), a blueprint targeting the alleviation of energy poverty, climate change, and protection of the environment (UNEP 2015).

Several countries in the world are generating substantial amounts of biomass on a daily basis. Currently, the total annual biomass production is estimated at around 146 billion metric tons, with natural vegetation being the primary source (Takki 2024). There are two main types of biomass found on Earth. These are dedicated crops and residues/waste. Specific examples of biomass include forest biomass, crop residues, livestock waste, sewage sludge, algae, municipal solid waste (MSW), and oil, starch and sugar crops (Kumar et al. 2018). Accumulation of waste biomass in the environment may cause severe adverse impacts on the natural ecosystem. Therefore, effective management of biomass could be an everlasting solution for sustainable utilization of natural resources and mitigates the overreliance on fossil fuels (Paul et al. 2024). Most products on the global market, such as vehicle fuels, fuel oil, and naphtha, are synthesized from non-renewable resources in a petrorefinery. Naphtha and natural gas are feedstocks for a diverse range of organic chemicals, polymers, and other commodities (Mailaram et al. 2021).

A new paradigm has recently evolved as a panacea to the supply of chemicals and fossil-based products (de Jong & Jungmeier 2015; Amoah et al. 2019). The aim is to produce biofuels, bioenergy, biochemicals, biopolymers, and other green

DOI: 10.1201/9781003650751-3

products from biomass resources. This approach is compatible with a petrorefinery, in which crude oil is replaced with biomass as a feedstock. The phenomenon is referred to as the biorefinery concept (Mailaram et al. 2021). A biorefinery concept is defined as an integrated system for converting biomass into a spectrum of energetic and other marketable products (Kamusoko & Mukumba 2025). Energetic products are composed of low-value biofuels (biogas, bioethanol, biodiesel, etc.) and bioenergy (electricity, heat and/or power). Bio-based products comprise high-value products such as biopharmaceuticals, biocosmetics, bionutrients, biochemicals, biomaterials, and biofertilizers (Budzianowski 2017). Biomass valorization in a biorefinery is sustainable and eco-friendly way to effectively manage the environment and promote a circular economy. The manufacture of bio-based products is a potential source of income for the chemical sector. An estimated revenue of about US$ 10–15 billion can be created from bio-based markets (de Jong & Jungmeier 2015).

In this millennium, a variety of schemes have been proposed to classify biorefineries. A simple and convenient method capitalizes on the feedstock and classifies biorefineries into lignocellulosic (LCF), algae, and waste (Ubando et al. 2022; Wagh et al. 2024). Full-scale adoption of biorefineries requires suitable technologies to convert biomass feedstocks into finished goods. These platforms fall into four main groups: biochemical (e.g. anaerobic digestion), physical/mechanical (e.g. particle size reduction), chemical (e.g. esterification), and thermochemical (e.g. gasification). Thermochemical and biochemical routes are core parts of biomass conversion. Feedstock pretreatment is an integral part of LCF biorefineries. It can be achieved through pre-processing recalcitrant biomass using physical and chemical methods (Calvo-Flores & Martin-Martinez 2022; Kamusoko & Mukumba 2025).

In this chapter, the principles of a biorefinery concept are discussed. Firstly, the chapter highlights the main drivers that serve as the foundation for the transition towards the biorefinery approach. It is well given that the function and expansion of biorefineries is driven by the availability of biomass feedstocks. Hence, biomass feedstocks are the epitome of a successful biorefinery project. This chapter underscores the different types of biomass feedstocks as potential resources for sustainable biorefineries. The classification of biorefineries on the basis of feedstock type is presented in this chapter. A biorefinery operates at the confluence of a variety of conversion processes. In this chapter, the main technological platforms for biorefineries and the resultant bio-based products are provided. The biorefinery's outlook is quite promising. Finally, the chapter delves into the outlook and provides guidance for future work towards large-scale utilization of biomass resources.

3.2 KEY DRIVERS FOR ADOPTING BIOREFINERIES

A transitional economic model for sustainable development of biorefineries is governed by a set of multifaceted factors. This section outlines the five critical factors, comprising environmental crisis, shortage of non-renewables sources, policy and regulatory issues, technological advancements, market trends and consumer behaviour (Wagh et al. 2024).

3.2.1 Environmental Concerns

One of the main reasons for the transformation from conventional fuels and chemicals to biorefineries is the detrimental environmental impacts caused by these products. Fossil products release ambient carbon and other GHG emissions that may contribute to global warming and climate change (Ghatak 2011). Climate change is growing at an alarming rate in the world. Over the 1995–2005 decade, a rise in GHG emissions from the burning of fossil products was reported. For example, CO_2 accumulated in the atmosphere at a rate of 1.9 ppm per annum, while CO_2 radioactive forcing accelerated by 20%. Emissions from petroleum and cement companies accelerated at an annual rate of 3% (Ghatak 2011). By 2040, GHG emissions are expected to increase to about 43 billion metric tons (Kamusoko et al. 2021).

3.2.2 Shortage of Non-Renewable Resources

Overdependence on finite resources such as petroleum and other non-renewable products has evolved as a key driving force for the development of biorefineries (Ghatak 2011; Wagh et al. 2024). More so, the relentless population growth has exerted pressure on natural resources and energy demand on the globe. By 2030, the world population is forecasted to rise to around 9 billion, bringing in 3 billion more middle-class users (Mohammad et al. 2024). Given that the world energy demand is growing at the rate of 1.6% per annum, it is anticipated to be approximately 700 EJ/y by 2030. Petroleum remains the dominant supplier of energy, contributing about 80% to the total world primary energy. Furthermore, the world relies on petroleum refineries for more than 90% of its organic chemicals (Ghatak 2011). It is against this background that there is likely to be a deficit in fossil reserves by 2070–2080 (Kamusoko et al. 2021). A circular bioeconomy that utilizes renewable ingredients, technologies, and products can provide an infinite alternative to oil and natural gas reserves.

3.2.3 Policy and Regulatory Issues

The biorefinery approach needs to be controlled and regulated by enacting robust strategies to fully adopt this sustainable technology. Several governments in the world are formulating policies and regulations to promote the biorefinery technology and a circular economy (Wagh et al. 2024). Policies such as subsidies, tax incentives, and regulatory frameworks for research and development (R & D) can create an enabling environment for the growth and expansion of a bio-based economy. As an international standard, governments are encouraged to develop a separate waste-to-energy policy to address renewable energy policy gaps (Kamusoko & Mukumba 2025).

3.2.4 Technological Advancements in Biorefineries

The quest to move towards a bio-based economy has been revolutionized by recent advancements in biorefinery technologies. Technological innovations can significantly improve the efficiency and sustainability of biorefinery systems. Innovations in biotechnology, nanotechnology, and data analytics have been modified to optimize the way raw materials are converted into bio-based products (Wagh et al. 2024). Advancements in technological aspects include hemicellulose integration and the utilization of sugars for multiple products, lignin valorization, development of efficient and reliable pretreatment technologies, and the search for highly efficient fermentation technologies (Yamakawa et al. 2018). As such, enzymatic hydrolysis has been introduced as a pre-processing step in some biorefinery facilities. It incorporates enzymes to degrade composite biomass into smaller usable molecules. Enzymatic hydrolysis is a low-cost and environmentally friendly technology compared to conventional methods (Wagh et al. 2024).

3.2.5 Market Trends and Consumer Behaviour

A dynamic shift with regard to biofuels, bioenergy, and bio-based products has been motivated by global market trends and consumer behaviour. Current market patterns demonstrate a switch in consumer perceptions towards more sustainable and safer products. There is a surging demand for bioproducts in the food, cosmetic, packaging, and construction industries. Most consumers are aware of global warming and its environmental impacts. This consumer behavioural change has stimulated policymakers and other stakeholders to invest in biorefinery enterprises and leverage them as an endeavour to promote socio-economic development (Wagh et al. 2024).

3.3 THE BIOREFINERY APPROACH

The biorefinery concept emerged in the late 1990s, motivated by conventional petroleum refineries (Takkellapati et al. 2018). Petroleum-based refineries fractionate crude oil into fuels, heat, power, chemicals, and materials (Ubando et al. 2022). A biorefinery converts biomass like agro-industrial residues, forestry waste, algae, etc., into a wide portfolio of products, including biofuels, food, feed, pharmaceuticals, biochemicals, and biomaterials (Figure 3.1). In this scenario, a network of dynamic facilities and processes are integrated in a sustainable biorefinery towards zero-waste circularity. Most biorefineries are designed to make fuel products such as bioethanol, biodiesel, or biogas. However, biofuel plants are not self-sustainable, and they are often short-changed without governmental intervention. Non-energetic bioproducts could be the next-generation carbon sinks and can have a huge market share in sustainable biorefineries (Sadhukhan et al. 2024). The main components of a biorefinery are upstream, midstream, and downstream processes (An et al. 2011; Aristizábal-Marulanda & Alzate 2019). Upstream involves the harvesting, collection and delivery of the raw material to the processing plant. Midstream covers the transformation of biomass into intermediate or final products. Downstream refers to extraction, purification, recovery, packaging, storage, and distribution of the final products to consumers (An et al. 2011). Biorefineries are driven by innovative technologies to transform biomass into multi-product streams. A biorefinery amalgamates physical, chemical, biochemical, and thermochemical routes to harness the chemical potential of biomass feedstocks. The carbon-neutrality of biomass may help to mitigate GHG emissions and promote sustainability and the circular economy (Sadhukhan et al. 2024).

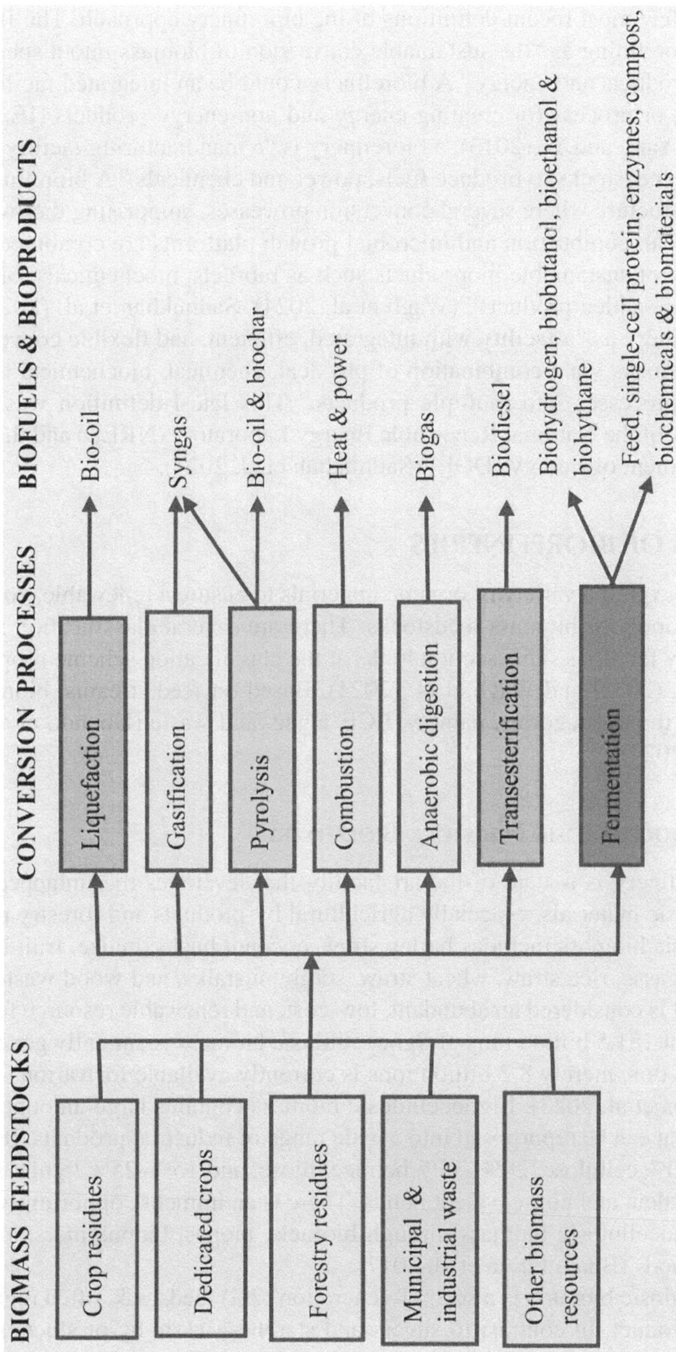

FIGURE 3.1 Conversion of biomass feedstocks into biofuels and bioproducts in a biorefinery.

People have mixed perceptions of the biorefinery concept. As a result, there is no single or mutually agreed definition of the biorefinery concept. This section will focus on the few most recent definitions of the biorefinery approach. The IEA Task 42 defined biorefining as "the sustainable conversion of biomass into a spectrum of marketable products and energy." A biorefinery could be an integrated facility, cluster, spectrum, or process for creating energy and non-energy products (IEA 2022). According to Yang and Yu (2013), a biorefinery is "a manufacturing facility that utilizes biomass feedstocks to produce fuels, power and chemicals." A biorefinery is "a type of infrastructure where several conversion processes, comprising thermochemical, biochemical, combustion and microbial growth platforms are combined to generate a variety of sustainable bioproducts such as biofuels, biochemicals, bioenergy and other value-added products" (Wagh et al. 2024). Sadhukhan et al. (2024) redefined a biorefinery as "a facility with integrated, efficient, and flexible conversion of biomass feedstocks via a combination of physical, chemical, biochemical and thermochemical processes into multiple products." This latest definition was derived from the work of the National Renewable Energy Laboratory (NREL) and the United States Department of Energy (DOE) (Sadhukhan et al. 2024).

3.4 TYPES OF BIOREFINERIES

Biorefineries exploit a variety of organic materials to ensure a renewable, consistent, and regular supply of biomass feedstocks. There are several classification schemes for biorefinery facilities. This section looks at the classification scheme proposed by Ubando et al. (2022) and Wagh et al. (2024). Based on feed streams, biorefineries may fall into three categories, namely, LCF, algae, and waste (Ubando et al. 2022; Wagh et al. (2024).

3.4.1 Lignocellulosic Feedstock Biorefinery

A LCF biorefinery is a state-of-the-art facility that leverages the untapped potential of cellulosic materials, especially agricultural by-products and forestry residues. Lignocellulosic biomass includes barley straw, coconut husks, maize, fruit bunches, sugarcane bagasse, rice straw, wheat straw, sorghum stalks, and wood waste (Wagh et al. 2024). It is considered an abundant, low-cost, and renewable resource for biorefineries. About 181.5 billion tons of lignocellulosic biomass is annually generated in the world. Of this, merely 8.2 billion tons is currently available for various applications (Mujtaba et al. 2023). Lignocellulosic biomass contains large amounts of carbohydrates that can be repurposed into a wide range of industrial products. It is made up of 35%–50% cellulose, 20%–35% hemicellulose, and 15%–25% lignin, strongly linked by covalent and non-covalent bonds. There is an immense opportunity to benefit from lignocellulosic biomass through biofuels, biogas, biopolymers, and other high-value goods (Bharathiraja et al. 2017).

Lignocellulosic biomass is a second-generation (2G) feedstock, often regarded as a waste by-product. In contrast to sugars and starches, waste by-products are non-edible resources that cannot threaten food security (de Jong & Jungmeier 2015). However, the economic viability and scalability of LCF biorefineries remain a challenge due to the heterogeneity of the feedstock. An extra pretreatment cost is required to realize LCF biorefinery as a business venture (Sadhukhan et al. 2018). Cellulolytic enzymes from bacteria and fungi can be deployed to hydrolyse complex compounds into simple constituents. Common enzymes comprise cellulases, xylanases, and ligninases that degrade cellulose, hemicellulose, and lignin, respectively. Lignin is a waste by-product of the pretreatment and bioprocessing of LCF. Valorization of lignin into gasoline fuel additives, adhesives, and bioplastics can reduce pretreatment costs and add value to LCF biorefineries (Maity 2015).

3.4.2 Algal Biorefinery

An algal biorefinery utilizes algae to produce green products, in the form of renewable fuels and chemicals. Algal biomass is a third-generation (3G) biorefinery feedstock (Shankar et al. 2022). Its popularity has been derived from minimal land use, rapid growth rate, and high biomass yield. Algae are subdivided into macroalgae and microalgae, depending on microbial size and nutrition (Ubando et al. 2022). Microalgae refers to microscopic, eukaryotic, and single-celled photoautotrophs that can survive in varied climatic and marine conditions. These microorganisms use light energy to convert CO_2 into cell biomass. Microalgae are reservoirs of various compounds such as lipids, carbohydrates, proteins, oil, carotenoids, and other useful pigments. Biocompounds have several nutritional and health benefits (Chandrasekhar et al. 2022; Shankar et al. 2022). Biorefining microalgae is a promising approach for producing chemicals, fuels, feed, and biogas. Industrialization of microalgal biorefineries is often limited by small biomass volume, economic concerns, uncompetitive prices, and restricted market share (Chandrasekhar et al. 2022).

Macroalgae are macroscopic, eukaryotic, and multicellular autotrophs that occupy a diverse range of aquatic and marine environments (Kostas et al. 2021). These macroorganisms are rich in lipids, proteins, carbohydrates, and photosynthetic pigments. Repurposing these components is a sustainable way to manufacture value-added bioproducts. Potential avenues for macroalgal biorefinery include food, biofuels, and biochemicals. However, macroalgae are overlooked as a bioresource for biorefineries. Several setbacks should be surpassed to recognize the full potential of macroalgae. Bioproducts that can be obtained from macroalgae need to be fully identified, whereas the techno-economic feasibility of macroalgal biorefineries still remains an issue to be addressed (Kostas et al. 2021).

3.4.3 Waste Biorefinery

Bioprocessing of waste streams from anthropogenic sources such as agricultural, municipal, forestry, and industrial activities is considered an attractive option to recover valuable products. Biomass, in the form of food waste, lignocellulosic material, paper waste, MSW, sewage sludge, and other organic materials, is a potential source of energy, fuels, and value-added products (Leong et al. 2021). The multiplicity of product streams includes butanol, biogas, methane, ethanol, hydrogen, biohythane, electricity, organic acids, biopolymers, proteins, enzymes, and bioplastics. Choosing a feedstock for waste biorefinery is influenced by the desired final products, feedstock availability, environmental concerns, and technological advancements (Ahmed et al. 2023). Waste biorefinery does not only generate bio-based goods but also addresses issues of waste management and promotes the development of the circular economy. This idea of sustainability resonates with many consumers in the circular bioeconomy based on the four Rs philosophy: reduce, reuse, recycle, and recover. Waste valorization is a potent strategy to create a carbon-neutral atmosphere and mitigate global climate change (Leong et al. 2021). Waste feedstocks are affordable, available throughout the year and do not compete with food or animal feed. However, a few waste materials are not easily biodegradable requiring a costly pre-processing step. Pretreatment is a critical phase towards valorization of biomass in waste biorefinery. The choice of pretreatment depends on waste composition and the resultant conversion technology (Ahmed et al. 2023).

3.5 BIOMASS FEEDSTOCKS FOR BIOREFINERIES

Biomass refers to any organic material that comes from plants and animals (Apotei 2019). It is a renewable feedstock that can be transformed into various energy and marketable products in a biorefinery. Biomass originates from biogenic sources, including agriculture, forestry, aquaculture, municipalities, and industries (Cherubini et al. 2009). Other sources of biomass are unicellular and multicellular organisms, and in particular, algae and fungi (Amoah et al. 2019). Consistent and infinite feedstock supply is an important criterion for sustainable biorefinery development. However, the availability of feedstock is influenced by agro-climatic conditions, geographical location, socio-economic issues, policies, and regulations.

Table 3.1 shows the different kinds of biomass feedstocks for biorefineries. A biorefinery can derive its feed from agriculture or forestry, in the form of dedicated crops. Dedicated crops comprise first-generation feedstocks that are mainly grown for food and feed and a few non-edible plants. Typical examples include woody plantations, grasses, herbs, and oil, starch and sugar crops (Ghatak 2011). The sustainability of feeding biorefineries with edible oils, sugars, and starches "versus" food security has been debated for quite some time. This issue can be resolved through the diversification of feedstock streams. Biomass residues or waste streams such as lignocellulosic material, non-edible oils and fats, food waste, etc., are suitable alternatives to dedicated crops in biorefinery systems (Ghatak 2011). These biomass residues are considered as 2G feedstocks. The main obstacle to feeding biorefineries with lignocellulosic residues is the slow hydrolysis rate. Hence, efficient pretreatment methods need to be developed to improve the sustainability and economic viability of lignocellulosic biorefineries. Recently, 3G feedstocks from algae and fourth-generation (4G) feedstocks from engineered microorganisms evolved as new feedstocks for integrated biorefineries (Kamusoko et al. 2021).

3.6 BIOREFINERY CONVERSION TECHNOLOGIES

The input of feedstocks in a biorefinery requires innovative technologies for sustainable conversion into valuable products. A full potential of biomass can be realized through microbial and enzymatic methods, advanced separation and purification

TABLE 3.1
Feedstocks for Biorefineries

Category	Feedstock Type	Example
Dedicated crops	Grasses and herbs	Switch grass, immature cereals, Alfalfa, *Miscanthus*
	Woody plantations	Hardwood, softwood
	Sugar crops	Sugarcane, sugar beet
	Starch crops	Corn, potato, wheat, cassava, sweet sorghum
	Oil crops	Soya, sunflower, rape, palm, Jatropha
	Marine biomass	Microalgae, macroalgae, seaweed
Residues/waste	Lignocellulosic biomass	Black liquor, paper mill waste sludge, timber house waste, saw mill waste, crop residues (wheat straw, rice straw, maize stover, etc.), bagasse
	Oils and fats	Fats from slaughterhouses, waste cooking oils
	Others	Slaughterhouse trimmings and bones, vegetable and fruit processing wastes, pith, poultry litter, animal farm wastes, molasses, municipal wastes

Source: Ghatak (2011).

techniques, and the transformative power of biotechnology (Wagh et al. 2024). Biomass conversion technologies are divided into four broad categories: mechanical/physical, chemical, biochemical, and thermochemical (Cherubini et al. 2009; Aristizábal-Marulanda & Alzate 2019). Chemical and physical methods are also involved in feedstock pretreatment, extraction, purification, and recovery of intermediates and products (Ghatak 2011).

3.6.1 Thermochemical Technologies

Thermochemical processes use high temperatures and pressures to convert biomass into energy products. Combustion is a simple and mature technology for biomass conversion. This process oxidizes biomass in the presence of oxygen, generating energy for cooking and heating (Jingura et al. 2013). The quest for combustion has been derived from its affordability, dependability, and ability to fit properly into pre-existing facilities. Residual lignin waste by-product was transformed into energy via combustion in the pulp and paper industry (Calvo-Flores & Martin-Martinez 2022). However, this method is not efficient as more energy is lost in open fires.

Carbonization is another ancient route for thermal degradation of biomass into charcoal-like biomaterials. This method occurs at a temperature varying from 500°C to 900°C under a limited supply of oxygen. Carbonization improves the carbon content per given volume of biomass (Calvo-Flores & Martin-Martinez 2022).

Advanced technologies such as gasification, pyrolysis, and hydrothermal liquefaction (HTL) are promising alternatives to traditional biomass use. Gasification converts woody material into syngas and other intermediates at high temperature (500°C–1000°C), pressure (0.98–2.94 atm), and partial oxygen atmosphere (Sher et al. 2025). Syngas is a mixture of combustible gases like hydrogen (H_2), carbon monoxide (CO), CO_2, water (H_2O), nitrogen gas (N_2), and methane (CH_4). It can be transformed into high-value products such as energy and biomaterials. Biomass gasification in combined heat and power (CHP) unit generates liquid biofuels, thermal energy, and electricity. This technology has been fully adopted at both pilot and commercial scales. Fly ash, volatile alkali metals, and tar are given off as by-products (Sher et al. 2025). Gasification generates high energy and heat compared to pyrolysis and HTL. It is considered the most efficient method for H_2 recovery from biomass (Handa & Rajamani 2023). Pyrolysis is a suitable thermochemical process for partial oxidation of solid biomass. It is performed at high temperature in the range of 300°C to 1000°C and mild pressure of 0.1–0.5 MPa. The resultant products are syngas, bio-oil, and biochar. Pyrolysis requires a catalyst to generate a pollutant-free gaseous mixture of H_2, CO and CO_2 (Speight 2020). HTL converts wet biomass from forestry and agriculture into biocrude oil. The reaction takes place under subcritical water and inert conditions, with or without a catalyst. Moderate temperatures and pressures in the range of 280°C–350°C and 5–21 Mpa, respectively, are more favourable for HTL (Shahbeik et al. 2024).

3.6.2 Biochemical Technologies

Bioconversion (e.g. anaerobic digestion (AD) and fermentation) capitalizes on enzymatic power to transform biomass into high-grade bioproducts. This reaction occurs under mild conditions, either aerobic, anaerobic or a combination of these conditions (Calvo-Flores & Martin-Martinez 2022). Fermentation is an old and low-cost technology to harness sugar in polysaccharides and transform it into alcohol and CO_2. Alcoholic beverages such as wine, beer, and cider have been traditionally produced via fermentation. Fermentation uses yeasts, fungi, bacteria, and other microbes to degrade biomass. Fungi (e.g. *Aspergillus, Rhizopus*, and *Penicillium*) and yeasts (e.g. *Saccharomyces, Candida*, and *Aureobasidium*) are active biomass fermenters. Bacteria like *Bacillus* and *Streptomyces* are also being explored for fermentation due to their ability to secrete thermostable enzymes (Handa & Rajamani 2023).

Nowadays, several biofuels and other bioproducts are widely manufactured using fermentation. A variety of foods, organic acids, and biopharmaceuticals can be derived from fermentation (Wagh et al. 2024). Fermentation of industrial waste, methanol, food waste, agro waste, and fruit waste is a potential source of single-cell protein (SCP). SCP can be used to formulate stock feed or as a food supplement. It is pertinent to select efficient and robust microbes that can tolerate harsh reactor conditions. Currently, researchers are focusing on engineering microbial strains to improve product yield and quality (Kamusoko & Mukumba 2025). Starch and cellulose can be actively fermented into bioethanol, a renewable vehicular fuel. Ethanolic fermentation is a three-step process involving pretreatment, saccharification, and fermentation. Pretreatment results in increased hydrolysis and fermentable sugar yield. In saccharification, complex polysaccharides are hydrolysed into monomeric units using cellulases and hemicellulases. Lastly, fermentation converts sugars from enzymatic saccharification into ethanol using microorganisms (Takkellapati et al. 2018). Biomass has been successfully converted to biobutanol via the acetone-butanol-ethanol (ABE) fermentation pathway using *Clostridium acetobutylicum* (Maity 2015).

AD is a well-established technology that occurs naturally in marshes, wetlands, landfills, and ruminant guts. It utilizes anaerobic microbes to stabilize waste, while producing bioenergy and a biofertilizer. AD microflora can be grouped into two categories, that is, methanogenic and non-methanogenic bacteria. Biofertilizers can replace chemical fertilizers in crop agriculture (Kamusoko & Mukumba 2025). Biomass contains huge amounts of organic matter that is amenable to AD. Biogas is the core product of AD. It is a combustible mixture of CH_4, CO_2, and other gases. Animal manure, sewage sludge, MSW, food waste, crop residues, and energy crops are biomass feedstocks for AD. The process involves three main steps: hydrolysis, fermentation, and methanogenesis. Hydrolysis degrades complex biomass compounds into fermentable substrates. Fermentation then breaks down hydrolytic products into alcohols, volatile fatty acids (VFAs), acetic acids, H_2, and CO_2. Metabolism of fermented gaseous mixture liberates biogas during methanogenesis (Handa & Rajamani 2023). Process parameters, including temperature, pH, agitation, inoculum size, and others, should be closely monitored for efficient gas production (Kamusoko et al. 2022). The biochemical route could be an attractive option for the production of various platform chemicals. For instance, engineered *Escherichia coli* was used to produce succinic acid from maize stover, attaining a titre of 112 g/L and 0.78 g/g glucose (John et al. 2025).

3.6.3 Other Conversion Processes

A myriad of chemical and physical methods is available to process biomass in biorefineries. These processes may not inevitably function in the realm of actual bioconversion, but for pretreatment and downstream modifications (Ghatak 2011). Pretreatment is a pre-extraction process performed to enhance the conversion of biomass into bio-based products. Although physical and chemical pretreatments are fast and effective, the methods are costly, consume more energy, and generate pollutants. All processes that use mechanical force to disintegrate biomass are physical methods (e.g. milling, grinding, ultrasound, and thermal treatment). Physical processes reduce feedstock particle size and increase the surface area for further bioprocessing (Kamusoko & Mukumba 2024). Biorefineries utilize many separation techniques to disorganize feedstocks into valuable products. For instance, detoxification can remove inhibitors while distillation can separate ethanol from other liquids in a corn-ethanol biorefinery (Huang et al. 2008). Cell disruption using mechanical forces and chemical agents can be applied for extracting cell inclusions and other resources from *E.coli* and *S. cerevisiae* (Lee et al. 2012). Acids, alkalis, and ionic liquids are all chemical agents for improving sugar yield in fermentation and AD (Kamusoko & Mukumba 2024).

Biodiesel is produced from biomass through transesterification. Transesterification is a chemical process for converting fats and oils into methyl esters and glycerol in the presence of a catalyst. Homogeneous alkaline catalysis is the most popular transesterification method for biodiesel production. Biodiesel is a liquid fuel that can replace petrodiesel in transportation. The glycerin by-product can be separated into products of economic importance (Wagh et al. 2024). Catalytic Fischer-Tropsch Synthesis is yet another method for chemical conversion. It converts syngas into fuels and chemical products (Ghatak 2011).

3.7 BIOREFINERY PRODUCTS

A biorefinery converts biomass into a multiplicity of energy and non-energy products (Ubando et al. 2022). Examples of biorefinery products include heat, power, biogas, bioethanol, biohydrogen, biobutanol, biohythane, food, feed, bioplastics, biocomposites, industrial enzymes, biopharmaceuticals, organic acids, solvents, dyes, and sorbents (Kamusoko & Mukumba 2025). These products can be categorized into biofuels, biochemicals, biomaterials, and bioenergy.

3.7.1 Biofuels and Bioenergy

Biorefinery-based technologies produce a gamut of advanced renewable fuels. Biofuels are emerging forms of solid, liquid, and gaseous fuels derived from biomass feedstocks. Currently, the world market is dominated by liquid biofuels such as biodiesel, bioethanol, and biobutanol (Mailaram et al. 2021). Biofuels are a source of bioenergy for heat and electricity generation. They can substitute fossil fuels in the transport sector. Biofuel consumption can reduce about 20%–60% GHG emissions (Handa & Rajamani 2023). Contrarily, biofuel production generates wastewater that may pollute the natural environment if improperly treated and disposed (Handa & Rajamani 2023).

3.7.1.1 Liquid Biofuels

Liquid biofuels can be blended with fossil fuels to improve engine performance and transportation efficiency. For example, biodiesel and bioethanol are blended with gasoline and petrodiesel up to a maximum of 15% and 20% (v/v), respectively (Mailaram et al. 2021). Biodiesel production can be dated back to as far as 1900 with the pioneering work of Rudolf Diesel (Mailaram et al. 2021). The world's biodiesel production has hit record levels and continues to grow, reaching approximately 45 million tons in 2019 (John et al. 2021). The European Union (E.U.) is the major producer of biodiesel, accounting for about 28% of the world's total production. Biodiesel reserves of the United States (U.S.) are around 24% of the world's biodiesel (Ali et al. 2022). Biodiesel is an alternative fuel for diesel engines. It is the main product from triglycerides of soya, canola, oilseed, cassava, and palm oil. Oils and fats from animals and oleaginous microorganisms are supplementary sources of biodiesel. Oleaginous microorganisms include bacteria, yeasts, and microalgae.

These organisms contain 30% lipids per cell weight. Oleaginous microbial systems are referred to as Single-Cell Oils (SCOs) that generate and store oil. Oleaginous species transform sugars, CO_2, and organic acids into SCOs. Biodiesel from SCOs is quite expensive, with an estimated total cost of US$ 1–80 per gallon (Amoah et al. 2019; Ali et al. 2022). Transesterification is the most suitable route for biodiesel production. The cultivation of oil crops requires vast amounts of land, limiting biodiesel production (Wagh et al. 2024). GHG emission savings for biodiesel are around 40%–65% (Laca et al. 2011). Biodiesel is either consumed as B100 or blended with fossil diesel. Biodiesel production generates significant amounts of glycerin as a co-product. Around 10% (w/w) glycerin is released along with biodiesel, accounting for 90% of the total glycerin. Glycerin can serve as the backbone for synthesis of value-added chemicals. For example, *E. coli* and an engineered strain of *C. acetobutylicum* can transform glycerin into 1,2-propanediol and 1,3-propanediol, respectively (Takkellapati et al. 2018).

Bioethanol is the world's promising biofuel that can substitute petrol in transportation. It is produced by microbial fermentation of sugar (beet and sugarcane) or saccharified starch (cassava, wheat, and corn). Cellulosic ethanol production is globally rising as a strategy to overcome food conflicts. As highlighted in Section 3.6, pretreatment and hydrolysis are critical to boost the release of fermentable sugars from complex biomass structure. The U.S. and Brazil are the leading suppliers of bioethanol, contributing about 85%–90% of the total global share (Amoah et al. 2019). Around 4.5 billion gallons of bioethanol are produced from about 12% of the U.S.'s corn yield. Brazil is actively involved in bioethanol production with 15 billion litres coming from sugarcane. This corresponds to around 33% of the country's gasoline demands (Handa & Rajamani 2023). Bioethanol production from the E.U. was estimated at 5.6 billion litres in 2021 (Calvo-Flores & Martin-Martinez 2022). Bioethanol is blended with gasoline to produce gasohol, a transportation fuel. It can be envisaged that bioethanol will overtake gasoline as a vehicular fuel in future. Bioethanol is a precursor and an organic solvent for the manufacture of biochemicals and biocomposites (Wagh et al. 2024; Kamusoko & Mukumba 2025). Bioethanol can reduce GHG emissions by 20%–90% over petroleum fuel competitors (Laca et al. 2011).

Biobutanol is another fuel suitable for use in internal combustion engines. It is an isomer of butanol generated by the ABE fermentation process. *C. acetobutylicum* is the most active bacteria involved in biobutanol fermentation. Other species consist of *C. eijerinckii*, *C. saccharoperbutylacetonicum*, *C. saccharobutylicum*, *C. sporogenes*, *C. pasteurianum*, *C. carboxidivorus*, *C. tetanomorphum*, and *C. aurantibutyricum* (Lin et al. 2023). The predominant isomers for biofuel production are *n*-butanol and isobutanol. The *t*-butanol cannot be used as a biofuel due to its persistence in the environment (Roberts & Patterson 2014). Biobutanol produces about 25% more energy than bioethanol; thus, it can be considered as a more powerful advanced transportation fuel. Its unique combination of quality and hydrophobicity makes it an ideal material to mix with gasoline or use in its pure state without engine adjustments (Mailaram et al. 2021). Upgrading biobutanol produces jet fuels (Yang & Yu 2013). Biobutanol is a highly flammable fuel, ignited by heat, flames or sparks. Its flash points range from 82°F to 98°F. Therefore, biobutanol must be safely stored underground in cool and well-ventilated facilities (Roberts & Patterson 2014). The scalability of biobutanol production has been hampered by high feedstock cost. This could be improved by using inexpensive and copious biomass feedstocks (Dlangamandla et al. 2025).

3.7.1.2 Gaseous Biofuels

The biofuel industry is gaining traction in the global market with its innovative gaseous products such as biomethane, biohythane, H_2, and syngas. In 2025, a surge by 8%–10% in the demand for H_2 was projected. H_2 is the resultant product of fermentation, mostly using photosynthetic bacteria and a few archaea (Wagh et al. 2024). Specific examples of H_2-producing microorganisms include *Clostridium*, *Enterobacter*, and *Bacillus*. Microbial growth can consume a portion of the substrate, thereby lowering the intended yield of H_2 (Amoah et al. 2019). Dark fermentation is a well-established process for H_2 production. It directly turns sunlight energy into H_2. H_2 is a clean, sustainable, and multipurpose fuel. It burns in fuel cells to release water in its oxidized form. H_2 is favoured to other biofuels due to its energy content of 120–142.9 MJ/kg and calorific value of 143 GJ/ton (Wagh et al. 2024). It is a renewable and carbon-free energy carrier for storage fuel cells, heat, power, and electricity generation. H_2 production liberates a series of intermediates and by-products, viz, organic acids, VFAs, and alcohols. These coproducts may help to improve the economic viability of the H_2 industry. They can be further polished into useful products, for instance, VFAs can be processed into biofuels and biopolymers (Kamusoko & Mukumba 2025).

Biogas is a product of the AD process using bacteria consortia under anaerobic conditions (Liu et al. 2011). It contains 45%–75% CH_4 by volume. Europe is the major producer of biogas, contributing around 50% to the world's biogas mix (IEA 2025). Biogas can be directly consumed, utilized for co-generation of heat and electricity, or upgraded to 97% CH_4 for use in transportation. Biogas purification involves dehydration, desulfurization, and extraction of CO_2, halides, and heavy metals (Liu et al. 2011). A few biogas plants have been commercialized in the transport sector across the world. Sweden, France, Denmark, Iceland, and Swiss are the major players in the biogas industry for transportation. The biogas fuel cell is a clean, competent, and silent device for electric power production (Liu et al. 2011). Production of I MJ of biogas electricity infers that about 0.4–0.9 MJ of petroleum is consumed. This can mitigate GHG emissions by approximately 75%–90%. Other environmental impacts such as acidification and eutrophication can be saved by around 95% (Laca et al. 2011). By-products of biogas production can be converted into several beneficial products. For example, a nutrient-rich digestate waste is an excellent resource for biofertilizer production. Pyrolysis of the digestate produces biochar that can be used as a soil conditioner or biosorbent. Upgrading biogas to CH_4 releases biogenic CO_2 with many industrial applications, especially in the food and beverage industry (IEA 2025).

Biohythane is novel fuel product from anaerobic fermentation of biowaste. It is a gaseous mixture of 10%–15% H_2 and 50%–60% CH_4 (Kannah et al. 2020). The gas is generated by combining dark fermentation and AD in a conventional two-stage thermophilic bioreactor. The hybrid process is performed by CH_4 and H_2-producing bacteria. Single-stage fermentation is a new alternative to the two-stage approach. It saves energy, labour, and capital requirements (Kamusoko & Mukumba 2025). The chemical composition of biohythane can outweigh the quality and economic value of conventional fuels. Biohythane is a high market and commercial value fuel for internal combustion engines and motor vehicles. It has higher quality and requires lower energy for ignition than traditional fuels (Kannah et al. 2020). Biohythane consumption can lower GHG emissions and reduce environmental pollution (Kamusoko & Mukumba 2025).

Syngas is the most profound product from biomass gasification. It is mainly a blend of CO and H_2. The gaseous mix is a source of power for electricity generation. It is an infinite fuel that can substitute fossil fuels. Besides, it is a platform chemical for the production of industrial chemicals and energy-producing compounds. Higher hydrocarbons can be made from syngas via the Fischer-Tropsch trajectory. Syngas is also a building block for methanol and dimethyl ether production (Usmani et al. 2021).

3.7.1.3 Solid Biofuels

Solid biofuels come from renewable materials like plant biomass, animal waste, and municipal waste. These fuels are a source of energy for cooking, heating, and power generation. Solid biofuels (e.g. wood chips, wood pellets, and fuelwood) constitute about 86% of the bioenergy required for cooking and heating (Manandhar et al. 2022). They can be used in their processed, semi-processed, or unprocessed form. Unprocessed biomass covers traditional fuels such as fuelwood, dung, charcoal, and crop residues. Traditional biomass is directly burnt in open fires to release thermal energy for domestic use. Using traditional biomass is quite a common practice amongst rural and peri-urban households in developing countries. About 65% of the total bioenergy consumption in the world is unprocessed biomass (Manandhar et al. 2022). However, burning of unprocessed biomass is often associated with socio-economic drawbacks. For example, the open fire method is an inefficient method with low conversion efficiency. Health-related issues caused by smoke from open fire burning can adversely affect the environment (Jingura et al. 2013). Semi-processed biomass comprises forestry processing waste (e.g. wood chips and saw dust) and MSW. This biomass is partially processed to improve conversion efficiency. Solid biomass can be upgraded to improve density, energy content, size, composition, and calorific value (Manandhar et al. 2022). This solid material is referred to as processed biomass. Pyrolysis, carbonization, gasification, and densification are mature technologies for upgrading biomass into advanced solid fuels (Jingura et al. 2013). Briquettes and pellets are the most common types of densified biofuels. Upgraded biomass forms include biochar, torrefied pellets, and torrefied biomass. Solid biofuels are regarded as carbon-neutral fuels. During their combustion, the volume of carbon emitted into the atmosphere equals the amount of CO_2 absorbed by photoautotrophic organisms (Manandhar et al. 2022).

3.7.2 Bio-Based Products

3.7.2.1 Biochemicals

Many substances that are part of our daily lives are synthesized in the chemical industry. A large portion of the chemicals arise from petroleum-based refineries. More than 80 million tons of chemicals worth about US$ 2 trillion are annually produced from petrorefinery (Yang & Yu 2013). The current mantra is to explore production of sustainable chemicals in contemporary biorefineries. The novelty of biochemicals has been a momentous breakthrough in the history of biorefineries. Biochemicals have gained popularity due their ability to substitute petrochemicals (John et al. 2025). A wide array of chemical products is produced in biorefineries through microbial fermentation, direct extraction, and biotransformation (Bhatia 2023). According to the U.S. Department of Energy (DOE), the ten most industrially important biochemicals are succinic acid, furanics, 3-hydroxypropionic acid/aldehyde, glycerol and derivatives, sorbitol, xylitol, levulinic acid, biohydrocarbons, lactic acid, and ethanol (Budzianowski 2017). Biochemicals are a major class of value-added products that can fed into a biorefinery. They can act as precursors or building blocks for the synthesis of complex compounds or biosolvents. Bio-based chemicals with more than one functional group (e.g. carboxylic acid, succinic acid, and alcohol) can be used as platform chemicals (Calvo-Flores & Martin-Martinez 2022).

Chemicals, including diols, diamines, and dicarboxylic acids can serve as monomeric units for synthesizing biopolymers and bioplastics. Biopolymers, for example, polylactic acid have been synthesized from lactic acid and polyhydroxybutyrate at industrial scale (Yang & Yu 2013). A plethora of biochemicals and fuels have also been commercialized from chemical conversion of biomass. These comprise lactic acid, succinic acid, 3-hydroxypropionic acid (HPA), levulinic acid, sorbitol, and xylitol (Budzianowski 2017). Dehydration of biomass sugars using acid catalysts produces furan derivatives like hydroxymethylfurfural (HMF). HMF is an excellent platform material for liquid alkane. Liquid alkane is a suitable fuel for diesel and jet engines (Yang & Yu 2013; Mujtaba et al. 2023). Other biochemicals such as glucaric acid, furans, adipic acid, isopropanol, and 1,5-diaminopentane are close to commercialization (Budzianowski 2017). Lactic acid is a precursor for bioplastics in the food and pharmaceutical sectors. The 1,3-propanediol serves as a platform chemical in polymer and cosmetic industries (John et al. 2025). Itaconic acid is an organic chemical with wide range of uses in the plastic, textile, paint, and pharmaceutical industries. It is industrially manufactured from microbial fermentation of biomass using *A. terreus* (Bhatia 2023).

3.7.2.2 Biopolymers

Biopolymers are renewable materials that can replace traditional polymers in packaging, film, and fibre industries. They are widely produced from first-generation feedstocks such as corn, rice, sugar beet, sugarcane, wheat, and sweet potatoes (Laca et al. 2011). Biopolymers resemble petrochemical polymers in terms of physical and mechanical properties. However, they are more superior than synthetic polymers due to their biodegradability and eco-friendly nature. Biopolymers of commercial importance include thermoplastic starch (TPS), polylactide (PLA), poly-β-hydroxybutyric acid (PHB), and polyhydroxyalkanoates (PHAs) (Laca et al. 2011). PHB is a storage biopolymer that is synthesized by cyanobacteria and eukaryotic algae such as *Phaeodactylum tricornutum* and *Chlamydomonas reinhardtii*. Whole algae can be utilized as a filler agent that can be combined with petroplastics to fabricate new bioplastics (Amoah et al. 2019). Biopolymers are struggling in the global market because of expensive raw materials and processing technologies. Efforts are underway to produce bioplastics from lignocellulosic biomass, including coffee husks, corn stover, pineapple leaf, spent coffee ground, banana peels, corn cob, rice husks, sugarcane bagasse, apple pomace, mango peel, grape pomace, and wheat straw. For example, itaconic acid has been extracted from corn stover, wheat chaff, bamboo, and wheat bran and explored for its potential to produce biopolymers. However, the extraction of fermentable sugars from non-edible materials is a costly process, limiting their utility as biopolymer resources (Wagh et al. 2024). PHAs can be fabricated via fermentation of agro- and forestry waste such as rye grass, wheat straw, rice straw, corn stover, and wood hydrolysates using strains of *B. megaterium, Ralstonia eutropha, Pseudomonas resinovarans, P. chlororaphis, Burkholderia sacchari, B. cepacia*, and *Paracoccus* sp., (Usmani et al. 2021).

3.7.2.3 Biopharmaceuticals

In the past years, chemical synthesis has been the most consistent route for the production of active pharmaceutical ingredients (APIs). Conventional routes exacerbate GHG emissions and release industrial solvents that may degrade the environment. This has motivated pharmaceutical producers to shift their attention towards green pharmaceuticals (Kumar et al. 2022). The term "biopharmaceuticals" refers to any pharmaceutical product that can be derived from biological resources. Biopharmaceuticals are manufactured through the fermentation of cells in stainless steel bioreactors. Fed-batch and perfusion cell culture are the most suitable processes for the fermentation of pharmaceutical products (Handa & Rajamani 2023). The global market demand for biopharmaceuticals is intensifying at a tremendous pace. In 2017, microbial-based biopharmaceuticals generated a total revenue of US$ 100 billion. Biomass feedstocks such as starch, cellulose, hemicellulose, pectin, and lignin are usually exploited for the synthesis of microbial-based pharmaceuticals. Microorganisms such as bacteria, yeast, and fungi are active biopharmaceutical producers. Agricultural and forestry waste have recently emerged as potential biopharma factories for the synthesis of high-value pharmaceutical products (Kumar et al. 2022). For example, olive pomace oil has been successfully fermented by *S. clavuligerus* to produce clavulanic acid. Clavulanic acid can replace glycerin in the manufacture of bacterial therapeutics. Dairy effluents and lignocellulosic waste have been examined for succinic acid production. Succinic acid is an API with multiple uses in the pharmaceutical industry. Other APIs of medical importance are shikimic acid, erythritol, cephalosporin, rifampicin, pregabalin, ectoine, etc. (Kumar et al. 2022).

3.7.2.4 Bio-Based Fertilizers

Valorization of organic waste into biofertilizers and additives is a sustainable and environmentally benign way to boost the agricultural economy. AD is a promising technology to alleviate energy poverty while producing fertilizing nutrients. Digestate waste from AD is a nutrient-rich resource for biofertilizer production. The nitrogen, phosphorous, and potassium content of digestate is comparable to chemical fertilizers. However, technologies still need further development to recover inorganic nutrients from biorefineries (Calvo-Flores & Martin-Martinez 2022). Composting is a mature process for the degradation and stabilization of organic matter. In this process, aerobes turn solid organic matter into a compost or biofertilizer, releasing heat and CO_2. Microbial population dynamics, pH, temperature, aeration, moisture, and biodegradable organic fraction should be frequently monitored as they affect composting efficiency. Compost requires less water and energy than synthetic fertilizers (Kannah et al. 2020). Vermicomposting is a suitable technology to recuperate nutrients from different types of biomass waste. Here, the collaborative efforts of microorganisms and earthworms convert organic waste into a vermicompost. Food scraps, yard waste, animal manure, paper waste, forestry residues, and agricultural waste are common feedstocks for vermicompost. Earthworm gut acts as a factory for decomposing organic matter while enriching the microbial flora. A vermicompost could be a futuristic source of plant nutrients. It contains humic acids that can promote plant growth and avert pathogenic outbreaks (Nascimento-Gonçalves et al. 2024). Vermicomposting of kitchen paper waste, kitchen yard waste, and cattle manure using red wriggler (*Eisenia fetida*) increased leaf area and biomass of radish, marigold, and upland cress. Consistency and quality of vermicompost are affected by feedstock composition, residence time, and various maturity parameters (Warman & AngLopez 2010).

3.7.2.5 Single-Cell Protein

SCP offers a low-cost, novel, and sustainable option for supplementing proteins in human, fish, and animal diets. It refers to edible, purified, and dried cell biomass or total protein obtained from microorganisms, including algae, yeasts, moulds, and bacteria (Kannah et al. 2020, Alokika et al. 2021). An ideal SCP microorganism should possess unique properties such as rapid growth, low nucleic acid content, good biological values, non-toxic, and non-pathogenic (Chaudhary et al. 2023). SCP is a potential substitute for conventional proteins like meat and soy. Microbial protein can be produced on complex substrates such as waste and by-products from food and agricultural activities, often with high carbon content (Bertasini et al. 2022). Biomass feedstocks for SCP include food waste, bagasse, rice polishings, citrus waste, soy molasses, yam peels (Kannah et al. 2020), glycerol-rich waste, wheat straw, rice bran, wheat bran, and other organic materials (Alokika et al. 2021). Natural gas, petroleum by-products, ethanol, methanol, n-alkanes, and acetic acid are high-energy sources of SCP (Bratosin et al. 2021). SCP production is more eco-friendly and mildly affected by agro-climatic conditions and saves water and land compared to agricultural protein (Spalvins et al. 2018). Its main problem is high amounts of toxic nucleic acids (3%–12%) that need to be removed during downstream processing (Chaudhary et al. 2023). This is an obstacle in the large-scale cultivation of SCP. The protein content of SCP depends on the type of microbial strain. Microalgae, bacteria, protists, and yeast contain 60%–70%, 30%–80%, 10%–20%, and 30%–50% w/w protein, respectively. Amongst them, yeasts are the best source of SCP due to their fast growth in a variety of sugary substrates (Bertasini et al. 2022). The cultivation of *S. cerevisiae* on food processing waste and digestate produced SCP with 28% w/w protein. This product could be used as feed for fish and non-ruminants (Bertasini et al. 2022).

3.7.2.6 Industrial Enzymes

In the 1960s, industrial enzymes were extracted from conventional sources like plant and animal tissues and exudates. Microbial fermentation became the primary source of enzymes in the early 1980s. In recent years, the bulk of commercial enzymes comes from microorganisms (Raina et al. 2022). Microbially derived enzymes are more preferred to plant and animal enzymes. This can be ascribed to their improved catalytic activity, stability, and simple production and optimization. Bacteria and fungi are the main producers of industrial enzymes, contributing 24% and 60% of the enzymes, respectively. Yeasts and *Streptomyces* are additional sources of microbial enzymes (Raina et al. 2022). Agro-industrial biomass is a valuable resource for producing industrial enzymes, including pectinases, lipases, amylases, proteases, tannases, xylanases, etc. (Kannah et al., 2020). This biomass is enriched in sugars, minerals, and proteins that promote the growth of microorganisms. Agricultural biomass (e.g. rice bran, wheat bran, wheat straw, sugarcane bagasse, and corn cob) is a plentiful, affordable, and renewable resource for commercial enzymes (Bharathiraja et al. 2017). Sugarcane bagasse has been commercially exploited for the production of high-value enzymes using solid-state and submerged fermentations (Alokika et al. 2021). Microbial enzymes are regarded as green substitutes for traditional catalysts in several industries, including pharmaceuticals, food and beverages, biofuels, textiles, paper, detergents, leather, and so on (Gopalan & Nampoothiri 2016; Raina et al. 2022). Amylases and pectinases are exploited in the food and beverage industry, cellulases in biofuel, leather and textile production, and tannases in the remediation of tannery waste (Panda et al. 2016; Bharathiraja et al. 2017). The food and beverage sectors are the largest consumers of industrial enzymes. European and American companies such as Novozymes, DSM, and DuPont are the major players in the industrial enzyme revolution. There is growing rivalry from Asian stakeholders, especially the Chinese and Japanese companies (Gopalan & Nampoothiri 2016).

3.8 OUTLOOK

A biorefinery is a cutting-edge technology for biomass valorization into multiple product streams. The prospect of biorefineries replacing petrorefineries is promising, with facilities at various stages of development worldwide. It is no doubt that biorefineries will substitute traditional fossil refineries in the long run. The conception is hinged on the conversion of biomass feedstocks into a diversity of clean fuels and chemicals, minimizing reliance on fossil products. The potential benefits of biorefineries over petrorefineries cannot be overemphasized. Yet a few hurdles need to be overcome towards maturation and full-scale adoption of the biorefinery concept. Apart from conventional biorefineries, the technology is still under R & D, pilot or small-scale demonstration.

Numerous strategies are essential for developing biorefineries that can compete in the market with petrorefineries. This calls for the development of integrated biorefineries targeted at producing a wide range of high-value products. Widening the product base offers an economically sound and more sustainable biorefinery environment. Multipurpose biorefinery facilities should be coupled with modern technologies to effectively manage and enhance the operational system. Computing technologies such as artificial intelligence, machine learning, and process modelling are innovative tools to optimize the economic viability of biorefineries (Mohammad et al. 2024). Either genetic or metabolic engineering of microbial cell factories is key to establishing a thriving biorefinery. A new paradigm is to find efficient, overproducing, and novel strains capable of metabolizing a broad spectrum of biomass feedstocks. This may help to create new products and increase yields of existing ones. Biomass feedstocks are morphologically diverse due to varying geographical conditions. As such, the global-scale development of biorefineries should be reinforced by local resources and adapted to the local environment. One of the key pre-requisites is to isolate and develop microbial cell cultures that can easily acclimate to the local feedstocks and conditions (Amoah et al. 2019).

LCF biorefinery seems to offer an attractive option for addressing energy and environmental challenges. Low-cost biowaste can be recycled, thereby promoting sustainability. However, feeding biorefineries with LCF can be a daunting task due to economic and technical hurdles. Lignocellulosic biomass is resistant to biodegradation, restraining the scalability of LCF biorefineries. There is a need for innovative pretreatment solutions to surpass this technological barrier. Research efforts have been primarily confined to physical and chemical methods. These technologies are expensive, use more energy, and may result in further pollution. It is obligatory to educate the global community on clean technologies, focusing on economically sound options for biomass utilization. Developing cost-effective and eco-friendly methods that utilize microbes and their enzymes could provide a sustainable and clean route to enhance biorefinery business models (Alokika et al. 2021).

Some countries, for example, Brazil, the U.S., and the E.U., have launched biorefineries at commercial scale to attain a zero-waste status quo. This owes to governmental initiatives through increased policy frameworks, regulations, subsidies, and incentives that can promote the green biorefinery platforms. Collaborative efforts from multiple stakeholders foster a conducive environment that is open for business development and promotes the growth of biorefineries. The converse is true for the developing world, where a few biorefineries have been industrialized. Most research findings for biomass valorization are restricted to small-scale setups. Small-scale findings require further translation to pilot and demonstration phases towards full-scale valorization. Securing funding opportunities for upgrading biorefineries is an impediment in most developing countries. Developing countries could benefit from a cross-sectoral approach, which encourages the sharing of knowledge, infrastructure, and resources to achieve sustainable development goals. Policy formulations should be coupled with a full analysis of the life cycle and environmental impact to ensure a healthy and viable biorefinery ecosystem.

REFERENCES

Ahmed, S. F., M. Kabir, A. Mehjabin, F. T. Z. Oishi, S. Ahmed, S. Mannan, M. Mofijur, F. Almomani, I. A. Badruddin, & S. Kamangar. 2023. Waste biorefinery to produce renewable energy: Bioconversion process and circular bioeconomy. *Energy Rep.* 10: 3073–3091.

Ali, S. S., R. Al-Tohamy, T. M. Mohamed, Y. A.-G. Mahmoud, H. A. Ruiz, L. Sun, & J. Sun. 2022. Could termites be hiding a goldmine of obscure yet promising yeasts for energy crisis solutions based on aromatic wastes? A critical state-of-the-art review. *Biotechnol. Biofuels* 15: 35.

Alokika, A., A. Kumar, V. Kumar, & B. Singh. 2021. Cellulosic and hemicellulosic fractions of sugarcane bagasse: Potential, challenges and future perspective. *Int. J. Biol. Macromol.* 169: 564–582.

Amoah, J., P. Kahar, C. Ogino, & A. Kondo. 2019. Bioenergy and biorefinery: Feedstock, biotechnological conversion and products. *Biotechnol. J.* 14: e1800494.

An, H., W. E. Wilhelm, & S. W. Searcy. 2011. Biofuel and petroleum-based fuel supply chain research: A literature review. *Biomass Bioenergy* 35: 3763–3774.

Antar, M., D. Lyu, M. Nazari, A. Shah, X. Zhou, & D. L. Smith. 2021. Biomass for a sustainable bioeconomy: An overview of world biomass production and utilization. *Renew. Sust. Energy Rev.* 139: 110691.

Apotei, K. G. 2019. Biomass as an alternative source of energy in Ghana. PhD diss, Vaasan Ammattikorkeakoulu Univ. Appl. Sci.

Aristizábal-Marulanda, V., & C. A. C. Alzate. 2019. Methods for designing and assessing biorefineries: Review. *Biofuels Bioprod. Bioref.* 13: 789–808.

Bertasini, D., R. L. Binati, D. Bolzonella, & F. Battista. 2022. Single-cell proteins production from food processing effluents and digestate. *Chemosphere* 296: 134076.

Bharathiraja, S., J. Suriya, M. Krishnan, P. Manivasagan, & S.-K. Kim. 2017. Production of enzymes from agricultural wastes and their potential industrial applications. In *Advances in Food and Nutrition Research* (vol. 80), ed. S.-K. Kim, & F. Toldra, 125–148. Cambridge, MA: Elsevier.

Bhatia, S. K. 2023. Biorefinery for the sustainable biochemicals production: Process design and technological advances. *Sustainability* 15: 7973.

Bratosin, B. C., S. Darjan, & D. C. Vodnar. 2021. Single-cell protein: A potential substitute in human and animal nutrition. *Sustainability* 13: 9284.

Budzianowski, W. M. 2017. High-value low-volume bioproducts coupled to bioenergies with potential to enhance business development of sustainable biorefineries. *Renew. Sust. Energy Rev.* 70: 793–804.

Calvo-Flores, F. G., & F. J. Martin-Martinez. 2022. Biorefineries: Achievements and challenges for a bio-based economy. *Front. Chem.* 10: 973417.

Chandrasekhar, K., T. Raj, S. V. Ramanaiah, G. Kumar, J. R. Banu, S. Varjani, P. Sharma, A. Pandey, S. Kumar, & S.-H. Kim. 2022. Algae biorefinery: A promising approach to promote microalgae industry and waste utilization. *J. Biotechnol.* 345: 1–16.

Chaudhary, V., P. Kajla, D. Verma, T. P. Singh, A. Kothakota, V. A. Prasath, G. Jeevarathinam, M. Kumar, S. Ramniwas, S. Rustagi, & R. Pandiselvam. 2023. Valorization of dairy wastes into wonder products by the novel use of microbial cell factories. *Trends Food Sci. Technol.* 142: 104221.

Cherubini, F., G. Jungmeier, M. Wellisch, T. Willke, I. Skiadas, R. V. Ree, & E. de Jong. 2009. Toward a common classification approach for biorefinery systems. *Biofuels Bioprod. Bioref.* 3: 534–546.

de Jong, E., & G. Jungmeier. 2015. Biorefinery concepts in comparison to petrochemical refineries. In *Industrial Biorefineries & White Biotechnology*, ed. A. Pandey, R. Höfer, M. Taherzadeh, K. M. Nampoothiri, & C. Larroche, 3–33. Amsterdam, Oxford, Waltham, MA: Elsevier.

Dlangamandla, N., J. Zininga, D. Osipov, M. Semenova, O. Sinitsyna, & K. Permaul. 2025. Biobutanol production from cellulose-rich sugarcane bagasse and hard aspen wood residue. *Waste Manag. Bull.* 3: 100223.

Ghatak, H. R. 2011. Biorefineries from the perspective of sustainability: Feedstocks, products, and processes. *Renew. Sust. Energy Rev.* 15: 4042–4052.

Gopalan, N., & K. M. Nampoothiri. 2016. Biotechnological production of enzymes using agro-industrial wastes: Economic considerations, commercialization potential, and future prospects. In *Agro-Industrial Wastes as Feedstock for Enzyme Production*, ed. G. S. Dhillon, & S. Kaur, 313–330. London: Academic Press.

Handa, A., & P. Rajamani. 2023. Waste management and environment. In *Basic Biotechniques for Bioprocess and Bioentrepreneurship*, ed. A. K. Bhatt, R. K. Bhatia, & T. C. Bhalla, 391–413. London: Academic Press.

Huang, H., S. Ramaswamy, U. W. Tschirner, & B. V. Ramarao. 2008. A review of separation technologies in current and future biorefineries. *Sep. Purif. Technol.* 62: 1–21.

IEA. 2022. *Global Biorefinery Status Report 2022. IEA Bioenergy: Task 42 Biorefining in a Circular Economy*. Paris: IEA Bioenergy.

IEA. 2025. *Outlook for Biogas and Biomethane. A Global Geospatial Assessment*. Paris: IEA Bioenergy.

Jingura, R. M., D. Musademba, & R. Kamusoko. 2013. A review of the state of biomass energy technologies in Zimbabwe. *Renew. Sustain. Energy Rev.* 26: 652–659.

John, M., M. O. Abdullah, T. Y. Hua, & C. Nolasco-Hipólito. 2021. Techno-economical and energy analysis of sunflower oil biodiesel synthesis assisted with waste ginger leaves derived catalysts. *Renew. Energy* 168: 815–828.

John, S. U., C. E. Onu, C. M. Ezechukwu, I. C. Nwokedi, & C. N. Onyenanu. 2025. Multi-product biorefineries for biofuels and value-added products: Advances and future perspectives. *Acad. Green Energy* 2: https://doi.org/10.20935/AcadEnergy7605

Kamusoko, R., R. M. Jingura, W. Parawira, & Z. Chikwambi. 2021. Strategies for valorization of crop residues into biofuels and other value-added products. *Biofuels Bioprod. Bioref.* 15: 1950–1964.

Kamusoko, R., R. M. Jingura, W. Parawira, & Z. Chikwambi. 2022. Biogas: Microbiological research to enhance efficiency and regulation. In *Handbook of Biofuels*, ed. S. Sahay, 485–497. London: Academic Press.

Kamusoko, R., & P. Mukumba. 2024. Potential of wheat straw for biogas production by anaerobic digestion in South Africa: A review. *Energies* 17: 4662.

Kamusoko, R., & P. Mukumba. 2025. Pineapple waste biorefinery: An integrated system for production of biogas and marketable products in South Africa. *Biomass* 5: 17.

Kannah, R. K., J. Merrylin, T. P. Devi, S. Kavitha, P. Sivashanmugam, G. Kumar, & J. R. Banu. 2020. Food waste valorization: Biofuels and value added product recovery. *Bioresour. Technol. Rep.* 11: 100524.

Khoaele, K. K., O. J. Gbadeyan, I. J. Mphahlele, B. Sithole, & V. Chunilall. 2024. Fundamentals of biomass waste valorization. In *Valorization of Biomass Waste Materials, A Biorefinery Concept*, ed. O. J. Gbadeyan, P. Sekoai, & V. Chunilall, 1–17. Newcastle: Cambridge Scholars Pub.

Kostas, E. T., J. M. M. Adams, H. A. Ruiz, G. Durán-Jiménez, & G. J. Lye. 2021. Macroalgal biorefinery concepts for the circular bioeconomy: A review on biotechnological developments and future perspectives. *Renew. Sustain. Energy Rev.* 151: 111553.

Kumar, S., K. Paritosh, N. Pareek, A. Chawade, & V. Vivekanand. 2018. Deconstruction of major Indian cereal crop residues through chemical pretreatment for improved biogas production: An overview. *Renew. Sustain. Energy Rev.* 90: 160.

Kumar, V., V. Bansal, A. Madhavan, M. Kumar, R. Sindhu, M. K. Awasthi, P. Binod, & S. Saran. 2022. Active pharmaceutical ingredient (API) chemicals: A critical review of current biotechnological approaches. *Bioengineered* 13: 4309–4327.

Laca, A., M. Herrero, & M. Díaz. 2011. Life cycle assessment in biotechnology. In *Comprehensive Biotechnology* (2nd ed.), ed. M. Moo-Young, 839–851. Amsterdam: Academic Press.

Lee, A. K., D. M. Lewis, & P. J. Ashman. 2012. Disruption of microalgal cells for the extraction of lipids for biofuels: Processes and specific energy requirements. *Biomass Bioenergy* 46: 89–101.

Leong, H. Y., C. K. Chang, K. S. Khoo, K. W. Chew, S. R. Chia, J. W. Lim, J.-S. Chang, & P. L. Show. 2021. Waste biorefinery towards a sustainable circular bioeconomy: A solution to global issues. *Biotechnol. Biofuels* 14: 87.

Lin, Z., W. Cong, & J. Zhang. 2023. Biobutanol production from acetone–butanol–ethanol fermentation: Developments and prospects. *Fermentation* 9: 847.

Liu, X., Z. Yan, & Z.-B. Yue. 2011. Biogas. In *Comprehensive Biotechnology* (2nd ed.), ed. M. Moo-Young, 99–114. Amsterdam: Academic Press.

Mailaram, S., P. Kumar, A. Kunamalla, P. Saklecha, & S. K. Maity. 2021. Biomass, biorefinery and biofuels. In *Sustainable Fuel Technologies Handbook*, ed. S. Dutta, & C. M. Hussain, 51–57. London: Academic Press.

Maity, S. K. 2015. Opportunities, recent trends and challenges of integrated biorefinery: Part II. *Renew. Sust. Energy Rev.* 43: 1446–1466.

Manandhar, A., S. H. Mousavi-Avval, J. Tatum, E. Shrestha, P. Nazemi, & A. Shah. 2022. Solid biofuels. In *Biomass, Biofuels, Biochemicals*, ed. G. S. Murthy, E. Gnansounou, S. K. Khanal, & A. Pandey, 343–370. Amsterdam: Elsevier.

Mohammad, R. E. E., S. S. Abdullahi, H. A. Muhammed, H. Musa, S. Habibu, A. H. Jagaba, & A. H. Birniwa. 2024. Recent technical and non-technical biorefinery development barriers and potential solutions for a sustainable environment: A mini review. *Case Stud. Chem. Environ. Eng.* 9: 100586.

Mujtaba, M., L. F. Fraceto, M. Fazeli, S. Mukherjee, S. M. Savassa, G. A. de Medeiros, A. d. E. S. Pereira, S. D. Mancini, J. Lipponen, & F. Vilaplana. 2023. Lignocellulosic biomass from agricultural waste to the circular economy: A review with focus on biofuels, biocomposites and bioplastics. *J. Clean. Prod.* 402: 136815.

Nascimento-Goncalves, E., T. Azevedo, H. Lopes, J. R. Sousa, P. A. Oliveira, M. Roboredo, A. M. Coimbra, & M. C. Morais. 2024. Vermicomposting as a valorization solution to the winery sector by-products. *Agronomy* 14: 1111.

Panda, S. K., S. S. Mishra, E. Kayitesi, & R. C. Ray. 2016. Microbial-processing of fruit and vegetable wastes for production of vital enzymes and organic acids: Biotechnology and scopes. *Environ. Res.* 146: 161–172.

Paul, N. M., T. H. Haritha, G. Rudran, & N. John. 2024. Biomass: Existing management and practices. In *Handbook of Biomass*, ed. S. Thomas, M. Hosur, D. Pasquini, & C. J. Chirayil, 1419–1447. Singapore: Springer.

Raina, D., V. Kumar, & S. Saran. 2022. A critical review on exploitation of agro-industrial biomass as substrates for the therapeutic microbial enzymes production and implemented protein purification techniques. *Chemosphere* 294: 133712.

Roberts, L. G., & T. J. Patterson. 2014. Biofuels. In *Encyclopedia of Toxicology*, ed. P. Wexler, 469–475. Amsterdam: Academic Press.

Sadhukhan, J., E. Martinez-Hernandez, M. A. A. Allieri, J. A. Z. Eguía-Lis, A. Castillo, D. Dominguillo, E. Torres-García, & J. Aburto. 2024. Strategic navigation of world-leading biorefineries and Mexico's policy landscape: A gateway to a sustainable circular bioeconomy. *J. Clean. Prod.* 434: 140386.

Sadhukhan, J., E. Martinez-Hernandez, R. J. Murphy, D. K. S. Ng, M. H. Hassim, K. S. Ng, W. Y. Kin, I. F. Md Jaye, M. Y. L. P. Hang, & V. Andiappan. 2018. Role of bioenergy, biorefinery and bioeconomy in sustainable development: Strategic pathways for Malaysia. *Renew. Sust. Energy Rev.* 81: 1966–1987.

Shahbeik, H., H. K. S. Panahi, M. Dehhaghi, G. J. Guillemin, A. Fallahi, H. Hosseinzadeh-Bandbafha, H. Amiri, M. Rehan, D. Raikwar, H. Latine, B. Pandalone, B. Khoshnevisan, C. Sonne, L. Vaccaro, A. Nizami, V. K. Gupta, S. S. Lam, J. Pan, R. Luque, B. Sels, W. Peng, M. Tabatabaei, & M. Aghbashlo. 2024. Biomass to biofuels using hydrothermal liquefaction: A comprehensive review. *Renew. Sust. Energy Rev.* 189: 113976.

Shankar, S., J. Singh, N. Chakravarty, A. Mathur, & R. P. Singh. 2022. Algal biorefinery: Challenges and opportunities. In *Production of Top 12 Biochemicals Selected by US DOE from Renewable Resources*, ed. A. K. Chandel, & F. Segato, 41–79. Amsterdam: Elsevier.

Sher, F., S. Hameed, N. S. Omerbegović, A. Chupin, I. Ul Hai, B. Wang, Y. H. Teoh, & M. J. Yildiz. 2025. Cutting-edge biomass gasification technologies for renewable energy generation and achieving net zero emissions. *Energy Convers. Manag.* 323: 119213.

Spalvins, K., L. Zihare, & D. Blumberga. 2018. Single-cell protein production from waste biomass: Comparison of various industrial by-products. *Energy Procedia* 147: 409–418.

Speight, J. G. 2020. Production of fuels from non-fossil fuel feedstocks. In *The Refinery of the Future* (2nd ed.), ed. J. G. Speight, 391–426. Cambridge, MA: Elsevier.

Takkellapati, S., T. Li, & M. A. Gonzalez. 2018. An overview of biorefinery derived platform chemicals from a cellulose and hemicellulose biorefinery. *Clean Technol. Environ. Policy* 20: 1615–1630.

Takki, K. 2024. *Biomass to Account for 18% of the Energy Market*. Helsinki: Statzon. https://statzon.com/insights/the-global-biomass-market (accessed 14 July 2025).

Tiseo, I. 2025. *Greenhouse Gas Emissions Worldwide – Statistics & Facts*. Statista. https://www.statista.com/topics/5770/global-greenhouse-gas-emissions/#topicOverview (accessed 17 July 2025).

Ubando, A. T., E. A. S. Ng, W. Chen, A. B. Culaba, & E. E. Kwon. 2022. Life cycle assessment of microalgal biorefinery: A state-of-the-art review. *Bioresour. Technol.* 360: 127615.

UNEP. 2015. *Transforming Our World: The 2030 Agenda for Sustainable Development*, A/RES/70/1. United Nations. https://www.refworld.org/legal/resolution/unga/2015/en/111816 (accessed 28 July 2025).

Usmani, Z., M. Sharma, A. K. Awasthi, N. Sivakumar, T. Lukk, L. Pecoraro, V. K. Thakur, D. Roberts, J. Newbold, & V. K. Gupta. 2021. Bioprocessing of waste biomass for sustainable product development and minimizing environmental impact. *Bioresour. Technol.* 322: 124548.

Wagh, M. S., S. Sowjanya, P. C. Nath, A. Chakraborty, R. Amrit, B. Mishra, A. K. Mishra, & Y. K. Mohanta. 2024. Valorization of agro-industrial wastes: Circular bioeconomy and biorefinery process – A sustainable symphony. *Process Saf. Environ. Prot.* 183: 708–725.

Warman, P. R., & M. J. AngLopez. 2010. Vermicompost derived from different feedstocks as a plant growth medium. *Bioresour. Technol.* 101: 4479–4483.

Yamakawa, C. K., F. Qin, & S. I. Mussatto. 2018. Advances and opportunities in biomass conversion technologies and biorefineries for the development of a bio-based economy. *Biomass Bioenergy* 119: 54–60.

Yang, S., & M. Yu. 2013. Integrated biorefinery for sustainable production of fuels, chemicals, and polymers. In *Bioprocessing Technologies in Biorefinery for Sustainable Production of Fuels, Chemicals, and Polymers*, ed. S. Yang, H. A. El-Enshasy, & N. Thongchul, 1–26. Hoboken, NJ: John Wiley & Sons.

4 The Biorefinery as a Business Model

A French Retrospective from a Territorial Value Point of View

Ion Lucian Ceapraz and Claude Chammaa

4.1 INTRODUCTION: FROM THE BIOECONOMY TO THE BIOREFINERY

Today, finding solutions to mitigate gas emissions that contribute to climate change is becoming urgent. To address this, various actions have been implemented, primarily through the Conference of the Parties (COP), which aims to limit the increase in global temperature to less than two degrees and fight climate change by reducing greenhouse gas emissions. The goal is to establish a common framework and secure funding for the most vulnerable countries to achieve net-zero carbon emissions by 2050.

Europe has expanded its biobased economy by more than 40% in less than a decade, largely due to the implementation of national bioeconomy strategies. To address this challenge, the biorefinery has been developed, which turns these problems into opportunities by converting waste into energy (Kleinschmit et al. 2014). Biorefineries are part of the Sustainable Development Goals (SDGs), helping to reduce fossil energy demand in favor of sustainable energy sources. The shift toward a bioeconomy primarily relies on the development of biorefineries (Kleinschmit et al. 2014).

Several definitions of the biorefinery are used in the scientific literature, depending on the conceptual context. From an economic perspective, a biorefinery is defined as "the sustainable processing of biomass into a range of marketable products (food, feed, materials, and chemicals) and energy (fuels, power, and heat)" (International Energy Agency (IEA) 2012). The IEA explains the complexity of a biorefinery using the Biorefinery Complexity Index, which helps decision-makers and investors develop strategies by considering technological and economic risks. Such investment decisions are also exposed to behavioral biases, which need to be managed to improve investors' decision-making (Chammaa & Haddad 2025).

The bioeconomic transition based on renewable resources is urgently needed to reduce global carbon emissions. Several conditions are crucial for an effective shift toward a carbon-free future. This transition requires "strong partnerships and

DOI: 10.1201/9781003650751-4

sustainable goals between different sectors, raw materials, and countries" (Christensen 2019). Thus, according to Roy and Tu (2022), the bioeconomic transition "has been incorporated into the national sustainable development strategies of many countries," and the concept of biorefinery has been considered as one of the vital elements of these strategies.

A biorefinery, as a comprehensive system, is based on a set of industrial, economic, political, and other logics, which include "processes for converting different types of biomasses using various conversion technologies, enabling the generation of bioenergy and various biobased products" (Brunnhofer et al. 2020). Using biomass is considered "the most important pillar for transforming the traditional bioeconomic model into a biobased economy" (Solarte-Toro & Alzate 2021). According to these authors, biomass is a vital element, but not enough to ensure the process's sustainability. Therefore, other key factors "related to the consumption of natural resources, waste, job creation, impact on the local community, economic growth, and regional income" must also be considered to ensure the sustainability of the biorefinery ecosystem (Solarte-Toro & Alzate 2021). Since one of the main reasons for establishing a biorefinery is its sustainability, we also include environmental, social, and economic prospects, which are very important and should "be valued throughout the value chain and life cycle of biorefineries."

The sustainability to be achieved must also be accompanied by a bioeconomic transition, which must be carried out on several criteria (Agridees 2025):

- Upstream, consistent public policies linked to the availability of biomass,
- Downstream, incentives to consume biobased products by highlighting the role of biomass in agricultural areas, and
- At the intermediate level, ensuring that innovation and research and development (R&D) play a priority role to remain competitive with petroleum-based products.

Concerning this transition, there is a spatial side to this value chain. Thus, "the territorial nature of the bioeconomy" is essential in the agricultural sector. Thus, the bioeconomy is "rooted in territories, defined as living areas, economic areas, with homogeneous agronomic and pedoclimatic conditions" (Agridees 2025). Within the framework of the bioeconomy, a diverse range of biorefineries is being developed (Laibach et al. 2019), with characteristics and regulations that must be translated into commercialization and technology transfer to markets facing specific challenges (Golombiewski et al. 2015). In this sense, new business models are needed to enable these technologies to bring added value and incorporate biobased products into the market.

4.2 EMPIRICAL DEFINITIONS: BIOECONOMY, BIOREFINERIES, AND BUSINESS MODELS

There is a lack of business models in the bioeconomy that address, for example, circularity (Salvador et al. 2023). Several types of business models have been described by these authors based on a literature and field review methodology: "optimizing resource efficiency and use; establishing biorefineries; value recovery from waste;

resource exchange; innovation towards bio- and renewable resources; valuing the local economy; service- and result-oriented value offers."

The literature on business models that characterize the transition to the bioeconomy and to new biobased technologies has been limited (Golombiewski et al. 2015; Van Lancker et al. 2016), especially in relation to "a complex knowledge base, commercialization issues, and fragmented policy programs." However, according to Bröring and Vanacker (2022), this literature is beginning to develop gradually.

Various factors can hinder the implementation of successful business models, such as skills originating from unrelated sectors (Reim et al. 2019). Other factors limiting the creation of business models include the ability of companies to adapt these skills (Cohen & Levinthal 1990) and even regulations on the adoption of new technologies. There are various types of business models in the scientific literature. Some explain that business models are generally ideas of how a company conducts its business. Furthermore, Teece (2010) emphasizes that a business model "describes how the company will convert its resources and capabilities into economic value."

Other scientific works describe the business model as a series of elements: "the value proposition (product/service offering, customer segments, and customer relationships), activities, resources, partners, distribution channels (i.e. value creation and delivery), and cost structure as well as the revenue model (i.e. value capture)" (Osterwalder & Pigneur 2010).

There are also other definitions of business models which propose a definition encompassing three elements, including "value proposition, value creation and delivery, and value capture."

4.3 WHY IS THE TERRITORIAL APPROACH IMPORTANT FOR A BUSINESS MODEL?

Åckerman (2021) highlights an important link between our business model in general and the innovation business model in particular, as well as regional development. Biorefinery as a concept, system, or production process is a decisive factor that can support regional development and reindustrialization, while reducing dependence on fossil resources (Bauer et al. 2017).

The implementation of the economic strategy in the EU began in 2012 with an update in 2018. The main objectives of this strategy, as outlined by the European Commission, are to ensure the long-term competitiveness of the European bioeconomy, to increase the circular and efficient use of resources, to secure a competitive and sustainable biomass supply, and to position the EU in a rapidly expanding market. Before proceeding further in the chapter, we will define territorial biorefineries.

The concept of territorial biorefinery is highlighted in our paper through several characteristics that will be developed, also in connection with the concept of territorial value. A new business model that is linked to its territory and involves territorial alignment, governance of its ecosystem, and coordination of stakeholders through a local network gives meaning to the territorial business model.

According to Ubando et al. (2020), "the biorefinery acts as a strategic mechanism for the realization of a circular economy" by considering the economic, environmental, and social aspects of the industrial world. The authors point out that the concept of biorefinery was introduced by Cherubini (2010) to demonstrate that it is used to "maximize the use of biomass and minimize waste and emissions associated with the conversion of bioenergy products." Cherubini (2010) defines a biorefinery as "a facility (or network of facilities) integrating biomass conversion processes and equipment to produce biofuels, electricity, and chemicals for transportation from biomass."

One of the most interesting definitions is given by the IEA, which describes the biorefinery as "the sustainable processing of biomass into a spectrum of marketable products (food, feed, materials, and chemicals) and energy (fuels, power, and heat)." Other definitions emphasize the biorefinery as a concept, a facility, a process, a plant, or even a cluster of facilities. Moreover, the Industries and Agro-Resources (IAR) competitiveness cluster describes the following definition of a biorefinery:

> a biorefinery is an industrial complex, located on the same site, which turns agricultural and forest biomass into a variety of biobased products (food, feed, chemicals, biomolecules, and agro-materials) and bioenergy (biofuels, electricity, and heat) as part of a sustainable strategy.

Before embarking on any conceptual approach to frame the concept of territorial biorefineries, we must clarify its definition. According to Ceapraz et al. (2016), territorial or territorialized biorefineries differ from other biorefineries in the following ways: their capacity for territorial anchoring and integration of stakeholders in the value-chain based on proximity but also in relation to resource needs (Colletis et al. 1999) and complementarity between the stakeholders involved.

The fact that the bioeconomy proposes different business models compared to the fossil fuel economy is "based on different geographical conditions and many disconnected stakeholders" (Wohlfahrt et al. 2019). The territorial bioeconomy introduces "space" as a new variable, as stakeholders in charge of different projects must manage multiple scales, ranging from local to international geographic scales (Wohlfahrt et al. 2019). Moreover, "a territorial bioeconomy system is a territory characterized by the presence of one or several biobased value chains (e.g. bioenergy plants and biorefineries), aimed at replacing non-renewable resources and partially dependent on local biomass feedstocks" (Wohlfahrt et al. 2019). The concept of the territorial biorefinery and its conceptualization in scientific literature have received little attention. This concept requires some theoretical clarification. According to Ceapraz et al. (2016), the territorial anchor of a biorefinery can initially be approached by looking back at the role of the territory in its development. There are two types of territorial anchoring for the bioeconomy, namely, "sectoral bioeconomy" and "territorial bioeconomy." Territorial bioeconomy is characterized by adaptation to local specificities in terms of product supply (differentiation) and demand (short and local supply chains). The sectoral bioeconomy, on the other hand, refers to an industrial framework and economies of scale (Benoit 2021; Callois 2022). Without providing a historical review of the literature, the work of Pecqueur (2005) aligns more closely with our vision in clarifying the status of exogenous or endogenous territories in relation to resource use and, implicitly, in value creation. The terms active or passive territoriality, or constructed and given territory, are also used. A simple definition of territory by Colletis (2010) highlights this distinction. Territory is thus "a space of potential resources to be identified, even an active entity that can develop its potential in its local resources, understood in the broad sense, that is, its activities, its actors, its networks, etc." This link between economic activity and territory is therefore part of the concept of territorial anchorage.

4.4 METHODOLOGICAL AND CONCEPTUAL FRAMEWORK

The concept of a "territorial business model" for biorefineries is introduced as a new approach to considering value. The business model, as defined by Richardson (2008), comprises three functions, such as value proposition, value generation, and value capture. According to Bröring and Vanacker (2022), the economic value of the business model has gradually been enriched with the addition of environmental value. In our chapter, we focus mostly on the territorial value of business models with additional "building blocks" observed in the works of Osterwalder & Pigneur (2010). We will describe this value according to our own methodology. This territorial value is described as very important in scientific literature, especially in sustainable business models (Maillefert & Robert 2017). According to these authors, the territorial dimension is "a key element in the creation of corporate value and can take on a territorialized value" with several dimensions. Other works in the literature confirm that the logic of territorialization in sustainable business models brings value to the company from a relational perspective and therefore implicitly through proximity approaches.

In our case, we consider that territorial value is composed of several attributes, conceptually constructed around the territorial anchoring (both exogenous and endogenous development of the territory) and the governance of the territory.

4.4.1 The First Approach of the Conceptual Framework: The Territorial Anchoring

Territorial anchoring links economic activity to the territory, seen as "the result of a joint construction, a collective learning process based on the coproduction of resources" (Zimmermann 2005). It represents a more settled and sustainable form of relationship with the territory, complementing the notion of localization with that of territorialization.

4.4.1.1 Exogenous Development of the Territory: Localization

Firstly, the localization of economic activity implies a significant role for productive activities according to the attributes and characteristics of a territory. Territories and their attributes, being immobile by definition, allow companies to define their spatial location based on real or perceived territorial advantages (Ceapraz et al. 2016).

Location factors depend on each company or regional project and can be considered countless in terms of choice. Space is considered exogenous and location "results from the specific characteristics and attributes of the region in question" (Lauriol et al. 2008). For example, Chan (2001) proposes several types of location determinants for a company, including quantifiable factors (such as transportation costs, personnel costs, energy costs, and land availability) and non-quantifiable factors (such as government stability, competition, customer/worker preferences, pollution, and others). In the case of biorefineries, their optimal location is part of companies' long-term resolutions and is considered one of the most important strategic decisions.

For example, according to Cerca et al. (2022), to describe the strategic decisions in a biorefinery, we must consider several aspects that "might include feedstock availability and sourcing, possible suppliers, the optimal location, its capacity, and markets that can be accessed" (Zahraa et al. 2020). Pettersson and Harvey (2013) describe the various criteria considered important in the location of biorefineries, which include feedstock supply, market access, economies of scale, industrial integration, local use of infrastructure and know-how, and financial capital.

Other authors, such as Lopez-Díaz et al. (2017), consider that there is a plurality of studies on the location of biorefineries that have linked the upstream and downstream supply chain.

Location decisions are crucial and give rise to Decision Support Tools (DSTs) that can assist in making informed location decisions. This is the case with Operations Research, also known as the Facility Location Problem (FLP) (Serrano-Hernandez et al. 2017). Piedrahita-Rodríguez et al. (2023) discuss the growing interest in the location of biorefineries, which affects all aspects of sustainability (Pettersson & Harvey 2013). Among the most important factors mentioned are the supply of raw materials, the market for biobased products, and industry expertise. Optimal location analysis is considered appropriate, when sustainability objectives are not affected (Sharara et al. 2020). Thus, location depends on the objectives set by the company, which can be either objective or subjective. For example, let us consider the economic aspect as an objective that the company sets for itself. The best location is one that offers very low economic/logistical costs (transport costs, labor or raw material costs, etc.). In environmental terms, too, the objectives are to find the lowest cost in relation to the location with the lowest possible carbon footprint.

4.4.1.2 Endogenous Development of the Territory: Activation of Resources, Coordination, and Proximities

Secondly, the territorialization as a complementary attribute of territorial anchoring, besides that of localization, can be seen as "a spatialized construct, whose logic of constitution is based on the intentional combination of individual and/or collective actions and the mobilization of specific resources" (Kotbi & Sauvee 2010). The field is no longer merely a passive support in location choices, but a part of the active construction of comparative advantages by private and institutional actors. We talk about "the idea of a common construction, the idea of collective learning based on the coproduction of resources" (Zimmermann 2005). It is therefore a question of activating these territorial resources or building them: "territorial resources do not pre-exist the territory, but are built with and within the territory" (Colletis & Pecqueur 2018). Thus, these resources might refer directly to the many economic theories of value. In our case, the focus is on the territory, which allows for a connection between these territorial resources and their value. According to Lauriol et al. (2008), the question to be addressed is that of the spatial specificity of these resources and their construction dynamics. This gives rise to the problem of coordinating these resources and the spatial organization necessary for their structuring, mobilization, and enhancement within the company. Lamara (2009) even refers to endogenous development of the territory as a "social coconstruction" resulting from "coordination between actors" with a view to "building territorial resources" through their activation.

Thus, territorial anchoring, which also involves the coordination of economic actors, requires differentiation between territories, bearing in mind that their mobilization in the use of resources involves the construction of comparative advantages (Guibert 2024). Therefore, these dynamics of resource building that generate comparative advantages (Colletis et al. 1999) involve not only choosing the optimal location but also coordinating economic and institutional actors.

In France, the group "Dynamiques de Proximité" believes that proximity changes the way we view the theoretical frameworks and applied tools of regional science. The proximity groups follow four major phases associated with the development of the concept of proximity, that is, birth, establishment, maturity, and the post-period that began after 2014. Strong collective cohesion, common characteristics, and simple organization mark the origin of this concept. Establishment is marked by the consolidation of the concept of proximity, characterized by an explosion of collaborations and productions. The influence of external authors, organizational evolution, and the role of key intermediaries (Rallet & Torre, 2004) ensures internal cohesion and external connections. Maturity is characterized by member autonomy and greater dissemination of the concept of proximity, and finally, the post-period, which is the current period. According to Talbot (2008), there are "two essential coordination mechanisms, namely, spatial and non-spatial proximity," which are broken down into several approaches. While spatial proximity refers to geographical space, this approach focuses on both geographical and organizational proximity. In contrast, non-spatial proximity is divided into three categories, including geographical, institutional, and organizational.

The concept of proximity sheds lights more specifically on the type of coordination that exists between stakeholders and becomes a "major element in the analysis of these coordination issues in territorial organization" (Lauriol et al. 2008). Proximity, thus, represents "a conception of economic and social reality that is essentially relational", based on "economic or geographical separation of agents (individual or collective), who hold different resources and, who have close or distant interactions in the resolution of an economic problem". Thus, two significant trends can be described as geographical proximity and organizational proximity.

Geographical proximity "corresponds to the quest of actors seeking to get closer to other economic or social actors, to natural or artificial resources, to places or technical objects." (Torre 2018). Thus, geographical proximity enables several economic agents to share the same space, facilitating links between them through either joint coordination or exchanges that reduce uncertainty (Talbot 2008).

Organizational proximity is a relational proximity and represents an organization's ability "to get its members to interact" (Rallet & Torre 2004). This interaction between members can be direct (organizational proximity) or indirect (institutional proximity). In the case of organizational proximity, we refer to links between "agents who participate in a specific activity within a particular structure" (Kirat & Lung 1995). On the other hand, for institutional proximity, a distinction must be made between two approaches: that of norms imposed on actors and that of institutions that depend on the actors themselves (Pecqueur 2005). This organizational proximity, achieved through the involvement of stakeholders from the same companies or clusters, increases coordination, exchange, and trust in situ.

Institutional proximity is considered a resource for mediation (language, social norms, etc.). It represents individuals who belong to mutual categories of rules and norms. This refers not only to organizations but also to behaviors. Institutional proximity presupposes the existence of institutions at both formal and informal levels, enabling the sharing of laws, regulations, cultural norms, and other elements. Its basis is a political framework that serves as a resource for all the projects involved (Talbot 2008). Thus, the smooth running of all types of coordination between economic agents and state institutions requires this kind of proximity (Talbot 2008).

4.4.2 The Second Approach of the Conceptual Framework: The Territorial Governance

A second important approach is that of territorial governance, which is closely linked to the contribution of proximity described above and considers the decision-making factors and the management of activities following the allocation of resources. It is defined as "an institutional and organizational process that attempts to make different modes of coordination between geographically close entities compatible" (Gilly & Wallet 2001). Governance focuses on several objectives, including the coordination of actors, strategic decisions on resource allocation, and the control of activities, such as conflict regulation (Lauriol et al. 2008). Based on these objectives, the territorial governance process enables the development of territorial development projects that utilize asymmetric resources and appropriate tools (Torre 2018). According to Lamara (2009), territorial governance considers "an interconnection between actors" that may or may not be sustainable and may also be territorial or non-spatial.

4.5 THE BIOREFINERIES IN FRANCE

The concept of biorefinery was extended in the 2010s to include the recovery of residual biomass and was termed the environmental biorefinery. The aim is to combine the environmental benefits of the biorefinery as an alternative to the petrorefinery with the added value of utilizing residual bioresources. Biorefineries in France vary in size, with large units that utilize biomass that differs slightly from one another, located in river or seaports and connected to global flows. There are also small biorefineries situated in rural areas with an ecosystem strongly connected to the rural world. According to Ceapraz et al. (2016), the first model "is based on threshold effects, specialization, and economies of scale", while the second one is related to "a more diverse and more thorough valuation of various biomasses of agricultural origin." We will focus on some examples to test our approaches.

4.5.1 Pomacle-Bazancourt Biorefinery

The first example we consider is that of the Pomacle Bazancourt biorefinery, located 15 km from Reims and spanning 250 ha, utilizing 1 million tons of wheat and 2.5 million tons of beets as raw materials. The biorefinery was founded in 1953 by a farmers' cooperative, which initially purchased a distillery. Since then, it has undergone several stages of transformation, from 1992 to 2009.

This biorefinery serves as a platform that brings together various stakeholders in the bioeconomy, characterized by a high degree of diversification in the agricultural and non-agricultural raw materials used and their subsequent transformation into high-value-added biobased products. It is an example of an integrated biorefinery. It operates with strong local roots based on the involvement of local farmers and their cooperatives and an entire ecosystem of more than 10 processing companies specializing in wheat and beet that deliver their products to a multitude of industries. This upstream-downstream coordination fosters proximity between players in the sector and ensures regular traceability of raw material supplies, while promoting efficient processing of biomass from technological, economic, and environmental perspectives.

In line with the theoretical and conceptual framework defined above, we will explain the operational determinants of the territorial value of the biorefinery, specifically the optimal location of the biorefinery and the coordination of economic actors through proximity, to achieve proper territorial governance. Several territorial values result from the combination of all these factors.

4.5.1.1 Optimal Location for Pomacle-Bazancourt

For the optimal location of Pomacle-Bazancourt, many important factors should be considered. The first factor relates to local resources and low supply costs, characterized by the abundant and consistent availability of raw materials, short collection distances, and stable short-term supply transactions. A second important factor is the importance of infrastructure and logistics costs, considering multimodal accessibility by both road and rail as well as proximity to the Parisian market and ease of export via ports (Le Havre, Rouen, and Antwerp). For the distribution of coproducts, there is accessibility by outlet, for example, for biofuels, there is a distribution chain with oil ports. For the food and feed industry, there is a large regional consumer base, and for more complex chemical industry products, there is a market for innovative products thanks to industrial and academic synergies and partnerships. This last criterion is a result of the importance of this biorefinery as a cluster of innovation and expertise.

The importance of the Pomacle-Bazancourt site as an integrated cluster-type biorefinery includes agglomeration effects due to the diversity of companies located there. There are approximately 10 players in the field of industrial production and academic researchers (with more than 50 specialists in bioeconomics and biotechnology). There is real synergy between these players through the transfer of know-how and the pooling of industrial utilities. The important role of institutional support provided by local authorities as well as the link with regional and European funding must also be considered. Finally, it is essential to emphasize the role of the IAR competitiveness cluster as a steering and demonstration hub for enhancing the integration of R&D and industry links.

4.5.1.2 Coordination of Actors and Proximities for Pomacle-Bazancourt

Concerning the nature of the proximities at the Pomacle-Bazancourt site, a brief description of all the proximities observed on site will be provided.

Firstly, geographical proximity is evident in the very short distances between the biorefineries and the biomass supply areas (beet and cereals). This means that the costs and risks associated with supply sources are reduced not only by shorter distances but also by the proximity of agricultural areas, which provide a regular and consistent supply of contractual volumes from cooperatives. Also, in terms of distance, it is worth mentioning the geographical location of the biorefinery downstream in the value chain for the distribution of coproducts (nearby ports, proximity to the Paris market). Regarding the biorefinery cluster, which brings together a group of players on-site, the physical proximity of all these entities, along with face-to-face contacts, facilitates the process of industrial symbiosis. This process yields advantages resulting from the cluster agglomeration through spillover effects.

Secondly, organizational proximity involves sharing the biorefinery value chain with a group of industrial and academic players located on the same site. One example is the vertical integration of certain specialized companies that control the value chain advantage. One example is the Cristal Union and Vivescia cooperatives, which offer vertical integration of their activities in terms of biomass supply. Proximity also facilitates the reduction of transaction costs and contractualization with farmers and cooperatives through better coordination, which facilitates exchange and trust. The agglomeration effects specific to this cluster enable, thanks to this proximity, the improvement of a shared technical and scientific understanding at the interpersonal level.

However, institutional proximity is primarily achieved through membership in the IAR competitiveness cluster and through top-down regional or national bioeconomy funding policies. Another example is standards and regulations (e.g. certification or labelling) within a legal framework linked to the national bioeconomy strategy. In addition, the primary academic player, the European Centre for Biotechnology and Bioeconomy (ECBB), is located in the same cluster and provides a major scientific framework for the smooth running of the Pomacle-Bazancourt biorefinery and the implementation of this institutional proximity.

4.5.1.3 The Territorial Governance for Pomacle-Bazancourt

Firstly, territorial governance for the Pomacle-Bazancourt biorefinery is multi-channel and multi-stakeholder. It employs a bottom-up approach, with strong involvement from local cooperatives (Cristal Union and Vivescia), as well as contributions from biomass suppliers. These primarily economic decisions enable territorial value to be captured, particularly upstream in the sector. Secondly, top-down governance decisions are also significant but involve other types of stakeholders at a more institutional level. These decisions concern European funding, the region's involvement, and inter-municipal cooperation in relation to the contractualization of the bioeconomy strategy, the site's sustainability, and the role of the Bioeconomy ZAC (joint development zone) as a priority axis of the Grand Reims development and decarbonization strategy.

Secondly, the site is considered a typical bioeconomy cluster, where there is real synergy between industrial, research, and institutional actors. This is a typical triple/quadruple helix coordination and, therefore, a true innovation ecosystem. Thirdly, the governance of the site, due to the arrival of new players and increasing synergies, allows this integrated biorefinery to evolve toward adaptive governance (Allais et al.

2021). With its numerous interdependent facilities, the biorefinery site facilitates the adoption of operational governance stemming from industrial symbiosis, which ensures a genuine regional industrial ecology (see the third approach below).

4.5.1.4 Territorial Values for Pomacle-Bazancourt

Firstly, the territorial value of location factors is reflected in qualitative and quantitative indicators. Concerning local resources, real added value is provided by contracts with cooperatives and farmers to secure inputs in terms of raw material supplies (beets and corn). It is also worth noting that securing these contracts has a long-term impact, providing stable income for farmers. There is also sectoral diversification within the cluster, which combines agricultural and industrial activities. In terms of figures, we can also add a significant boost to employment, with approximately 1,000 employees at the biorefinery site, which spans 250 ha. In terms of logistics infrastructure, there are collective costs on the site, but also a reduction in transport costs, which are made possible by a multimodal transport architecture close to markets (the Paris basin) and ports. In terms of logistics, the cluster is anchored in an agricultural region, which increases the maintenance and upkeep of transport infrastructure that serves other industries. The cluster nature of this integrated biorefinery enables real industrial synergy among the various industrial and academic players located in the same place, resulting in cost reduction, easier transfer, and sharing of knowledge and innovation, and attractiveness to other economic players. This cluster thus facilitates spillover effects in terms of knowledge and information exchange, enabling the region's real economic diversification. This cluster maintains an international image as a major player in the bioeconomy, a guarantee of a region that innovates due to its strategic location and the abundant presence of local resources. All the players involved are supported by public funding, which gives the region an image not only of innovation but also of enabling the region's real economic diversification and resilience, thanks to all these cooperation levers.

Secondly, the territorial value created by the various proximities combines both job creation through the location of multiple actors in the cluster and a set of added values through the coordination of activities across the entire value chain. This added value is primarily a result of low transportation costs compared to biomass supply basins and outlets. Secondly, there is a reduction in transaction costs, along with a decrease in the risks associated with contractualization. However, the exchange of know-how, the transfer of knowledge, and the creation of innovation between all these actors, supported by an institutional framework, means that local collectives and the IAR competitiveness cluster produce multiple spillovers in this territory.

Thirdly, the territorial governance of the biorefinery enables the territory to create multiple territorial values. Economically speaking, we are referring to the value captured across the entire value chain. For example, upstream in the chain, in relation to biomass supply, cooperatives play a crucial role in ensuring fair and stable income distribution among farmers. Within the cluster, the presence of multiple players enables cost reduction through the pooling of resources and knowledge transfer, which is linked to high productivity and investment. The emergence of new players promotes diversification in terms of both ownership and biobased products/biomaterials on the market.

On a social level, employees benefit from stability and increased incomes through training offered via apprenticeships as well as the diversity of products and markets, which impacts the structure of employment. Environmentally, industrial symbiosis leads to real savings in both economic and logistical terms, resulting in increased system stability (less CO_2 and waste). Institutionally, partnerships on-site with a rich diversity of players reinforce the image of the biorefinery as a significant hub of the French bioeconomy.

4.5.2 La Mède Biorefinery

We will now move to the opposite side of France to illustrate the diversity of regions hosting biorefineries. The La Mède biorefinery is located in Bouches-du-Rhône, near Marseille. It is considered one of the first oil refineries to be converted into a biorefinery, boasting one of its kind biofuel processing facilities in the country and one of Europe's largest biorefineries. Its main activity is the production of renewable diesel. The complex also features AdBlue production, a product added to diesel cars to reduce emissions of pollutants as well as a solar plant.

This former petroleum refinery was established in 1935 and was reconverted by Total Energies in 2015. Authorized in 2019, the La Mède biorefinery produces 500,000 metric tons of renewable diesel annually. It can process various certified sustainable feedstocks, including vegetable oils, used cooking oils, and animal fats sourced from the circular economy. As of January 1, 2023, palm oil is no longer processed at La Mède.

As in the case of the Pomacle-Bazancourt biorefinery, we will now describe the most important factors for the optimal location of the La Mède biorefinery, followed by those relating to coordination and proximity as well as territorial governance. These factors translate into territorial values that express the true potential of this biorefinery.

4.5.2.1 Optimal Location for La Mède

The crucial importance of the La Mède biorefinery lies in its proximity to the Port of Marseille, which facilitates the transport of raw materials and finished products, while reducing costs. Having transformed into an existing refinery is an important factor in reducing costs. Additionally, Marseille is a city on the Mediterranean, which facilitates the transportation of finished products to all the countries bordering the Mediterranean by reducing transportation costs.

The optimal location of La Mède, in addition to its highly diversified production of renewable diesel, includes fuel for aviation and the eco-friendly diesel additive (AdBlue) for terrestrial transportation, positioning the site to serve multiple market segments, from passenger vehicles to heavy transport and aviation. This versatility strengthens its role as a leader in distributing sustainable energy products.

4.5.2.2 Coordination of Actors and Proximities for La Mède

One of the most crucial considerations is the geographical proximity. Indeed, the La Mède location has recently modified its production procedure to convert all used cooking oil into jet fuel. It will produce 15,000 tons of this fuel per year. It should be noted that it has already signed a contract with Air France for its flights departing from Marseille airport. This new fuel does not require any modifications to existing aircraft engines. It will reduce CO_2 emissions by up to 90%, which is a crucial step in the sustainable development process for the aviation sector.

Its conversion into a biorefinery, therefore, secures part of France's energy production capacity in a green growth optic, as such a facility was considered beneficial not only for the company itself but also for local communities and the country. Finally, the organizational and institutional proximity that was evident during the transition phase led local trade unions to welcome the takeover, as it helped preserve 250 out of the 500 direct jobs at the refinery. They also managed to reach a compromise with Total Energies to avoid using palm oil, choosing instead to use cooking oils and animal fats.

4.5.2.3 The Territorial Governance for La Mède

The governance of the La Mède biorefinery is top-down, as it was established by acquiring an existing refinery, a strategic decision that leveraged an existing asset to reduce costs and capitalize on the former biorefinery's strategic location. This process was governed by French regulations on renewable energies and actively supported by local authorities. However, it has received criticism and objections from trade unions because it primarily aimed to process palm oil, which has attracted intense criticism, particularly regarding deforestation and environmental concerns. Environmental non-governmental organizations put pressure on the biorefinery to change its supply model to the same materials as the Total Energies, namely, used oils, animal fats, etc.

Ultimately, this governance model has been marked by numerous conflicts between Total Energies and both farmers and environmental associations, which has inevitably led to the need to negotiate with the French government. This suggests that it is based primarily on conflict rather than cooperation, highlighting the difficulty of reconciling the biorefinery strategy with environmental concerns and regional expectations.

Finally, La Mède's governance may be called "adaptive," since it changes over time as industrial, institutional, and social actors engage with one another. The project has evolved due to the pressures from numerous stakeholders. Now, decisions are continually being changed to find a balance between economic needs, regulatory frameworks, and social acceptance.

4.5.2.4 Territorial Values for La Mède

Firstly, as already mentioned in the case of Pomacle-Bazancourt biorefinery, the territorial value of location factors is reflected in both qualitative and quantitative indicators.

Since this biorefinery took over a former refinery, it has successfully maintained industrial activity on a site that was initially designed for this purpose. This has allowed the preservation of an industrial activity on a site that was previously developed for this purpose. This biorefinery provides on-site industrial training for local machine maintenance professionals, offering a deeper understanding of how biorefineries operate. This training enables participants to gain hands-on experience directly on site and apply the concepts they have learned. As a result, the R&D department benefits from stronger development, since those involved in research are directly engaged in the operational environment.

Secondly, the La Mède biorefinery combines several advantages in terms of territorial value resulting from the combination of various proximities. The biorefinery is located in a highly integrated ecosystem, which includes a support system that reduces costs and provides close links to petrochemical refineries and Marseille-Provence Airport. Thirdly, the territorial value of governance focuses on a multi-level and multi-stakeholder perspective. This involves several industrial players (e.g. Total Energies and Marseille-Provence Airport), institutional and environmental stakeholders, as well as different spatial scales ranging from the local territory to regional communities and the EU level. The value here is reflected in the use of waste and residues within an industrial and territorial ecology approach, which incorporates a legitimate circular economy, thereby significantly reducing the risk of conflicts of use. Local anchoring thus enables the integration of local actors through partnerships and collective actions that foster exchange between industry, R&D, and institutions.

4.6 THE TERRITORIAL BUSINESS PLAN FOR A BIOREFINERY

For Lange (2022), the biobased business models started with the "biomass to biofuel" biorefinery. Chronologically, the first biorefineries on the market had complex business models, as they were the result of partnerships between several experienced biobased enterprises, which later evolved into mergers or acquisitions (Lange 2022). Secondly, biofuel biorefineries were evaluated on the commercial and sustainability dimensions in relation to "energy content and not the biomass structures" (Lange 2022). Lastly, more recent biorefineries are described as having multiple business models, as they primarily optimize the valuation of all biomass components and thus propose higher-value products.

As outlined in the conceptual framework, several business models related to concepts of sustainability or circularity and linked to environmental, social, and economic challenges were recently developed within the transition to a circular economy. The development of business model terminology is expanding, with a multitude of definitions, and applies to the entire spectrum of industrial value chains.

One particular methodology of detailing business models by their components was inspired by the works of Osterwalder and Pigneur (2010), who identified several "building blocks" of a business model: "the value proposition, customer segments, customer relationships, channels, key partners, key activities, key resources, cost structure and revenue streams." In considering this methodology, we outline that "the value proposition" of this concept is the most important part of this model. It is described as the benefits and motivations offered to the customer, that is, the products and services introduced by the biorefinery to different markets. For our case, the value proposition is, as mentioned before, "a territorial value proposition" which was detailed already from a conceptual point of view and correlated to the concept of territorial biorefinery. This "territorial value proposition" was already described for both biorefineries, Pomacle-Bazancourt and La Mède, in the Sections 4.5.1.4 and 4.5.2.4. We continue focusing here on describing the other blocks of the business model.

Firstly, in the case of Pomacle-Bazancourt biorefinery, here are the most important characteristics of these building blocks.

4.6.1 The Territorial Value Proposition of Pomacle-Bazancourt

This biorefinery generates regional value from a cluster/hub of sustainable and circular activities rooted in the region, led by a group of agricultural, industrial, and research stakeholders. Its operation, based on the in situ transformation of local agricultural resources into high-value-added biobased products, contributes to the creation of regional jobs by shortening value chains and accelerating innovation. The importance of regional anchoring through cooperative governance ensures the stable supply of biomass and promotes sustainable rural income. The industrial symbiosis specific to this cluster of activities, therefore, generates high environmental, social, and economic value. The entire production of high value-added biobased products is accompanied by a reduction in carbon footprint and waste, while promoting resilient local growth.

4.6.2 The Customer Segments

There are numerous potential customers from various industrial sectors. We have identified customers from the following industries. In the energy and biofuel industries as well as in the transport and logistics sector, numerous companies purchase biofuel to reduce CO_2 emissions. In the agri-food industry, there are buyers for a range of products, including sucrose, glucose, starch, and gluten. In the animal feed industry, there are buyers for beet pulp, wheat derivatives, and alfalfa pellets. For the chemical and cosmetics industry, there are buyers for intermediate products or ingredients of organic origin. One sector considered innovative is biotech and biomanufacturing, which benefits from collaboration with the cluster's R&D companies in various fields, such as process development and industrial demonstrations. Other customers include public institutions seeking an institutional relationship with the biorefinery in support of the national energy transition strategy and environmental organizations seeking to recover waste.

4.6.3 Customer Relationships

There are many different types of partnerships, which can take the form of long-term contracts with transport or logistics companies for the supply of biofuel. At the institutional level, partnerships are established through collaborative agreements with organizations for environmental certification and CO_2 emissions reduction. Partnerships can also be established in connection with EU funding, supply chain traceability, and public–private partnerships that strengthen the development of innovation. This institutional framework is based on industrial and territorial symbiosis, promoting interaction and trust among local companies located in the area.

One of the most important relationships at the industrial level is through cooperatives, which involve customers who purchase basic ingredients for various industries.

Another type of relationship is emerging at the level of open innovation platforms, which serve as a foundation for launching pre-industrial projects and mitigating investment risks.

Finally, in connection with R&D, there is a symbiosis between future customers and the world of research via the ECBB through the creation of industrial chairs or start-ups.

4.6.4 The Value Chains

The main value chains include the industrial, commercial, and consumer retail sectors as well as open-innovation and R&D chains, which prioritize ARD and ECBB. The IAR competitiveness cluster also contributes to enriching the networking chain, thereby developing a large network of members. Another interesting chain is that of distribution logistics to consumers with an international reach. Finally, it is imperative to note the importance of the communication chain in promoting the sustainability and circularity of the cluster.

4.6.5 Key Partners

Several key partners are involved in managing the cluster. These include (a) the agricultural cooperatives Cristal Union and Vivescia, (b) industrial partners ADM Bazancourt, Givaudan, Active Beauty, and Air Liquide, (c) R&D players, such as the ARD platform, (d) academic partners (ECBB with AgroParisTech, Centrale Supélec, University of Reims Champagne-Ardenne (URCA), and Neoma Business School chairs), (e) public authorities, (f) partners in the animal feed industry, and (g) partners in technology consortia, such as IIFP Energies nouvelles (FPEN, National Institute for Agriculture, Food and Environment (INRAE), and ARD.

4.6.6 Key Activities

The key activities from upstream to downstream in the value chain are as follows: (a) biomass supply activities carried out by the Cristal Union and Vivescia cooperatives; (b) agro-industrial processing activities; (c) industrial symbiosis activities through the exchange of utilities, by-products, and efficient joint resource management; (d) R&D activities and management of open innovation platforms and academic activities; (e) technology transfer activities; and (f) activities related to the network and local communities involved in the regional bioeconomy to attract new investments.

4.6.7 Key Resources

The key resources mobilized in the biorefinery are represented by regional biomass, co-located industries on the site, shared infrastructure that serves as a basis for industrial symbiosis, R&D platforms, on-site academic centers, and agricultural cooperatives that manage procurement.

4.6.8 Cost Structure

The costs incurred in the context of the biorefinery are multiple and concern: (a) the purchase of biomass, (b) energy and utility costs, (c) biomass processing and treatment costs, (d) costs related to industrial symbiosis through sharing and the relationships that develop within the cluster, (e) R&D costs, (f) costs to cover risks of all kinds, (g) costs related to investments, and (h) costs related to futures markets and, implicitly, the volatility of agricultural commodity prices.

4.6.9 Revenue Streams

The sale of the following products carries the most significant revenue: (a) basic ingredients, (b) wheat-derived ingredients, (c) specialty ingredients, (d) coproducts for animal feed, (e) CO_2 removal through carbon credits, (f) maintenance services, and (g) marketing technologies. Secondly, in the case of La Mède biorefinery, here are the most important characteristics of these building blocks.

4.6.10 The Territorial Value Proposition of La Mède

The main value proposition of La Mède is the conversion of used cooking oils and fats into biofuels (renewable diesel and bio jet fuel). Other value propositions include the reuse of an existing refinery site, which preserves jobs and supports the local economy. A significant factor in this value proposition is the proximity of the Port of Marseille, which guarantees access to various Mediterranean countries. Not to mention the vocational training programs that improve skills and develop R&D. Finally, La Mède contributes to France's decarbonization by reducing CO_2 emissions through sustainable fuels.

4.6.11 The Customer Segments

The customers mainly consist of fuel distributors or companies wishing to transition to renewable energy sources as well as the aeronautics industry, including Marseille Airport, located nearby, and other European airports that use sustainable aviation fuel (SAF) fuel in the interest of sustainable development. Customers also include transport and logistics companies, public institutions in the region, particularly those located near biorefineries that support the SDGs, and finally, environmental organizations.

4.6.12 Customer Relationships

As we mentioned earlier, for Pomacle-Bazancourt, long-term contracts with logistics companies are a significant factor in the marketing and viability of biorefineries. In the case of Mède, in addition to contracts with logistics and transport companies, it has formed partnerships with airlines and municipalities for the distribution of biofuels.

In customer relationships, trust and transparency are essential for the sustainability of partnerships. This is certified by environmental certifications, such as International Sustainability and Carbon Certification (ISCC), and ensures the traceability of raw materials. Finally, this relationship is consolidated through a commitment to Corporate Social Responsibility (CSR), including the development of employment in the region and internal training for employees.

4.6.13 Key Partners

Several key partners are contributing to the development of the biorefinery, including (a) Total Energies, (b) Marseille Airport, (c) the Port of Marseille, (d) research institutions, (e) construction companies, (f) public institutions, and (g) the EU, and finally, training institutions.

4.6.14 Key Activities

The key activities for La Mède are listed as follows: (a) advanced biofuel production from used oils and residues (renewable diesel, SAF), (b) provision of basic materials and recycling within the framework of the circular economy, (c) collaboration and cluster management with institutional and industrial partners, (d) R&D and innovation to improve process efficiency and diversify long-lasting materials, (e) industrial training programs for professionals and biorefinery technicians, and (f) environmental and quality management to ensure regulatory compliance and reduce emissions.

4.6.15 Key Resources

The key resources of La Mède are primarily its strategic geographic location (Port of Marseille), its human capital (the maintenance staff, professionals, and research experts), and its retransformed infrastructure (an initial refinery transformed into a biorefinery), Total Energies' know-how, institutional support, and new laws that support sustainable production. Finally, we can add to all of the above financial investments in technological development.

4.6.16 Cost Structure

The costs engaged in the biorefineries, particularly La Mède, are mainly: (a) maintenance costs of the site, (b) recycling costs (residues and used oil), (c) training costs investments, (d) environmental and compliance costs (like certifications and waste treatment), (e) logistics costs, administrative, financial and partnership contracts costs, and finally, (f) R&D innovation costs.

4.6.17 Revenue Streams

The revenue streams are mostly provided by (a) sales of SAF to industrial clients and airports, (b) revenues from byproducts and energy recovery, (c) public and EU funding for decarbonization, innovation, and circular economy projects, (d) training services and technical expertise, and finally, (e) export revenues through the Port of Marseille to Mediterranean and international markets.

4.7 CONCLUSION

Several types of business plans can be highlighted, depending on the conceptual approach, methodology used, and the number of empirical cases that can be identified. In our case, a comparative study between two large French biorefineries enabled us to highlight several key pillars of analysis, identifying commonalities and differences. Our study prioritized the territorial values of these biorefineries based on two main conceptual approaches, that is, the territorial anchoring and the territorial governance. These territorial values were transcribed to ultimately propose a comprehensive business plan that includes higher-value products and market opportunities for each of these biorefineries.

The two biorefineries, La Mède and Pomacle-Bazancourt are two complementary examples of biomass processing in France. The former focuses on large-scale biofuel production in industrial integration, while the latter specializes in biochemical processing and innovation in a rural setting.

The Pomacle-Bazancourt biorefinery is a well-developed bioeconomic agro-industrial hub with strong roots in regional and national agriculture. It focuses primarily on biochemistry, green polymers, and biomaterials. It produces biomaterials and green polymers, which gives it a leading position in the bioeconomy value chain. The circular use of biomass demonstrates the extent to which rural ecosystems are evolving into new low-carbon industrial hubs, with a focus on sustainable development.

On the other hand, the La Mède biorefinery symbolizes the new model of circular industrial conversion, transforming a former fossil fuel refinery into a biorefinery dedicated to renewable fuels. Its unique features include its logistics, the transformation of used oils (value creation), and its participation in SDGs efforts in France and the EU, particularly in air transport and logistics. As a result, it is helping to accelerate the energy transition by demonstrating that, starting from what already exists, it is possible to move forward and create new opportunities in support of a sustainable industrial strategy.

Most products derived from biomass, such as bioplastics, biosurfactants, green hydrogen, and organic acids, are highly competitive. These products enhance productivity and profitability, while strengthening industrial and territorial resilience by diversifying raw materials, absorbing CO_2, and integrating circular economy cycles.

In the future, biorefineries will primarily rely on their ability to expand beyond biofuels to become biohubs, combining biotechnology, the circular economy, and digital optimization, particularly with the emergence of artificial intelligence (AI). Cross-sector partnerships and flexible raw material strategies enable biorefineries to become regional and international pillars of the low-carbon economy by integrating R&D ecosystems, delivering both economic value and global environmental regeneration.

REFERENCES

Åckerman, N. 2021. *What is a Business Model and What's in it for Regional Development?*, CRA Working Paper 2021:01. Gothenburg: Center for Regional Analysis, University of Gothenburg.

Agridees. 2025. *La bioéconomie territoriale: Vers de nouveaux modèles agricoles et industriels*. Paris: Agridees.

Allais, F., H. Lescieux-Katir, & J.-M. Chauvet. 2021. The continuous evolution of the Bazancourt-Pomacle site rooted in the commitment and vision of pioneering farmers: When reality shapes the biorefinery concept. *EFB Bioecon. J.* 1: 100007.

Bauer, F., L. Coenen, T. Hansen, K. McCormick, & Y.V. Palgan. 2017. Technological innovation systems for biorefineries – A review of the literature. *Biofuels Bioprod. Biorefin.* 11: 534–548.

Benoit, S. 2021. Bioéconomie et diversité des ancrages territoriaux. *Écon. Rur.* 376: 77–91.

Bröring, S., & A. Vanacker. 2022. Designing business models for the bioeconomy: What are the major challenges? *EFB Bioecon. J.* 2: 100033.

Brunnhofer, M., G. Németh, J.P. Schöggl, T. Stern, & A. Posch. 2020. The biorefinery transition in the European pulp and paper industry – A three-phase Delphi study including a SWOT-AHP analysis. *For. Policy Econ.* 110: 101882.

Callois, J.-M. 2022. *Le renouveau des territoires par la bioéconomie: Les ressources du vivant au cœur d'une nouvelle économie*. Versailles: Quae.

Ceapraz, I.L., G. Kotbi, & L. Sauvée. 2016. The territorial biorefinery as a new business model. *Biobased Appl. Econ.* 5: 47–62.

Cerca, M., A. Sosa, E. Gusciute, & F. Murphy. 2022. Strategic planning of biobased supply chains: Unlocking bottlenecks and incorporating social sustainability into biorefinery systems. *Sustain. Prod. Consum.* 34: 219–232.

Chammaa, C., & L. Haddad. 2025. Artificial intelligence and behavioral finance biases: Improving Lebanese investors' decision-making. *J. Decis. Syst.* 34: 2537076.

Chan, Y. 2001. *Location Theory and Decision Analysis*. Cincinnati, OH: South-Western College Publishing.

Cherubini, F. 2010. The biorefinery concept: Using biomass instead of oil for producing energy and chemicals. *Energy Convers. Manag.* 51: 1412–1421.

Christensen, L.D. 2019. *When Bioeconomy Policy Objectives (Fail To) Overlap: Rethinking the Analysis of Necessity to Detect Causal Interdependencies among Sustainable Development Goals in the Nordic Bioeconomy Strategy*. COMPASSS Working Paper 93:1–23. Aarhus: Aarhus Universitet.

Cohen, W.M., & D.A. Levinthal. 1990. Absorptive capacity: A new perspective on learning and innovation. *Admin. Sci. Q.* 35: 128–152.

Colletis, G. 2010. Coévolution des territoires et de la technologie: Une perspective institutionnaliste. *Rev. Écon. Rég. Urbaine* 2: 235–249.

Colletis, G., J.-P. Gilly, I. Leroux, B. Pecqueur, J. Perrat, F. Rychen, & J.-B. Zimmermann. 1999. Construction territoriale et dynamiques économiques. *Sci. de la Soc.* 48: 25–47.

Colletis, G., & B. Pecqueur. 2018. Révélation des ressources spécifiques territoriales et inégalités de développement: Le rôle de la proximité géographique. *Rev. Écon. Rég. Urbaine* 5: 993–1011.

Gilly, J.-P., & F. Wallet. 2001. Forms of proximity, local governance and the dynamics of local economic spaces: The case of industrial conversion processes. *Int. J. Urban Reg. Res.* 25: 553–570.

Golombiewski, B., N. Sick, & S. Bröring. 2015. The emerging research landscape on bioeconomy: What has been done so far and what is essential from a technology and innovation management perspective? *Innov. Food Sci. Emer. Technol.* 29: 308–317.

Guibert, C. 2024. *Ancrage territorial et durabilité des filières élevage sous SISQ en région Occitanie*. PhD diss, Toulouse: Université de Toulouse.

IEA. 2012. *Biobased Chemicals: Value Added Products from Biorefineries*. Paris: IEA.

Kirat, T., & Y. Lung. 1995. Innovations et proximites: le territoire, lieu de deploiement des processus d'apprentissage'. In *Coordination iconomique et apprentissage des firmes*, ed. N. Lazaric, & J.M. Monnier, 206–227. Paris: Economica

Kleinschmit, D., B.H. Lindstad, B.J. Thorsen, A. Toppinen, A. Roos, & S. Baardsen. 2014. Shades of green: A social scientific view on bioeconomy in the forest sector. *Scand. J. For. Res.* 29: 402–410.

Kotbi, G., & L. Sauvee. 2010. La place du territoire dans les choix stratégiques des groupes sucriers français: Enjeux et perspectives du changement institutionnel et concurrentiel. *Identité, Qualité et Compétitivité Territoriale, Colloque Joint ASRDLF – AISRe*, Sep 2010, Aoste, Italie.

Laibach, N., J. Börner, & S. Bröring. 2019. Exploring the future of the bioeconomy: An expert-based scoping study examining key enabling technology fields with potential to foster the transition toward a biobased economy. *Technol. Soc.* 58: 101119.

Lamara, H. 2009. Les deux piliers de la construction territoriale: Coordination des acteurs et ressources territoriales. *Dév. Durable Territ.* doi:10.4000/developpementdurable.8208

Lange, L. 2022. Business models, including higher value products for the new circular, resource-efficient biobased industry. *Front. Sustain.* 3: 789435.

Lauriol, J., V. Perret, & F. Tannery. 2008. L'espace et le territoire dans l'agenda de recherche en stratégie. *Rev. Fr. Gest.* 184: 181–198.

Lopez-Díaz, D.C., L.F. Lira-Barragán, E. Rubio-Castro, J.M. Ponce-Ortega, & M.M. El-Halwagi. 2017. Optimal location of biorefineries considering sustainable integration with the environment. *Renew. Energy* 100: 65–77.

Maillefert, M., & I. Robert. 2017. Nouveaux modèles économiques et création de valeur territoriale autour de l'économie circulaire, de l'économie de la fonctionnalité et de l'écologie industrielle. *Rev. Écon. Rég. Urbaine* 5: 905–934.

Osterwalder, A., & Y. Pigneur. 2010. *Business Model Generation*. Hoboken, NJ: Wiley.

Pecqueur, B. 2005. Le développement territorial: Une nouvelle approche des processus de développement pour les économies du Sud. In *Le territoire est mort: Vive les territoires! Une refabrication au nom du développement*, ed. B. Antheaume, & F. Giraut, 295–316. Paris: IRD.

Pettersson, K., & S. Harvey. 2013. Optimal location of biorefineries. In *Systems Perspectives on Biorefineries*, ed. B. Sandén, & K. Pettersson, 29–56. Gothenburg: Chalmers University of Technology.

Piedrahita-Rodríguez, S., S.C. Urrea, D.A. Escobar García, M. Ortiz-Sánchez, J.C. Solarte-Toro, & C.A.C. Alzate. 2023. Life cycle assessment and potential geolocation of a multi-feedstock biorefinery: Integration of the avocado and plantain value chains in rural zones. *Bioresour. Technol. Rep.* 21: 101318.

Rallet, A., & A. Torre. 2004. Proximité et localization. *Écon. Rur.* 280: 25–41.

Reim, W., V. Parida, & D. R. Sjödin. 2019. Circular business models for the bioeconomy: A review and new directions for future research. *Sustainability* 11: 2558.

Richardson, J. 2008. The business model: An integrative framework for strategy execution. *Strateg. Change* 17: 133–144.

Roy, B.B., & Q. Tu. 2022. A review of system dynamics modeling for the sustainability assessment of biorefineries. *J. Ind. Ecol.* 26: 1384–1399.

Salvador, R., M.V. Barros, M. Pieroni, D.A.L. Silva, F. Freire, & A.C. De Francisco. 2023. Overarching business models for a circular bioeconomy: Systematizing archetypes. *Sustain. Prod. Consum.* 43: 349–362.

Serrano-Hernandez, A., J. Faulin, J.M. Pintor, & J. Belloso. 2017. Determining an optimal area to locate a biorefinery under economic and environmental criteria. *Transp. Res. Proc.* 22: 95–104.

Sharara, M., K. Sahoo, A.D. Reddy, & S. Kim. 2020. Sustainable feedstock for bioethanol production: Impact of spatial resolution on the design of a sustainable biomass supply chain. *Bioresour. Technol.* 295: 122896.

Solarte-Toro, J.C., & C.A.C. Alzate. 2021. Biorefineries as the base for accomplishing the sustainable development goals (SDGs) and the transition to bioeconomy: Technical aspects, challenges and perspectives. *Bioresour. Technol.* 340: 125626.

Talbot, D. 2008. Les institutions créatrices de proximités. *Rev. Écon. Rég. Urbaine* 3: 387–408.

Teece, D.J. 2010. Business models, business strategy and innovation. *Long Range Plann.* 43: 172–194.

Torre, A. 2018. Développement territorial et relations de proximité. *Rev. Écon. Rég. Urbaine* 5: 1043–1075.

Ubando, A.T., C.B. Felix, & W.-H. Chen. 2020. Biorefineries in circular bioeconomy: A comprehensive review. *Bioresour. Technol.* 299: 122585.

Van Lancker, J., E. Wauters, & G. Van Huylenbroeck. 2016. Managing innovation in the bioeconomy: An open innovation perspective. *Biomass Bioenergy* 90: 60–69.

Wohlfahrt, J., F. Ferchaud, B. Gabrielle, C. Godard, & B. Kurek. 2019. Characteristics of bioeconomy systems and sustainability issues at the territorial scale: A review. *J. Clean. Prod.* 232: 898–909.

Zahraa, S.M., N. Shiwakoti, & P. Stasinopoulos. 2020. Biomass supply chain environmental and socio-economic analysis: 40-years comprehensive review of methods, decision issues, sustainability challenges, and the way forward. *Biomass Bioenergy* 142: 105777.

Zimmermann, J.-B. 2005. Entreprises et territoires: Entre nomadisme et ancrage territorial. *Rev. IRES* 47: 21–36.

5 Life Cycle Assessment of Biomass Feedstocks for Biorefineries

Stanley C. Msiska, Denzel C. Makepa, Richard K. Padi, and Annie Chimphango

5.1 INTRODUCTION

The urgent global transition toward low-carbon, resource-efficient production systems has positioned bio-based industries as a cornerstone of sustainable development strategies. In sub-Saharan Africa (SSA), where energy poverty, agricultural inefficiencies, and rural unemployment persist, the bioeconomy offers a unique opportunity to link agricultural transformation with clean energy access and environmental stewardship (Jha & Schmidt 2021). However, the majority of biorefinery developments in SSA have remained conceptually centralized and feedstock-specific, relying on imported technologies and mono-feedstock models that are poorly aligned with the fragmented, smallholder-based farming systems that dominate the region. This disconnection underscores the need for context-adapted biorefinery frameworks that are both feedstock-integrative and product-flexible, capable of valorizing locally available residues into multiple energy and value-added products (Katakojwala & Mohan 2021).

The SSA, with Malawi as a representative case, offers a particularly relevant context for advancing such innovations. The country's agricultural sector is characterized by smallholder intercropping systems where maize and pigeon pea dominate as food and cash crops (Ngwira et al. 2020; Silberg et al. 2020; Chaima et al. 2023). These systems generate abundant crop residues such as maize stover, pigeon pea stalks, and leaves that are largely underutilized or openly burned, contributing to greenhouse gas emissions and soil nutrient depletion (Gondwe et al. 2017). Yet despite this availability, Malawi's energy mix remains heavily dependent on imported fossil fuels and traditional biomass, while fertilizer imports exacerbate trade imbalances and limit soil productivity recovery (Ngwira et al. 2020). The valorization of such residues through intercropped mixed-feedstock (IMF) biorefineries presents a strategic pathway for achieving energy sovereignty, waste valorization, and circular agricultural growth (Gondwe et al. 2017; Msiska et al. 2024).

 DOI: 10.1201/9781003650751-5

Globally, the evolution of biorefineries has progressed from single-feedstock and single-product systems to integrated facilities that utilize multiple feedstocks and produce several products. This shift reflects the growing need for operational resilience, economic robustness, and environmental efficiency. Furthermore, studies by Cherubini and Jungmeier (2010) and Juneja et al. (2019) highlighted the critical role of residue streams in enhancing the overall sustainability of biorefineries. Studies from Europe, Asia, and Latin America have shown that feedstock diversification, particularly through residue blending, enhances process stability and reduces supply risk (Berchem et al. 2020). However, these approaches have rarely been contextualized for low-input, rainfed systems typical of SSA, where biomass yields, logistics, and market access differ markedly from industrial contexts (Whalley et al. 2017). Moreover, most life cycle assessment (LCA) studies on bio-based production have focused on single-feedstock scenarios (e.g. maize stover, sugarcane bagasse, and rice straw), overlooking intercropped residue systems that could deliver synergistic environmental and agronomic benefits (Liu et al. 2021; Gaffey et al. 2024). This gap limits the empirical understanding of how feedstock integration and product complexity jointly influence environmental performance in smallholder-driven biorefinery systems.

Previous LCAs of bioenergy and biochemical production in SSA have largely focused on macro-scale or hypothetical systems, often neglecting field-level residue dynamics and intercropping practices (Jha & Schmidt 2021). Few studies have incorporated the coproduction of high-value chemicals such as lactic acid alongside traditional bioenergy outputs, despite their growing relevance for rural bioeconomy diversification. Similarly, while ethanol production from maize residues has been extensively modeled, its integration with biogas and combined heat and power (CHP) in SSA contexts remains underexplored. This study, therefore, advances the literature by developing an assessment framework tailored to Malawi's smallholder agricultural systems, integrating residue availability, conversion efficiencies, and energy balances within a coherent techno-environmental context.

The significance of this research extends beyond local applicability. The comparative evaluation of scenarios S1–S6 offers globally relevant insights into how feedstock diversity and product integration modulate environmental sustainability in decentralized biorefineries. For instance, mixed-feedstock configurations (S4–S6) are hypothesized to achieve substantial global warming potential (GWP) reductions, up to 61% compared with single-feedstock analogues, through enhanced process synergy and nutrient recycling (reflected in the results). Conversely, the transition from ethanol-based to lactic-acid-based production (S3 and S6) is expected to increase environmental burden due to higher energy and chemical requirements. These contrasts provide crucial evidence for policymakers and investors seeking to balance rural economic empowerment with environmental integrity in resource-constrained conditions.

Against this backdrop, the present study develops and evaluates six biorefinery scenarios structured to reflect two contrasting feedstock systems and three levels of product diversification. Scenarios S1–S3 represent monocrop-based systems using single feedstocks, producing: S1, electricity via CHP; S2, ethanol with electricity and biogas coproducts; and S3, lactic acid alongside electricity and biogas. Scenarios S4–S6 mirror these configurations but replace single feedstocks with IMF systems to capture the agronomic and environmental synergies of integrated residue utilization: S4, IMF producing electricity; S5, IMF producing ethanol with electricity and biogas coproducts; and S6, IMF producing lactic acid alongside electricity and biogas. This systematic design enables a two-dimensional comparative analysis: (i) assessing the effect of feedstock integration (single vs mixed residues) and (ii) evaluating the environmental trade-offs of expanding product portfolios from energy-only to chemical coproduction systems. The six-scenario framework thus provides a comprehensive lens for quantifying how incremental technological and feedstock diversification shifts environmental performance within the same smallholder context.

The conceptual underpinning of this scenario structure is rooted in both biophysical and systemic approaches. From a biophysical perspective, the inclusion of pigeon pea residues in IMF systems is assessed for the potential to improve carbon, to improve the carbon-to-nitrogen balance of the feedstock mixture, enhancing digestion efficiency and biogas yield while enabling nutrient recycling through organic fertilizer substitution (Msiska et al. 2024). From a systemic perspective, the coproduction of electricity, ethanol, and lactic acid is assessed for their potential in aligning biorefinery systems with circular economy principles by maximizing resource recovery and product value (Parada et al. 2017). However, increasing product diversification also raises energy and material demands, potentially amplifying environmental burdens in categories such as GWP, human toxicity, and fossil depletion, a phenomenon documented in recent LCA literature (Osman et al. 2024; Zhang et al. 2024). Evaluating these countervailing effects is therefore essential to identifying optimal configurations for sustainable deployment.

In this study, a cradle-to-gate LCA is employed to quantify environmental impacts across the six scenarios using process inventory data derived from Aspen Plus simulations (Althaus et al. 2007; Liu et al. 2021). The analysis follows ISO 14040/44 standards (ISO 2006) and applies the ReCiPe 2016 midpoint method (Huijbregts et al. 2017) to capture six key impact categories, including GWP, terrestrial acidification, freshwater eutrophication potential, water use, land-use, and fossil depletion potential. The cradle-to-gate boundary excludes downstream product use and disposal phases, focusing instead on feedstock cultivation, collection, transport, conversion, and internal energy flows within the biorefinery (Lee & Inaba 2004; Ubando et al. 2022).

The study aligns directly with regional and global sustainability goals. It contributes to SDG 2 (Zero Hunger) by supporting soil fertility restoration through organic fertilizer substitution; SDG 7 (Affordable and Clean Energy) by enhancing local renewable energy generation; SDG 12 (Responsible Consumption and Production) by valorizing agricultural residues; and SDG 13 (Climate Action) by reducing carbon footprints of rural production systems. At the same time, it supports Malawi's National Energy Policy (2018) and the African Union's Agenda 2063 (AU 2015) by promoting local energy sovereignty and bioindustrialization.

In essence, this study fills a critical knowledge and implementation gap at the intersection of bioenergy, circular agriculture, and environmental sustainability in SSA. By systematically varying both feedstock configuration and product scope, it identifies the environmental implications of scaling from simple electricity generation to multifunctional biorefineries within smallholder agricultural contexts. The findings not only quantify the environmental trade-offs of IMF biorefineries in Malawi but also establish a replicable methodological template for similar agro-ecological contexts in developing regions.

5.2 LIFE CYCLE ASSESSMENT METHODOLOGY

LCA is a robust and standardized methodology used to evaluate the environmental impacts associated with all stages of a product's life, from raw material acquisition through production, use, and end-of-life disposal (Lee & Inaba 2004; Ubando et al. 2022). In the context of biorefineries in this study, LCA provides a systematic framework for assessing the sustainability of biomass feedstock options and biorefinery pathways, enabling informed decisions that balance environmental performance, resource efficiency, and socio-economic viability. To support a structured and transparent evaluation of feedstock sustainability, Figure 5.1 presents the standardized LCA framework as defined by ISO 14040 (ISO 2006). This framework outlines the four key phases: (i) goal and scope definition, (ii) inventory analysis, (iii) impact assessment, and (iv) interpretation, providing a consistent basis for comparing biomass feedstocks and guiding decision-making in biorefinery development.

5.2.1 Goal and Scope

The primary goal of this LCA was to quantify and compare the environmental sustainability of six biorefinery configurations (Scenarios S1–S6) designed to process agricultural residues from maize and pigeon pea production in Malawi as a case study for SSA. The assessment aimed to identify how feedstock type (single versus intercropped mixed residues) and product focus (electricity-only, ethanol-energy coproduction, and lactic acid-energy coproduction) influence the overall environmental

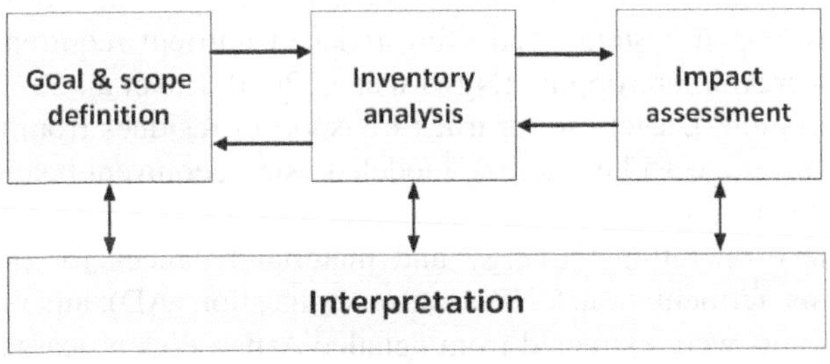

FIGURE 5.1 Standardized life cycle assessment framework based on ISO 14040.

(Adapted from ISO 2006.)

performance of integrated biorefinery systems. Scenarios S1–S3 represent single-feedstock systems based on maize stover, whereas S4–S6 substitute these with IMF systems comprising maize stover and pigeon pea residues.

A cradle-to-gate system boundary was adopted, encompassing upstream agricultural residue generation, feedstock collection and transportation, and biorefinery conversion processes up to the production of ethanol, lactic acid, electricity, biogas, and organic fertilizer (Figure 5.2). The use phase and end-of-life of biorefinery products were excluded as the focus was on production-stage environmental trade-offs rather than downstream utilization. The LCA results were used to compare environmental burdens across all six scenarios, providing insights into how integrated mixed feedstocks and diversified product portfolios can enhance sustainability in SSA's smallholder farming systems.

5.2.2 Functional Unit and Data Inventory

The functional unit (FU) was defined as 1 t of dry feedstock processed at the biorefinery gate encompassing either single maize stover residues (S1–S3) or intercropped mixed residues (S4–S6). This feedstock-based FU provides a consistent reference for comparing the environmental profiles of systems with multiple coproducts and conversion routes. It ensures that emissions, inputs, and outputs are normalized on a uniform physical basis, enabling transparent cross-scenario comparisons of energy- and product-oriented systems.

Economic allocation was applied at the farming stage to partition environmental burdens between grains (the primary economic output) and residues (secondary byproducts used as feedstocks). Beyond the field boundary, all impacts were fully attributed to the residues as they are transported and processed industrially, consistent with ISO 14044 guidelines and similar bioenergy LCA studies (Lee & Inaba 2004; Ngwira et al. 2020; Ubando et al. 2022).

The life cycle inventory covered three main stages (Liu et al. 2021; Ubando et al. 2022):

1. Farming: Fertilizer application, field emissions (N_2O, NH_3, NO_X), and residue management under monocropping and intercropping conditions. Fertilizer inputs were 92 kg N/ha, 20 kg P_2O_5/ha, and 10 kg K_2O/ha for monocropping systems, and 46 kg N/ha, 10 kg P_2O_5/ha, and 5 kg K_2O/ha for intercropped systems, reflecting reduced nutrient requirements under legume-cereal intercropping (Ngwira et al. 2020; Liu et al. 2021).
2. Transportation: Diesel use for truck transport of residues from field to biorefinery (average 15 km radius), modeled using ecoinvent heavy-duty road transport datasets.
3. Biorefinery operations: Energy and material balances for pretreatment, hydrolysis, fermentation, CHP, anaerobic digestion (AD), and product purification units were extracted from detailed Aspen Plus process simulations for each scenario. These balances provided the basis for quantifying inputs (chemicals, utilities, and water) and outputs (products, coproducts, emissions, and waste streams).

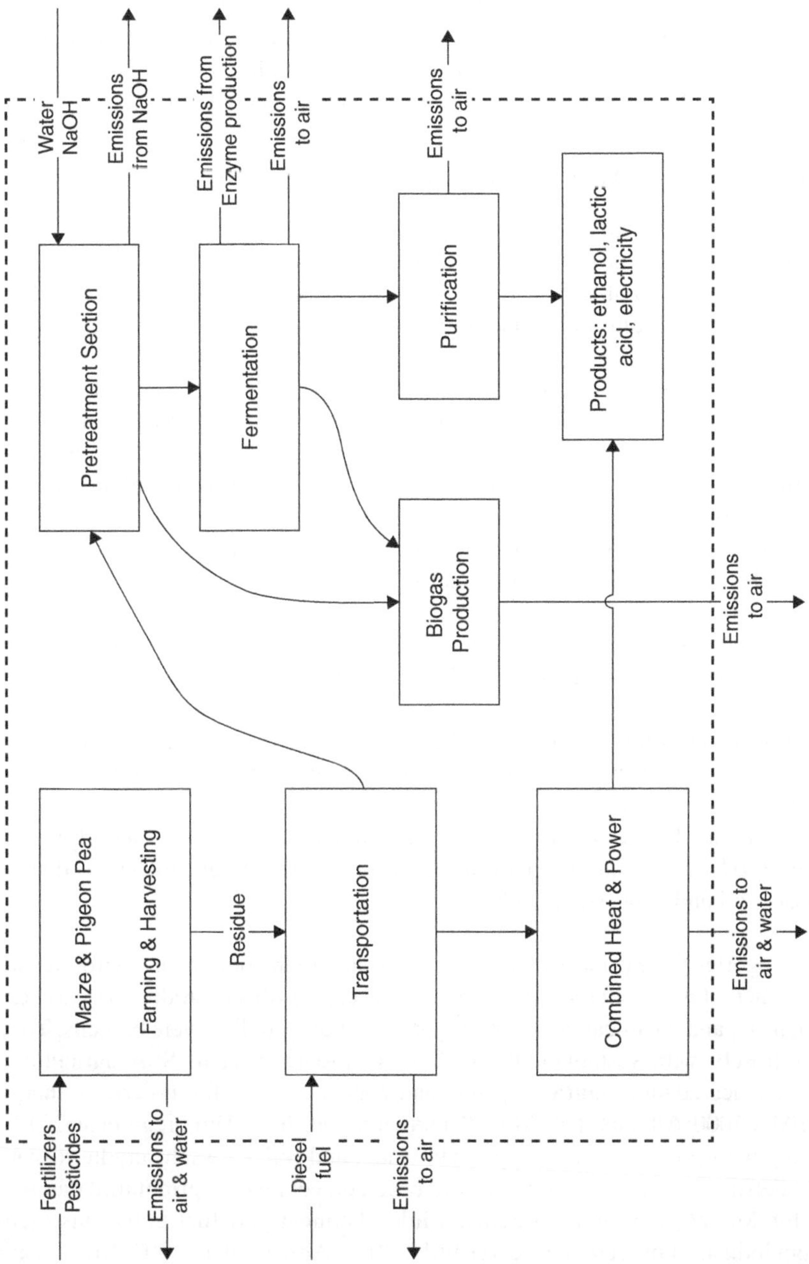

FIGURE 5.2 Illustration of the cradle-to-gate boundary for life cycle analysis of the intercropped mixed feedstock biorefinery products. **(Adapted from Msiska et al. 2024.)**

Process inventories incorporated the use of NaOH, HCl, NH_3, SO_2, $(NH_4)_2HPO_4$, and glucose, all modeled using corresponding industrial datasets in ecoinvent v 3.9.1 (EFMA 2000; Althaus et al. 2007; Brinkmann et al. 2014). Field and process emissions were modeled according to Intergovernmental Panel on Climate Change, European Monitoring and Evaluation Program, and the European Environment Agency guidelines embedded in ecoinvent, ensuring methodological consistency and transparency (EFMA 2000; Althaus et al. 2007; Brinkmann et al. 2014).

5.2.3 Environmental Impact Assessment and Interpretation

The environmental assessment was conducted using SimaPro v 9.6, implementing the ReCiPe 2016 midpoint (H/A) method (Huijbregts et al. 2017). Among the 18 midpoint categories available, six impact indicators were emphasized due to their direct relevance to agricultural residue utilization and biorefinery operations:

- GWP (kg CO_2 eq/t) reflects climate change impacts from CO_2, CH_4, and N_2O emissions associated with fertilizer use, transport, and process energy.
- Terrestrial ecotoxicity, TE (kg 1,4-DCB eq/t), captures toxic effects of chemical use and heavy metals in soil linked to fertilizer and process effluent emissions.
- Ionizing radiation, IR (kBq Co-60 eq/t), represents radiological impacts from electricity generation, especially where grid electricity is supplemented during processing.
- Fossil resource depletion, FRD (kg oil eq/t), quantifies depletion of fossil energy resources, primarily through diesel and auxiliary energy inputs in transport and processing.
- Freshwater eutrophication, FWE (kg P eq/ton), accounts for phosphorus runoff from fertilizer application and nutrient discharges from biorefinery wastewater streams.
- Terrestrial acidification, TA (kg SO_2 eq/t), reflects emissions of acidifying gases (NH_3, NO_x, and SO_2) from agricultural and energy processes that impact soil and ecosystem health.

These categories were prioritized because they capture the dominant environmental trade-offs inherent to biomass-based systems where fertilizer production, process energy intensity, and chemical usage are major contributors to life cycle burdens. They also align closely with sustainability challenges most relevant to SSA agriculture, including soil degradation, nutrient losses, and dependence on fossil-derived energy inputs (EFMA 2000; Althaus et al. 2007; Brinkmann et al. 2014; Huijbregts et al. 2017).

Modeling incorporates emissions into air, water, and soil across the cradle-to-gate boundary. Fertilizer-derived emissions were based on ecoinvent agricultural datasets adjusted for Malawian residue systems, while chemical and fuel emissions were drawn from industrial process datasets (EFMA 2000; Althaus et al. 2007; Brinkmann et al. 2014; Huijbregts et al. 2017). Energy-related emissions from CHP and AD were calculated from process-specific combustion and digestion efficiencies obtained from simulation results.

5.3 RESULTS AND DISCUSSION

5.3.1 Environmental Performance Trends Across Scenarios

The LCA results reveal consistent patterns across all six midpoint impact categories, GWP, TE, IR, FRD, FWE, and TA (Figures 5.3 and 5.4). Two key trends emerge: (i) environmental burdens increase with product-system complexity, shifting from electricity-only systems to ethanol-electricity-biogas, and further to lactic acid-electricity-biogas systems; and (ii) substituting single monocrop feedstocks with IMFs

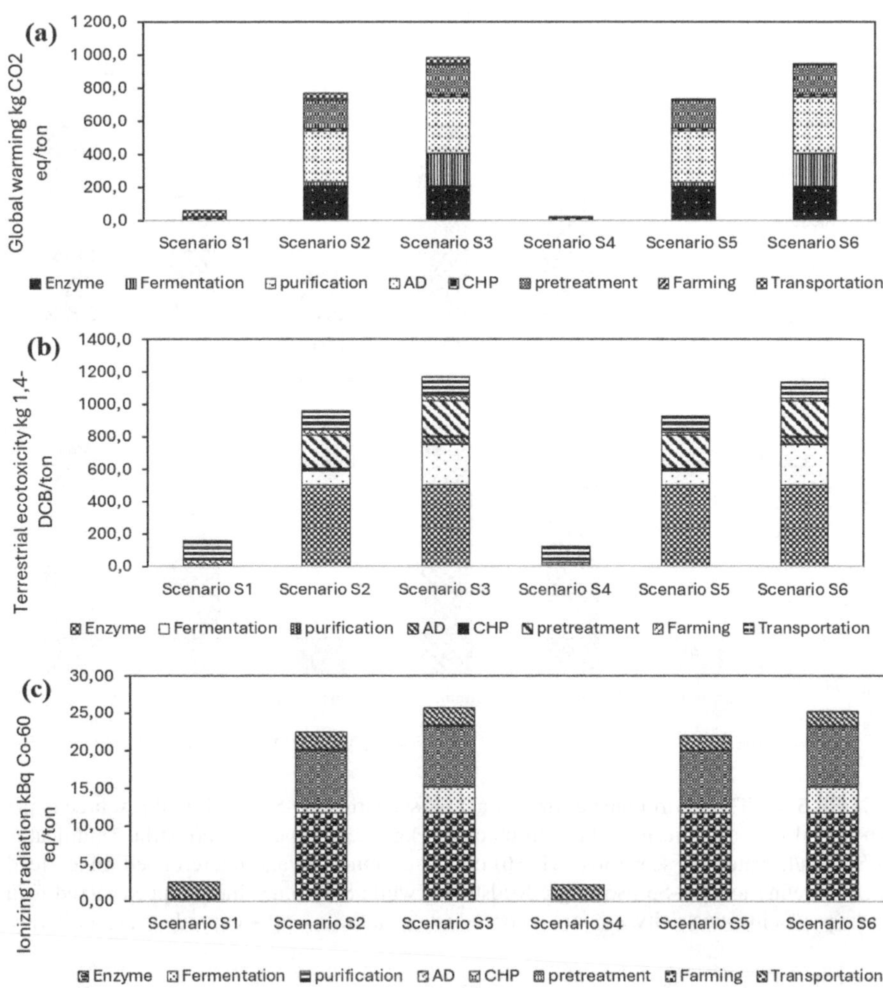

FIGURE 5.3 The environmental impact across scenarios S1–S6: (a) global warming impact kg CO_2 eq/t, (b) Terrestrial ecotoxicity kg 1,4-DCB/t, and (c) Ionizing radiation kBq Co-60 eq/t. Note: Six scenarios (S1–S6) compare monocrop and intercropped mixed feedstock biorefineries: S1–S3 use single feedstocks, while S4–S6 use intercropped mixed feedstocks producing electricity, ethanol + coproducts, and lactic acid + coproducts, respectively.

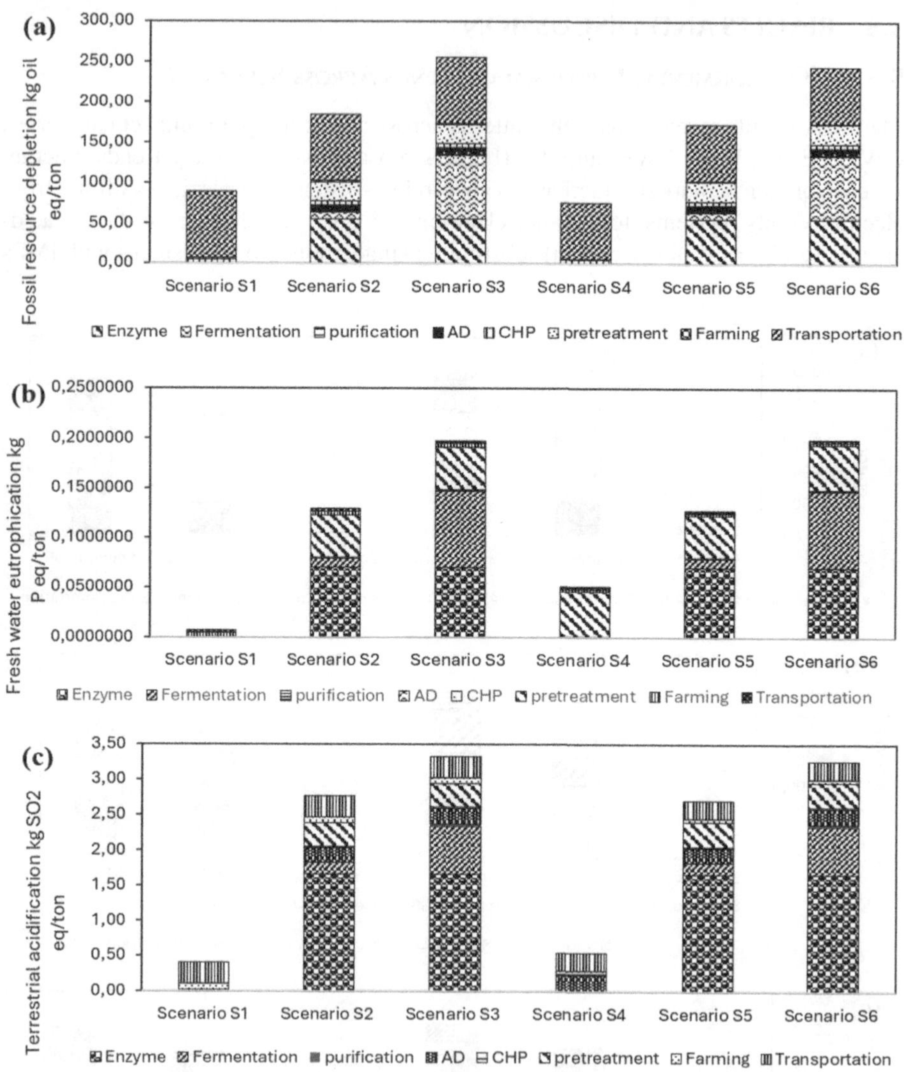

FIGURE 5.4 The environmental impact across scenarios S1–S6: (a) Fossil resource depletion kg oil eq/t, (b) Fresh water eutrophication kg P eq/t, and (c) Terrestrial acidification kg SO_2 eq/t. Note: Six scenarios (S1–S6) compare monocrop and intercropped mixed feedstock biorefineries: S1–S3 use single feedstocks, while S4–S6 use intercropped mixed feedstocks producing electricity, ethanol + coproducts, and lactic acid + coproducts, respectively.

substantially lowers most impact magnitudes, confirming the synergistic advantages of integrated residue utilization. For example, GWP emissions rise sharply with increasing system complexity. The single-feedstock electricity pathway (S1) records 59.6 kg CO_2 eq/t, which escalates to 770.0 kg CO_2 eq/t in the ethanol-biogas configuration (S2) and 982.8 kg CO_2 eq/t in the lactic acid-biogas system (S3) (Figure 5.3a). Introducing

IMFs (S4–S6) reduces these to 23.0, 733.5, and 946.3 kg CO_2 eq/t, respectively, corresponding to reductions of 61%, 4.7%, and 3.7%. These findings align with Canter et al. (2015) who observed that integrating CHP systems into maize-based biorefineries can reduce greenhouse gas emissions by up to 74%. However, our results show that such reductions are most pronounced in energy-centric systems, while chemically intensive pathways like lactic acid production offer limited mitigation.

A similar trend is observed for TE, which rises from 157.4 kg 1,4-DCB eq/t in S1 to 959.5 and 1,169.8 kg 1,4-DCB eq/t in S2 and S3, respectively (Figure 5.3b). The introduction of IMFs decreases TE by 22%, 3.6%, and 2.9% to 123.0, 925.2, and 1,135.5 kg 1,4-DCB eq/t, respectively. This pattern is consistent with Muñoz et al. (2013) who found that maize stover ethanol systems exhibit relatively low ecotoxicity, but fermentation and purification steps in more complex systems significantly elevate chemical burdens. The higher TE in ethanol and lactic acid systems is attributed to the chemical and nutrient inputs required during fermentation, neutralization, and downstream purification. The larger reduction in S1 vs. S4 emphasizes the upstream benefit of reduced fertilizer and pesticide use in IMF systems.

For IR, impacts are modest but follow the same direction, increasing from 2.56 (S1) to 22.43 and 25.68 kBq Co-60 eq/t (S2, S3) (Figure 5.3b). The IMF cases record 2.13, 21.99, and 25.24 kBq Co-60 eq/t, equivalent to reductions of 16.8%, 2.0%, and 1.7%, respectively. These improvements reflect lower embodied energy in cultivation and preprocessing, which reduces indirect radiological burdens from grid electricity inputs, a trend noted in regional LCA studies of low-input cropping systems.

FRD mirrors the GWP pattern (Figure 5.4a). S1 records 88.74 kg oil eq/t rising to 184.47 (S2) and 225.81 kg oil eq/t (S3). IMF substitution lowers these to 75.41, 172.46, and 243.81 kg oil eq/t (Figure 5.4a). The first two IMF cases show clear reductions (−15% and −7%), whereas the lactic acid pathway (S6) shows an 8% increase, reflecting the heavy fossil-derived reagent demand (e.g. triethylamine and calcium carbonate) and high-temperature energy use in lactic acid purification. The increase is consistent with findings by Manandhar and Shah (2023) who report that lactic acid production from lignocellulosic biomass incurs high fossil resource demand due to reagent use (e.g. triethylamine and calcium carbonate) and energy-intensive purification.

FWE increases with production intensity, from 0.0069 in S1 to 0.1290 (S2) and 0.1973 kg P eq/t (S3) (Figure 5.4b). With IMFs, the values become 0.0511, 0.1277, and 0.1989 kg P eq/t, representing a 26% reduction for electricity-only systems but negligible (<1.5%) changes for the more complex pathways. The greater improvement in S1 vs. S4 stems from reduced fertilizer use via biological nitrogen fixation, while limited gains in S5–S6 indicate that wastewater nutrient loads dominate eutrophication in advanced systems.

TA rises with product complexity from 0.40 in S1 to 2.76 in S2 and 3.32 kg SO_2 eq/t in S3. IMF cases record 0.54, 2.70, and 3.26 kg SO_2 eq/t (Figure 5.4c).

The terrestrial acidification (TA) results show a contrasting pattern when comparing single-feedstock scenarios (S1–S3) with their corresponding IMF scenarios (S4–S6). Notably, TA increases by 35% from S1 to S4, while modest declines of 2%–3% occur from S2 to S5 and S3 to S6. This contrast reflects the interplay of pathway-specific mechanisms. In the electricity-only pathway (S1 → S4), fertilizer use in the monocrop system is moderate, and switching to IMF introduces legume residues (pigeon pea), which decompose and release NO_x, locally elevating acidification. Because chemical and thermal inputs in this pathway are low, the reduction in fertilizer-related acidifying emissions is insufficient to offset NO_x contributions, resulting in a net increase in TA. In the more chemically intensive pathways (S2 → S5 and S3 → S6), upstream inputs, including fertilizer production, reagents, and energy for fermentation and purification, dominate acidification potential. Switching to IMF reduces fertilizer demand and enhances nutrient recycling from legume residues, which outweighs the acidifying effects of NO_x emissions, leading to slight net declines in TA. This nuanced pattern aligns with Khoiri et al. (2025), who observed elevated NO_x emissions in legume-based intercropping systems, highlighting that intercropping mitigates upstream impacts while potentially introducing localized nitrogen-related effects. Overall, the TA patterns demonstrate that the net acidification outcome of IMF integration depends on the relative magnitude of upstream chemical/energy-related emissions versus nitrogen fluxes from legume decomposition, emphasizing the need to consider pathway-specific sensitivities when evaluating intercropping strategies.

Overall, the environmental burdens across the evaluated systems follow two distinct hierarchical patterns. Firstly, impacts increase progressively with product-system complexity following the order electricity < ethanol < lactic acid, indicating that as the product portfolio expands from energy-focused to more chemically intensive pathways, the environmental footprint correspondingly intensifies. Secondly, for any given product-system, impacts consistently decrease in the order monocrop > mixed feedstock, demonstrating that the integration of intercropped mixed residues substantially mitigates environmental pressures. This dual gradient underscores that while product diversification toward high-value chemicals elevates resource and energy demands, and feedstock diversification through residue integration remains a highly effective strategy for reducing overall environmental burdens and enhancing system sustainability.

5.3.2 Mechanisms for the Observed Trends

The mechanisms underlying the observed environmental trends stem from both agronomic and process configuration dynamics. Agronomically, intercropping maize with pigeon pea enhances nutrient cycling and soil fertility through biological nitrogen fixation, thereby reducing the reliance on synthetic fertilizers, a major contributor to GWP, FWE, and TE (Ngwira et al. 2020). For example, Sun et al. (2025) found that maize/pea intercropping increased symbiotic nitrogen fixation efficiency by 23% and improved overall nitrogen use efficiency and yield by 15%, demonstrating strong interspecific root interactions that underlie our observed fertilizer savings. Khoiri et al. (2025) showed that pigeon pea intercropping can alter soil microbial communities, enhancing nitrogen metabolism and ecosystem multifunctionality, which helps to explain enhanced soil function and reduced upstream inputs. The leguminous component also enriches soil organic matter, which improves carbon sequestration potential and soil structure. Building on these agronomic benefits, Renwick et al. (2020) highlighted that land-sparing effects of intercropping (achieving higher biomass yields per hectare) can mitigate indirect land-use change emissions, supporting the productivity and land-use outcomes in our IMF systems. Furthermore, the use of

IMFs can improve land-use efficiency by generating higher total biomass yields per hectare without requiring additional cultivated land, thus mitigating indirect land-use change emissions (Ngwira et al. 2020; Silberg et al. 2020). As a result, IMF configurations (S4–S6) consistently achieved lower GWP, FRD, and TE values due to reduced upstream fertilizer demand and optimized residue utilization.

In contrast, the elevated impacts associated with lactic acid systems (S3 and S6) are primarily driven by their chemically intensive operations. Lactic acid purification requires significant quantities of neutralizing agents and solvents, coupled with energy-intensive distillation, leading to substantial increases in GWP and FRD. These process "hotspots" are well-documented in literature, which highlights solvent recovery, renewable process heat integration, and substitution of fossil-based neutralizers with bio-based alternatives as key strategies for mitigation. Ögmundarson et al. (2020) identified that energy utilities and downstream separation stages contribute to the majority of environmental burdens in lactic acid systems derived from varied feedstocks. Manandhar and Shah (2023) corroborate these findings, detailing that neutralization and distillation are key hotspots, similar to our observed emission spikes in S3 and S6. To mitigate the elevated environmental impacts associated with lactic acid production, particularly those arising from chemically intensive purification processes, the literature proposes several process optimization strategies. Among these, the implementation of solvent recovery systems and the integration of renewable process heat have been identified as effective measures to reduce energy consumption and associated emissions (Ögmundarson et al. 2020). Building on this, advancements in downstream separation technologies such as aqueous two-phase extraction, membrane separation, and electrodialysis have shown considerable promise in minimizing both chemical usage and thermal energy demands, thereby improving the overall environmental performance of lactic acid biorefineries (Gugel et al. 2025). Further innovations, including emulsion liquid membrane techniques and nanofiltration, offer additional pathways to reduce reagent consumption and eliminate the need for high-temperature distillation, which is a major contributor to FRD and GWP (Kumar et al. 2019; Salatein et al. 2025). These findings underscore the critical role of process intensification and circular resource recovery in achieving sustainability targets, particularly for biorefineries aiming to produce high-value chemicals alongside energy carriers.

5.3.3 Comparative Environmental and Policy Implications

These results align closely with global evidence from LCA studies on residue-based and multi-product biorefineries. Analyses from East Africa, South Asia, and Latin America consistently report a 40%–65% reduction in greenhouse gas intensity and a 20%–30% decline in nutrient-related impacts when mixed or intercropped feedstocks are used (Oke et al. 2016; Patrizi et al. 2020). Similar findings from Europe and North America show that residue blending can enhance process stability and reduce supply risk, supporting the premise that feedstock diversification is a robust decarbonization lever (Oke et al. 2016; Patrizi et al. 2020).

Regionally, the implications for SSA are significant. Intercropping of maize with pigeon pea is widespread among smallholder farmers, making IMF adoption both agronomically compatible and socioeconomically feasible (Renwick et al. 2020; Frimpong et al. 2025). The demonstrated 61% reduction in GWP, alongside concurrent decreases in FRD, TE, and IR, shows that IMF-based biorefineries can strengthen rural energy systems, enhance fertilizer self-sufficiency, and improve soil resilience, aligning with SDGs 2 (Zero Hunger), 7 (Affordable and Clean Energy), 12 (Responsible Consumption and Production), and 13 (Climate Action).

At the policy level, these insights advocate for a dual approach, promoting feedstock diversification in the near term while investing in process innovation for long-term sustainability. Incentivizing smallholder participation in IMF value chains, expanding residue collection cooperatives, and integrating renewable heat systems into biorefineries would yield both environmental and economic dividends.

Globally, the study underscores that coupling feedstock diversification with careful product selection is pivotal for transitioning to sustainable bioeconomies. While feedstock strategies can deliver immediate reduction in GWP and FRD, true circularity and competitiveness in high-value bioproducts demand innovation across the entire value chain, from cultivation to conversion and product recovery.

5.4 CONCLUSIONS

This study demonstrates that the environmental sustainability of SSA biorefineries is strongly influenced by feedstock configuration and product-system complexity. Intercropped mixed feedstocks consistently outperform single monocrop residues, while product diversification from electricity-only to ethanol- and lactic acid pathways increases environmental burdens. Feedstock diversification mitigates upstream fertilizer- and energy-related emissions, whereas chemically intensive conversion steps remain key environmental hotspots. The work highlights the synergies between feedstock diversity, nutrient cycling, and conversion efficiency, providing region-specific LCA data for African systems that are often underrepresented in global databases. From a policy and development perspective, integrating maize–pigeon pea residues into decentralized or cooperative biorefineries can deliver climate, energy, and nutrient benefits, supporting SDGs on zero hunger, clean energy, responsible production, and climate action.

Globally, the findings emphasize that sustainable bioeconomy transitions require coupling feedstock innovation with low-impact process engineering, including renewable heat, solvent recovery, and greener catalysts. Future research should explore spatially explicit residue availability, process optimization, and water-energy-food nexus indicators to further enhance low-impact biorefinery design. Overall, IMFs provide a robust and regionally appropriate pathway for sustainable biorefinery development in Africa, demonstrating that feedstock diversification and process decarbonization must evolve together to realize low-impact bioeconomies.

Declaration of Generative AI and AI-Assisted Technologies in the Writing Process

The authors declare that they only used generative AI and AI-assisted technologies, for example, Grammarly and ChatGPT, for editing grammar and spelling.

ACKNOWLEDGMENTS

The authors would like to acknowledge the Malawi University of Science and Technology, Center for Agricultural Transformation, National Research Foundation, and Stellenbosch University for the financial support. The authors would like to acknowledge the technical staff in the Departments of Chemical Engineering and Wood Science at Stellenbosch University for their technical support.

REFERENCES

Althaus, H.-J., M. Chudacoff, R. Hischier, N. Jungbluth, M. Osses, & A. Primas. 2007. *Life cycle inventories of chemicals*. Ecoinvent report No. 8, v2.0. Dubendorf, CH: Swiss Center for Life Cycle Inventories. http://www.ecoinvent.org/

AU. 2015. *Agenda 2063: The Africa we want: Framework document*. Addis Ababa: African Union Commission.

Berchem, T., Q. Schmetz, T. Lepage, & A. Richel. 2020. Single and mixed feedstocks biorefining: Comparison of primary metabolites recovery and lignin recombination during an alkaline process. *Front. Chem.* 8: 517586.

Brinkmann, T., G.G. Santonja, F. Schorcht, S. Roudier, & L.D. Sancho. 2014. *Best available techniques reference document for the production of chlor-alkali*, Report number: EUR 26844 EN. Luxembourg: European Commission.

Canter, C.E., J.B. Dunn, J. Han, Z. Wang, & M. Wang. 2015. Policy implications of allocation methods in the life cycle analysis of integrated corn and corn stover ethanol production. *BioEnergy Res.* 9: 77–87.

Chaima, E., J. Lian, C. Ma, S. Kaunda, S. Kavwenje, M. Nkhata, S. Phiri, & P. Kumambala. 2023. Long-term electricity demand scenarios for Malawi's electric power system. *Energy Sustain. Dev.* 73: 23–38.

Cherubini, F., & G. Jungmeier. 2010. LCA of a biorefinery concept producing bioethanol, bioenergy, and chemicals from switchgrass. *Int. J. Life Cycle Assess.* 15: 53–66.

EFMA. 2000. *Best available techniques for pollution prevention and control in the European fertilizer industry booklet no. 2 of 8: Production of nitric acid*. Fertilizers Europe. https://www.fertilizerseurope.com/wp-content/uploads/2019/08/Booklet_2_final.pdf

Frimpong, F., E.O. Danquah, S.K. Amponsah, T. Frimpong, J. Adu, F.O. Danquah, N.E. Amengor, & P. Amankwaa-Yeboah. 2025. Diversifying cropping systems to enhance productivity using agroforestry trees: A case study of maize-pigeon pea intercropping in Ghana. *Explora Environ. Resour.* 2: 025130026.

Gaffey, J., M.N. Collins, & D. Styles. 2024. Review of methodological decisions in life cycle assessment (LCA) of biorefinery systems across feedstock categories. *J. Environ. Manage.* 358: 120813.

Gondwe, K.J., S.S. Chiotha, T. Mkandawire, X. Zhu, J. Painuly, & J. L. Taulo. 2017. Crop residues as a potential renewable energy source for Malawi's cement industry. *J. Energy South. Afr.* 28: 19–31.

Gugel, I., F. Marchetti, S. Costa, E. Baldini, S. Vertuani, & S. Manfredini. 2025. Efficient downstream processing of second-generation lactic acid from lignocellulosic waste using aqueous two-phase extraction. *Bioresour. Bioprocess.* 12: 1–13.

Huijbregts, M.A.J., Z.J.N. Steinmann, P.M.F. Elshout, G. Stam, F. Verones, M. Vieira, M. Zijp, A. Hollander, & R. van Zelm. 2017. ReCiPe2016: A harmonized life cycle impact assessment method at midpoint and endpoint level. *Int. J. Life Cycle Assess.* 22: 138–147.

ISO. 2006. *ISO 14040: 2006 Environmental management life cycle assessment – Principles and framework*. Geneva: ISO.

Jha, P., & S. Schmidt. 2021. State of biofuel development in sub-Saharan Africa: How far is sustainable? *Renew. Sustain. Energy Rev.* 150: 111432.

Juneja, A., G. Zhang, Y.S. Jin, & V. Singh. 2019. Bioprocessing and techno-economic feasibility analysis of simultaneous production of D-psicose and ethanol using engineered yeast strain KAM-2GD. *Bioresour. Technol.* 275: 27–34.

Katakojwala, R., & S.V. Mohan. 2021. A critical view on the environmental sustainability of biorefinery systems. *Curr. Opin. Green Sustain. Chem.* 27: 100392.

Khoiri, A.N., N.R. Costa, C.A.C. Crusciol, C.M. Pariz, C. Costa, J.C. Calonego, A.M. de Castilhos, D.M. de Souza, P.R. de Lima Meirelles, I.V. Cru, L.G. Moretti, J.W. Bossolani, & E.E. Kuramae. 2025. Pigeon pea-mediated soil microbial shifts improve agroecosystem multifunctionality in long-term maize-palisade grass intercropping. *Environ. Microbiome* 20: 1–19.

Kumar, A., A. Thakur, & P.S. Panesar. 2019. Lactic acid and its separation and purification techniques: A review. *Rev. Environ. Sci. Biotechnol.* 18: 823–853.

Lee, K.-M., & A. Inaba. 2004. *Life cycle assessment: Best practices of ISO 14040 Series.* Suwon: Center for Ecodesign and LCA (CEL), Ajou University.

Liu, F., X. Guo, Y. Wang, G. Chen, & L. Hou. 2021. Process simulation and economic and environmental evaluation of a corncob-based biorefinery system. *J. Clean. Prod.* 329: 129707.

Manandhar, A., & A. Shah. 2023. Techno-economic analysis of the production of lactic acid from lignocellulosic biomass. *Fermentation* 9: 641.

Msiska, S.C., R.K. Padi, & A. Chimphango. 2024. Comparative techno-economic and life cycle assessments of single and intercropped mixed feedstocks-based biorefineries for profitable and cleaner production of energy. *J. Clean. Prod.* 450: 141988.

Muñoz, I., K. Flury, N. Jungbluth, G. Rigarlsford, L.M. Canals, & H. King. 2013. Life cycle assessment of biobased ethanol produced from different agricultural feedstocks. *Int. J. Life Cycle Assess.* 19: 109–119.

Ngwira, A.R., V. Kabambe, P. Simwaka, K. Makoko, & K. Kamoyo. 2020. Productivity and profitability of maize-legume cropping systems under conservation agriculture among smallholder farmers in Malawi. *Acta Agric. Scand. B Soil Plant Sci.* 70: 241–251.

Ögmundarson, O., S. Sukumara, A. Laurent, & P. Fantke. 2020. Environmental hotspots of lactic acid production systems. *GCB Bioenergy* 12: 19–38.

Oke, M.A., M.S.M. Annuar, & K. Simarani. 2016. Mixed feedstock approach to lignocellulosic ethanol production – Prospects and limitations. *BioEnergy Res.* 9: 1189–1203.

Osman, A.I., B. Fang, Y. Zhang, Y. Liu, J. Yu, M. Farghali, A.K. Rashwan, Z. Chen, L. Chen, I. Ihara, D.W. Rooney, & P. Yap. 2024. Life cycle assessment and techno-economic analysis of sustainable bioenergy production: A review. *Environ. Chem. Lett.* 22: 1115–1154.

Parada, M.P., P. Osseweijer, & J.A.P. Duque. 2017. Sustainable biorefineries, an analysis of practices for incorporating sustainability in biorefinery design. *Ind. Crops Prod.* 106: 105–123.

Patrizi, N., M. Bruno, F. Saladini, M.L. Parisi, M.L. Parisi, R.M. Pulselli, A.B. Bjerre, & S. Bastianoni. 2020. Sustainability assessment of biorefinery systems based on two food residues in Africa. *Front. Sustain. Food Syst.* 4: 522614.

Renwick, L.L.R., A.A. Kimaro, J.M. Hafner, T.S. Rosenstock, & A.C.M. Gaudin. 2020. Maize-pigeon pea intercropping outperforms monocultures under drought. *Front. Sustain. Food Syst.* 4: 562663.

Salatein, N.M., A.M. Abdelghany, B.M. Elmowafy, R.K. Hassan, A.F. Omara, A.R. Mansour, & I.S. Fahim. 2025. Lactic acid separation technologies: Enhancing efficiency and purity using membrane separation technology (mini review). *Results Chem.* 15: 102280.

Silberg, T.R., R.B. Richardson, & M.C. Lopez. 2020. Maize farmer preferences for intercropping systems to reduce Striga in Malawi. *Food Sec.* 12: 269–283.

Sun, Y., Z. Wu, F. Hu, H. Fan, W. He, L. Zhao, C. Guo, X. Bao, Q. Chai, & C. Zhao. 2025. Maize and pea root interactions promote symbiotic nitrogen fixation, thereby accelerating nitrogen assimilation and partitioning in intercropped pea. *Agronomy* 15: 1615.

Ubando, A.T., E.A.S. Ng, W.H. Chen, A.B. Culaba, & E.E. Kwon. 2022. Life cycle assessment of microalgal biorefinery: A state-of-the-art review. *Bioresour. Technol.* 360: 127615.

Whalley, S., S.J.W. Klein, S. Klein, & B. Jeffrey. 2017. Economic analysis of woody biomass supply chain in Maine. *Biomass Bioenergy* 96: 38–49.

Zhang, Z., Z. Chen, J. Zhang, Y. Liu, L. Chen, M. Yang, A.I. Osman, M. Farghali, E. Liu, D. Hassan, I. Ihara, K. Lu, D.W. Rooney, & P. Yap. 2024. Municipal solid waste management challenges in developing regions: A comprehensive review and future perspectives for Asia and Africa. *Sci. Total Environ.* 930: 172794.

6 Advances in Biohydrogen and Bioelectricity Production

KeChrist Obileke, Patrick Mukumba, and Kaitano Dzinavatonga

6.1 INTRODUCTION

Generally, waste is known to pose a serious environmental and public health concern and, at the same time, to offer significant promise for the production of clean energy (Logan & Rabaey 2012). Conversely, as an alternative to treating these wastes, there is a need to harness the energy potential of waste streams through a cost-effective mechanism (Moqsud et al. 2013; Moqsud 2021). One of these methods is the microbial fuel cells (MFCs), which has received extensive attention in recent times due to its use in microbial treatment of anaerobic digestion, in particular. Also, the technology has been shown to convert waste biomass into energy without emitting or contributing to environmental pollution (Li et al. 2014; Sevda et al. 2018; Moradian et al. 2021). Furthermore, Mohan et al. (2008) and Li et al. (2014) stated that the dark fermentation via anaerobic processes has been employed to generate clean biohydrogen ($BioH_2$). With these developments and contributions, both microbial treatment and dark fermentation via anaerobic processes are integrated to treat solid and liquid wastes using the MFC technology (Chandrasekhar et al. 2015).

The interconnection between $BioH_2$ and bioelectricity production is primarily enabled by the MFC technology. According to Kafaei et al. (2023), MFCs oxidize organic substrates via electrochemically active microorganisms and release electrons. These electrons are transferred to the anode chamber to produce bioelectricity, and the external circuit is then connected to the cathode chamber for H_2O_2 and H_2O production via oxygen reduction using a $2e^-$ or $4e^-$ pathway. This mechanism is based on a bioelectrochemical system, known for its green technologies that convert biomass from liquid and solid wastes (Palanisamy et al. 2019). The same process is referred to as electrohydrogenesis due to its shared biological basis.

Both $BioH_2$ and bioelectricity involve microbial metabolism, whereby organic matter is broken down into simple sugars through biological processes and then converted into electricity and protons, but differ in how they generate energy

 DOI: 10.1201/9781003650751-6

(Kundu et al. 2025; Logan et al. 2015). An in-depth discussion of the modes of operation for $BioH_2$ and bioelectricity production is presented later in this chapter. On the contrary, microbial electrolysis cells (MECs) are used to produce $BioH_2$ by applying an external voltage of 0.4–1.11 V. The pathway of $BioH_2$ production appears attractive due to its strong independence from external electricity supply. It is worth noting that $BioH_2$, as a source of bioelectricity, has been regarded as the fuel of the future due to its clean, renewable energy. Hence, it has a relatively high energy content per unit weight of 122–124 kJ/g (Kotay & Das 2008; Chandrasekhar et al. 2015).

It is imperative to establish that both MFC and MEC technologies are applicable in relation to $BioH_2$ and bioelectricity production. The mechanism proposed by Sołowski (2018) states that the common method for producing $BioH_2$ involves integrating MFCs and MECs, in which electricity produced by MFCs powers MECs for $BioH_2$ production. This is utilized in either solid or liquid waste streams. There has been progress in developing a single, continuous MEC for bioelectricity and $BioH_2$. For instance, Guadarrama-Pérez et al. (2019) examined the production of bioelectricity and $BioH_2$ from xylose using single and mixed cultures. It was observed that the system reported a maximum voltage of 671 mV at a resistance of 1,000 Ω, a power density of 67 mW/m^2, and a $BioH_2$ rate of 5.5 L H_2/L.d, respectively.

Another process for generating bioelectricity is electrogenesis. The process uses MFCs to generate bioelectricity from cellulose and other biodegradable organic matter (Logan & Regan 2006). In this case, the oxidation of organic matter outside the cell to the MFC anode results in the exoelectrogenic bacteria transferring electrons, while releasing protons into the solution (In-Sepo et al. 2006). The electrons, protons, and oxygen react at the cathode, thereby producing water. The anode potential can approach the maximum possible value based on the free energy of the substrate, as in the case of acetic acid at a neutral pH, approximately 0.3 V.

Having examined the technologies of both production processes, the chapter aims to provide an overview of the development, optimization, and scaling of biological processes that convert organic waste, biomass, and wastewater into clean energy. This is necessary as it forms another alternative to replacing fossil fuels, thereby supporting and encouraging the circular economy. The chapter also examines the rising conversion rates and lower costs enabled by recent advances in $BioH_2$ and bioelectricity production.

6.2 BIOHYDROGEN AND BIOELECTRICITY PRODUCTION

Recent advances in $BioH_2$ production must improve yield, efficiency, and cost-effectiveness through diverse biological and technological strategies. Strategies involve genetic modification of microorganisms, optimizing operational parameters (pH, temperature, feedstock and employing advanced technologies. $BioH_2$ provides a carbon-neutral, renewable energy solution that helps mitigate overreliance on fossil fuels (Arimi et al. 2015). One of the advantages of hydrogen is its lightness and abundance, which have enabled its use in various applications, such as transportation, industry, and heating (Lubbe et al. 2023). Furthermore, as the most promising future carrier of energy, hydrogen is considered environmentally neutral, when produced from water, although there seems to be an economic barrier during this process (Xie et al. 2016). Therefore, the phrase "biohydrogen" is used when hydrogen is produced from biological sources and means. As an energy source, $BioH_2$ contributes to waste management, wastewater treatment, and the processing of biomass residues as well as to the overall system efficiency, thereby creating job opportunities in most cases (Barghash et al. 2021; Sai et al. 2023). During $BioH_2$ production, one pathway is dark fermentation, which produces volatile fatty acids (VFAs) as a metabolic byproduct. The VFA's function is to facilitate the subsequent conversion of organic waste into biomethane.

The International Energy Agency (IEA) reported that approximately 80% of hydrogen is produced from fossil fuels. In 2021, about 47% of the global hydrogen production came from natural gas, a fossil fuel. In the same year, coal and oil accounted for around 27% and 22% of global hydrogen production, respectively. Based on these findings, there is a strong likelihood of significant environmental impacts from the generated emissions. For record purposes, the expected CO_2 emissions to produce 74 million tons of hydrogen are estimated at 830 million tons (IEA 2022). Therefore, to address dependence, $BioH_2$ production offers an effective and efficient way to provide a sustainable, renewable energy source.

The advances in bioelectricity production focus on improving MFC technology. This is the main focus of the production, thereby exploring fuel cells and the employment of organic materials for bioenergy. MFC is considered the primary technology for generating bioelectricity, as noted by Obileke et al. (2021). Obileke et al. (2021) defined MFCs as bio-electrochemical devices that convert chemical energy from organic materials into electrical energy. This takes place in the presence of microorganisms. For bioelectricity production, the use of organic material (wastewater) during MFC operation makes the technology an eco-friendly device (Xu et al. 2017).

The production of bioelectricity is facilitated by the formation of a biofilm by the electron-releasing microorganism. Also, removing organic carbon from wastewater serves the same purpose. In this case, the use of photosynthetic microorganisms at the anode of the MFC tends to reduce waste and generate power, as mentioned by Angelaalincy et al. (2018). This is further illustrated in Figure 6.1. Wang et al. (2018) stated that MFCs is an interesting technology for bioelectricity generation because they degrade antibiotics and alter their toxicity, thereby enabling conversion to bioelectricity. The process occurs through a biological mechanism. Furthermore, the ability of MFCs to generate electricity from various feedstocks through organic waste recycling has increased interest in and popularity of the technology. Specifically, animal excrement and leaf mold are expected to be the most suitable substrates for MFC-based electricity production (Gazali & Moqsud 2017). With the aid of the dual chamber of MFC, Ghazali et al. (2017) used empty fruit bunches of palm oil trees to generate electricity.

With reference to Sarma et al. (2018) and Dessie et al. (2021), the biodegradation of hazardous chemicals using a nanocomposite polymer-coated anode tends to

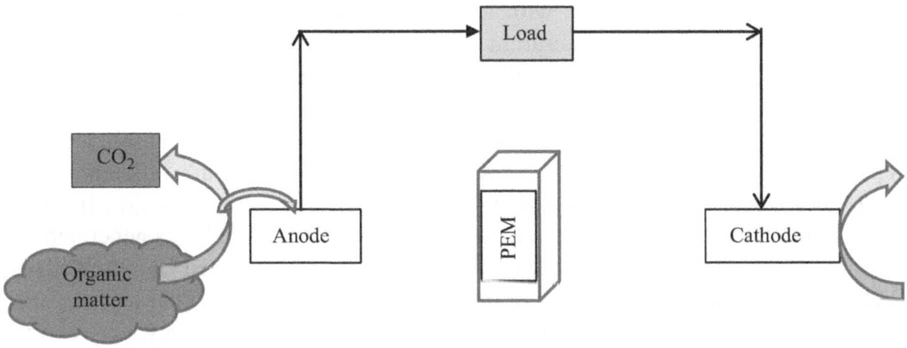

FIGURE 6.1 Schematic illustration of generating electricity via waste reduction in a microbial fuel cell using microorganisms.

enhance electricity output. In this case, given that algae are well-known for biofuel production, integrating them with cyanobacteria in MFCs can improve energy recovery. To improve MFC performance, Dessie and Tadesse (2022) noted that high specific surface area, low cost, strong electrical conductivity, and biocompatibility of the anodic material are factors.

6.3 MAJOR FEEDSTOCKS FOR BIOHYDROGEN AND BIOELECTRICITY PRODUCTION

BioH_2 and bioelectricity are produced using MEC or MFC technologies. These technologies are known to depend on organic-rich feedstock that is prone to microbial degradation. The primary and sustainable sources of feedstocks are agricultural waste, industrial effluents, and sewage (Ahmed et al. 2022). Solid particles in wastewater can reduce the efficiency of MFCs. According to Kim et al. (2013), lack of proper treatment or management is a major challenge to severe pollution in agro-industrial wastewater. On the other hand, agricultural residues are among the most prominent substrates for renewable energy production and a source of carbon. This is because of their availability as a cheap feedstock or substrate for renewable energy. In MFC, the microbial community is unable to directly utilize agricultural residue to produce bioelectricity. Therefore, to easily generate fermentable sugar hydrolysates, pretreatment is required.

6.3.1 Agro-Industrial Wastewater

Agro-industrial wastewater is considered a low-cost, profitable feedstock for BioH_2 and bioelectricity production. The feedstock has been used for BioH_2 production via microbial metabolism, resulting in increased yield through optimization of the carbon-to-nitrogen ratio. The palm oil mill effluent (POME) is a good example of agricultural industrial wastewater. Treatment of POME results in high-energy generation. Inoculating POME with anaerobic sludge can enhance a simple two-chambered MFC, thereby generating a higher power output of 45–304 mW/m^3 and a lower coulombic efficiency of 0.8%–45% COD removal (Cheng et al. 2010; Baranitharan et al. 2015).

Treatment of mustard tuber wastewater for its utilization in two-chambered MFC reported 246 mW/m^2 and 67%–85% COD removal and coulomb efficiency (Guo et al. 2013). Molasses contains a high COD concentration of 60–100 g/L. In MFCs using diluted molasses wastewater, a power density of 62 mW/m^2 was generated and 81% COD was removed. Treatment of sugar molasses generated bioelectricity of 0.18 W/m^2 from a MFC using *Brevibacillus borstelensis* STRII (Zhang et al. 2013; Hassan et al. 2019). The production of $BioH_2$ from molasses wastewater reported $BioH_2$ yields of 3–5 L/d, biogas and COD removal rates of 15%–19%, and a pH of 4.0–4.5 (Hong et al. 2010).

Brewery wastewater is another example of agro-industrial wastewater. Its high-strength wastewater is advantageous for energy generation and is mostly used in fermentation and saccharification processes (Lu et al. 2019). The negative effect of winery wastewater on MFC performance is due to its high polyphenolic content. Different winery waste types (e.g. white, red, etc.) have distinct characteristics that determine the MFC's power output. This is attributed to the microbial population present in the anode (Pandit et al. 2021). The wastewater is a highly potent substrate for $BioH_2$ production via anaerobic fermentation. It can also mitigate environmental risks due to its high organic load, up to 49 g COD/L (Ioannou et al. 2015).

The winery wastewater has a toxic effect on microorganisms, thereby reducing and inhibiting hydrogen productivity in many strains. Thus, its selection of the microbial culture is an important aspect for its valorization (Policastro et al. 2022). Other agricultural activities, effluents from food, plant, and animal processing industries, and waste from high-volume wastewater treatment plants also generate organic feedstocks. For example, sugar beet processing generated a power of 1.41 W/m^2 and 2.56 g COD/L under a retention time of 40 days in an MFC. Olive mill wastewater (OWM) is not a promising substrate for energy production in MFC technology. However, a combination of other wastewater with OMW generates high power density, with a removal of COD and low coulomb efficiency (Rahman et al. 2018; Roy & Pandit 2019).

6.3.2 Agricultural Residues

Agricultural residues are abundant, low-cost lignocellulosic feedstock for producing sustainable $BioH_2$ and bioelectricity. The advancement of pretreatment (ultrasound, torrefaction, etc.) has shown and proven to enhance the yield of $BioH_2$ from agricultural residues (Prado & Rial 2025). Examples of agro-waste include corn stover, rice husk, wheat straw, and sugarcane bagasse. The key technologies for this purpose are dark fermentation for $BioH_2$ and MFCs for bioelectricity. These processes turn waste into energy, thereby reducing emissions and burning of residues (Plakar et al. 2025). Wheat straw shows a high efficiency as a substrate for MFCs. The conversion of solid residue of wheat straw (cellulose 34%–40%; hemicellulose 21%–26%;

and lignin of 11%–23%) into carbohydrate-rich liquid makes it a good substrate for MFCs, thereby obtaining 123 mW/m^2 with 1 g/L substrate. Corn stover (70% cellulose and hemicellulose) for MFCs is subjected to cellulolytic enzymatic treatment or steam explosion to produce a sugar hydrolysate. The pretreatment of wheat straw and corn stover with hydrothermal acid, alkali, acid-heat, and alkali-heat destroys the structural complexity and improves the enzymatic saccharification potential (Zhang et al. 2020). Despite this, more improvements are needed for both corn stover and wheat straw to make them suitable for MFC technology (ElMekawy et al. 2014; Pandit et al. 2021).

Rice straw is a lignocellulosic biomass with varying organic carbon composition. The direct transformation of rice straw to electricity and hydrogen by a single yeast strain waste was studied by Moradian et al. (2024). The study revealed that the yeast MFCs exhibit a maximum power density of 28.56 mW/m^2 and simultaneously produce hydrogen gas at 4.9 L/m^3. It was also observed that the yeast strain secreted enzymes that degraded rice straw into sugar and organic acids. These are said to serve as fuel for the bioelectricity and BioH_2 production. Therefore, a carbohydrate-rich hydrolysate from rice straw was reported to achieve 400 mg COD/L removal, with a maximum energy output of 137 mW/m^2. In a two-chambered MFC without pretreatment, rice straw was applied to the anode side in the presence of a mixed culture, resulting in a 54.3% increase in energy generation. Power density of 190 W/m^3 was generated from non-pretreated rice straw in an MFC with a mixed culture containing bacteria capable of breaking down cellulose (Gurung & Oh 2015).

Cassava mill effluent is another example of agricultural residues for BioH_2 and bioelectricity production. In this case, cassava is rich in carbohydrates and has a high organic content of 10–16 g/L, making it suitable for MFC technology. In studies by Kaewkannetra et al. (2011) and Kim et al. (2013), COD removal of 88% and 1.7 W/m^2 power were achieved from the treatment of cassava mill effluent, due to its feasibility and biodegradability. Energy recovery of 22.19 W/m^3, within the pH range of 5–9, was reported in a single-chamber MFC (Kaewkannetra et al. 2011; Kim et al. 2013).

The electrogenic population in vegetable waste has made the feedstock suitable for BioH_2 and bioelectricity production. Vegetable waste is a rich, biodegradable resource for the production of BioH_2 and bioelectricity through anaerobic digestion and MFCs. It involves breaking down high-carbohydrate waste using microorganisms under optimized conditions, such as pH and temperature, to yield significant, clean energy (Kafaei et al. 2023). For BioH_2 production from vegetable waste, dark fermentation has been used, yielding a high BioH_2 production potential, whereas for bioelectricity, MFC is employed, achieving maximum power density of up to 1,540 mW/m^2. The MFC utilizes a slurry form of the waste, which is more suitable for hydrolysis. The waste increases the coulombic yield by 86.3%, while removing the COD. A microbial consortium (*Firmicutes, Proteobacteria*, and *Geobacter* sp.) is needed as a suitable electron donor for electron transfer. A current density of 314 mA/m^2 at a resistance of 123 Ω was obtained during the extraction of vegetable waste as a substrate to demonstrate a power density output in a dual-chambered MFC (Javed et al. 2017; Pandit et al. 2021).

Fruit waste accounts for a significant amount of the 1.3–2.5 billion tons of food waste annually. It is a highly biodegradable, carbohydrate-rich feedstock suitable for producing renewable energy in the form of $BioH_2$ and bioelectricity. The utilization of waste like peels and rotten fruit helps manage environmental waste and improve the circular economy (He et al. 2021). He et al. (2021) reported that fruit waste extract can increase microbial diversity and enhance power generation in MFCs. Findings from Miran et al. (2016) revealed that fruit waste can generate about 330 mV during conversion with reference to the biotransformation of effluent from fruit processing. Also, in a dual-chambered MFC with lemon effluent, an energy recovery of 0.99 A/m^2 and a coulomb efficiency of 32.3% were reported.

The plant and yard waste contain high concentrations of cellulose, hemicellulose, and lignin. They are highly promising, low-cost, and abundant feedstocks for producing $BioH_2$ and bioelectricity, offering a sustainable alternative to fossil fuels. These types of waste are broken down by microorganisms through biochemical processes (Shaikh et al. 2021). During microbial oxidation in MFCs, the pretreatment method readily degrades this type of waste. Findings from Zang et al. (2010) and Wang et al. (2013) using an air-cathode MFC indicate that 0.45 W/m^3 of power density can be generated with the use of hydrolysates from plant and yard-based waste. Hydrolysate generated from plant and yard waste achieved an energy output of 1.02 W/m^2, with COD removal and coulombic efficiencies of 76% and 69%, respectively, in an air-cathode MFC.

6.4 STUDIES ON BIOHYDROGEN AND BIOELECTRICITY PRODUCTION

For evaluating $BioH_2$ production, various substrates and feedstocks have been employed. The MEC is a biotechnological system for producing $BioH_2$. This process usually involves different types of wastewater and organic wastes, including sodium acetate, glucose, glycerol, domestic wastewater, effluent from sugar and food processing, and industrial wastewater. Muddasar et al. (2022) used cellulose substrate for the evaluation of the $BioH_2$ production. In the study conducted by Muddasar et al. (2022), the authors noted that the design configuration, type and nature of electrode materials, the substrate for maximum hydrogen production, and the coulombic efficiency usually affect the MECs. These factors also affect the MEC performance.

In another study, Soares et al. (2022) evaluated the potential of biomass wastes (rice bran, rice husk, brewer's spent grain, soybean waste, and wheat waste) as substrates for $BioH_2$ production. The substrates were treated by acid hydrolysis under different conditions of acid concentration, temperature, time, and moisture. Results reported reducing sugar concentrations in the range of 80.38–600.97 g/kg. The total sugar concentration, which comprises of cellobiose, glucose, xylose, and arabinose, gave rise to varying yields of 42.48–78.03 g/kg. A total of 96.14 million m^3 of $BioH_2$/year was produced from the biomass. The authors concluded that the biomass waste is a potential feedstock for $BioH_2$ production.

dos Santos et al. (2025) evaluated the production of $BioH_2$ from residual biomass of agricultural crops using rice straw, corn stover, wheat straw, and soybean straw. These are regarded as the main agricultural residues in Brazil, due to their abundance and diversity across the country. Findings revealed that corn and soybean straw had the highest potential for $BioH_2$ production, with 6.67 and 5.56 m^3/year, respectively. Their energy potential was estimated at 143.2 PJ/year. This shows that the biomass residues are capable of producing $BioH_2$, with corn straw as the most promising feedstock.

The use of wastewater as a substrate for $BioH_2$ production was reviewed by Thulasisingh et al. (2023). The substrate was used for its cost-effectiveness, an alternative to other sources of biomass waste. The use of wastewater for $BioH_2$ production is possible with the help of microorganisms. This occurs through light-independent (dark fermentation) and light-dependent (photofermentation) processes (Figure 6.2). Wastewater with high sugar content is preferred for $BioH_2$ production because it serves as a good carbon source. From the review, distillery waste is the most common source of organic wastewater, due to its high sugar content. To enhance the efficiency of wastewater for $BioH_2$ production, it is important to add trace elements (Na^+, Mg^+, Zn^+, etc.). One process for applying wastewater as a substrate for this purpose is pretreatment.

Albuquerque et al. (2024) conducted a review on the advances and perspectives in $BioH_2$ production from POME. The major takeaway message concerns the challenges and opportunities for POME in $BioH_2$ production. The first stage for $BioH_2$ production from POME is the pretreatment, mostly via anaerobic stabilization ponds. This technique is widely employed because of its low cost. Despite POME's useful properties for $BioH_2$, it has some limitations that affect its performance. These include long residence times, low efficiency, and a tendency to emit greenhouse gases such as methane and carbon dioxide. To address this problem, Mamimin et al. (2015) proposed combining POME treatment with energy generation. The main feature of POME for $BioH_2$ production is its high cellulose and hemicellulose content, which makes the effluent difficult to degrade. However, it increases the need for economic analysis, basically on the characteristics of the effluent. Table 6.1 shows some selected case studies for $BioH_2$ production using various biological routes.

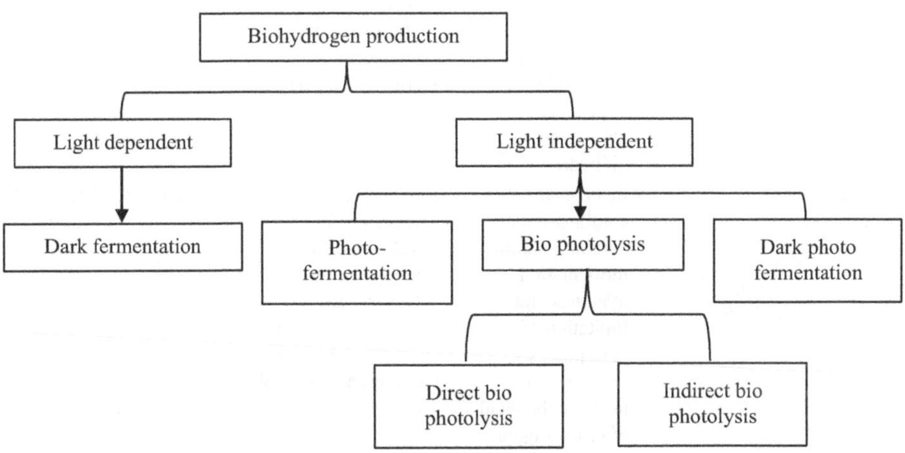

FIGURE 6.2 Schematic processes for the biohydrogen production.

(Adapted from Thulasisingh et al. 2023.)

TABLE 6.1
Illustration of Case Studies and Various Methods for Biohydrogen Production

Biohydrogen Route	Descriptor	Case Study	References
Dark fermentation	• The carbon-rich substrate is fermented by facultative and obligate anaerobic microorganisms to produce H_2 gas, volatile organic acid, and metabolites.	**Source**: Palm oil mill effluent **Conditions**: pH 6; 40°C; mixed bacterial culture; batch process **Yield**: 4.83 mol H_2/mol COD	Thulasisingh et al. (2023); Ren et al. (2006)
Photo fermentation	• This is a light-dependent reaction, in which photosynthetic bacteria convert volatile fatty acids to H_2 and CO_2 under anoxygenic condition to produce biohydrogen	**Source**: Dairy wastewater **Conditions**: pH: 7.0; 28°C; thermal sterilization pretreatment; *Rhodobacter sphaeroides* **Yield**: 3.2 L H_2/L	Thulasisingh et al. (2023); Seifert et al. (2010)
Biophotolysis	• Involves the use of blue-green algae and cyanobacteria to split the water into H_2 and O_2 in the presence of sunlight, to produce photosynthetic biohydrogen	**Source**: Pulp and paper wastewater **Conditions**: pH 7; 30°C; *Rhodobacter sphaeroides*; ultrasonication pretreatment **Yield**: 9.96 mL H_2/mL	Thulasisingh et al. (2023)
Dark and photo fermentation	• Involves the combination of dark and photo fermentation to address the limitation of both processes, with the aim to increase the yield of biohydrogen	**Source**: Starch wastewater **Conditions**: Cont inuous culture; *Clostridium* and *Rhodopseudomonas palustris*; pH 6.8 **Yield**: 0.24–1.43 mol H_2/mol	Nasr et al. (2015)

6.5 ADVANCES IN BIOHYDROGEN AND BIOELECTRICITY PRODUCTION

In recent years, advances have been made regarding the production of $BioH_2$ to improve its yield. Thulasisingh et al. (2023) noted that one challenge of such advances is the low efficiency of biological fermentation. This process involves integrating primary and secondary dark fermentation to improve hydrogen production. To improve this, it is necessary to identify and enrich the most effective microorganism. Hence, this is a function of the fermentation process.

In bioelectricity production, Obileke et al. (2021) published a mini-review on MFCs, a renewable energy technology for bioelectricity production. The review, which aims to establish a new and insightful perspective for further research and development, highlights that MFCs are beneficial for both waste treatment and energy production. The review progressively outlines the steps necessary for scaling up, while bearing in mind the reinforcement criteria for selecting microbes and their corresponding activities. In their review, Obileke et al. (2021) reported that higher power density can be achieved through suitable microorganism selection, the mode of operation, enhanced MFC designs, and the nature of the material. It was found that generating electricity via MFC results in 50% COD removal and a power density of 420–460 mW/m^2. Therefore, to optimize the efficiency and effectiveness of MFCs, different configurations and operating modes need to be implemented.

Having established that the MFC is the main technology for the production of bioelectricity, as mentioned in Obileke et al. (2021), the utilization of lignocellulosic biomass (LCB) as a potential feedstock plays an important role during the MFC process. This was evident in the study by Sani et al. (2021), who showed that biomass and wastewater were treated in an MFC. In the MFC technology, the anodic chamber contains microbial consortia that involve LCB to generate electricity. The first step of LCB in MFC is the pretreatment, utilizing various physical, chemical, and biological methods. Using physical processes, the LCB undergoes mechanical treatment to reduce particle size and increase the degree of polymerization, thereby improving the biodegradability of biomass. Within the MFC environment, LCBs are complex in terms of biodegradability. Based on this, various pre-treatment and extraction techniques are employed. Although the biodegradability of LCB is hindered by its complex structure and crystallinity, this issue is mitigated by its higher moisture content.

As mentioned earlier, LCB is a promising feedstock for MFCs for bioelectricity production. An example of the LCB is the lettuce waste, which was used in the study conducted by Rojas-Villacorta et al. (2023) as a substrate for MFCs. The aim of this study was to investigate vegetable waste as an alternative substrate to produce bioelectricity in a single-chamber MFC, using zinc and copper electrodes. The single-chamber MFC design contains a PVC chamber cut on the right side with a Zn electrode and a Cu electrode on the left. In the study, a statistical analysis was carried out on the continuous variables, voltage, and current, using Pearson linear regression. The equation from the Pearson linear regression was used to estimate the electrical resistance. In so doing, the coefficient of regression (R^2) was evaluated using the regression model. Findings revealed that the electrical current and voltage were 5.7 mA and 0.96 V, respectively, after 14 days. The pH and electrical conductivity of the substrate were 7.87 and 118.96 mS/cm/cm, respectively. When generating clean, sustainable energy, a voltage of 2.81 V was obtained from an LED lamp powered by a microbial combustion cell connected in series.

A review of advancements in bioelectricity generation using nanomaterial-modified anode electrodes in MFCs was conducted (Dessie & Tadesse 2022). Generally, nanotechnology in biochemical systems has been widely employed to recover bioelectricity from waste, making the procedure an appealing alternative to existing methods. The scope is to enhance the anode electrode by applying different conducting polymer nanocomposites based on metal oxides, thus improving the MFC's performance. It was revealed that interfacial bioelectrocatalysis between electroactive microorganisms and hierarchical porous nanocomposite materials addresses the cost-effective bioanode material challenge in MFCs. The review makes it clear that bioelectricity production in MFCs can occur in different ways, ranging from energy production to metal recovery to bioremediation. Therefore, with different waste streams processed by microorganisms in the anode chamber, bioelectricity can be produced at low cost using a suitable nanocomposite. This provides a sustainable source of energy and makes the process sustainable. With this, fossil fuel needs are reduced, thereby enabling green energy.

The operation of an MFC works on the principle of oxidizing microbes to produce waste material, thereby generating carbon dioxide, electrons, and protons at the anode. This can be witnessed in the study by Chakrawarti et al. (2024) on bioelectricity production from industrial and domestic wastes. The authors mentioned that the MFC converts domestic and industrial waste into electricity by the utilization of microbial catalyst, such as *Saccharomyces cerevisiae*, *Escherichia coli*, *Geobacter clostridium* sp-EG3, etc. During MFC operation, electrons are transferred through an external circuit, producing bioelectricity. The power density from MFCs is usually 1–2,000 mW. However, the industrial and sewage waste from the study reported bioelectricity production of up to 594 mW.

Pandit et al. (2021) examined agricultural waste and wastewater as feedstocks for bioelectricity generation using MFCs. Both feedstocks are said to be rich in organic substances, cellulose, hemicellulose, and lignin. As a result, they have a relatively high amount of energy. The authors' aim in conducting the study was to recover energy from agricultural waste for global energy security. To achieve this, various technologies and recovery methods were developed, such as the MFC. During the operation and design process of the MFC, the agricultural waste is placed in the anode compartment of the fuel cell. In agricultural waste, biodegradation is limited due to its complex structure and crystallinity. However, to address this challenge, the moisture content of the agricultural waste is an advantage. The application of various agricultural waste and fruit wastewater would break down the solid biomass through bacterial activity. The techno-economic evaluation reveals that the capital cost of the MFC is always higher than its operating cost compared to the conventional method. This high cost is attributed to the expensive electrode materials (catalyst and membrane) used in the MFC design.

Having looked at the application or involvement of microorganisms in the production and advancement of $BioH_2$ and bioelectricity. Tables 6.2 and 6.3 present a summary of these applications from selected literature and energy or power yield.

TABLE 6.2
Production of Biohydrogen from Different Biomass Waste

Feedstock	Microorganism(s)	Operating Parameters	Yield	Reference
Rice straw	*Thermotoga neapolitana*	pH 7.5, 75°C	68.2 L/kg	Nguyen et al. (2010)
Corn stover	Mixed culture (*Enterococcus, Sporanaerobacter, Para clostridium,* and *Clostridium_sensu_stricto_1*)	pH: 5.5, 40°C, municipal sewage sludge as inoculum	80.1 L/kg	Li et al. (2018)
Wheat straw	—	Mix of cow manure and sediment as inoculum, ozone pretreatment, 35°C, pH 6.0	87.4 L/kg	Wu et al. (2013)
Soybean straw	Mixed bacteria dominated by *Clostridium butyricum*	35°C, pH 7, chemical treatment	60.2 L/kg	Han et al. (2012)
Grape residues	*Clostridia* sp.	37°C, HRT-24 h, pH 5.7, OLR (14–28 g COD/L)	2.2×10^{-4} mol H_2/min	Guellout et al. (2022)
Winery wastewater	*Lactobacillus, Pectinatus*, and *Clostridium* species	37°C; HRT-5.5 h, pH 5.0–5.6, OLR-108.6 g COD/L	0.05 L H_2/L/d	Meija-Saucedo et al. (2022)
Cheese whey	*Rhodobacter sphaeroides*	OLR-70.4 kg COD/m^3/d, pH 4.25, 28°C, HRT-8 h	3.78×10^{-3} mol H_2/d	Rao & Basak (2021)
Food waste	*Enterococcus* and *Veillonella*	OLR-22 g COD/L/d, pH 5.5–7.0, 37°C, HRT-3 d	0.73 L H_2/L/d	Moreno-Andrade et al. (2023)
Potato waste	*Enterobacter aerogenes* MTCC2822	OLR-20 g COD/L/d, pH 7.0, 50°C, HRT-48 h	1.58 L H_2/L/d	Das & Basak (2024)
Brewery wastewater	*Clostridia* sp.	OLR-38.44 g COD/L/d, pH 4.13–4.72, 55°C, HRT-20 d	0.60–1.50 g/L/d	Sganzerla et al. (2023)
Cassava wastewater	*Eubacterium*	OLR-14 g COD/L/d pH 6, 30°C, HRT-4 d	0.7 L H2/L/d	Mari et al. (2020)

6.6 ADVANTAGES AND DISADVANTAGES OF BIOHYDROGEN AND BIOELECTRICITY

One of the recent advances in the application of BioH_2 and bioelectricity production lies in their advantages and disadvantages as presented in Table 6.4.

TABLE 6.3
Production of Bioelectricity Using Different Microorganisms

Feedstock	Microorganism(s)	Microbial Fuel Cell	Maximum Power (mW/m^3)	Reference
Sewage sludge	*Saccharomyces cerevisiae*	Single chambered	91	Nevin et al. (2008)
Glucose	*Escherichia coil*	Two chambered	16	Rahimnejad et al. (2009)
Lactate	*Geobacter* spp.	Two chambered	52	Jung & Regan (2007)
Acetate	*G. sulfurreducens*	Two chambered	48.4	Jung & Regan (2007)

TABLE 6.4
Advantages and Disadvantages of Biohydrogen and Bioelectricity Production

Advantages	Disadvantages
Biohydrogen	
• Environmentally friendly technology.	• Low substrate conversion efficiencies.
• It is possible to integrate biohydrogen production with anaerobic digestion to increase net energy and eco-friendliness.	• Inhibition by oxygen of hydrogenases, which are key enzymes in photobiological reaction.
• Biohydrogen production via biophotolysis by green microalgae can fix atmospheric nitrogen.	• Methanogens can compete with hydrogen-producing microbes, slowing down the reaction.
• The process can eliminate and reduce contaminants in wastewater.	• Continuous light requirements for photofermentation may lead to unfavorable thermodynamic conditions, suppressing biohydrogen production.
Bioelectricity	
• Energy-efficient process, which often occurs during non-dependence on light.	• Process instability and low-quality end-product, which makes purification difficult.
• No oxygen inhibition of the process due to the use of nitrogenases, which makes the process irreversible.	• Uses expensive bioreactors with low solar conversion efficiency and rate of production.
• Utilizes solar energy to produce hydrogen directly from water and sunlight, reducing greenhouse gas emissions.	• Low hydrogen productivity due to the sensitivity of hydrogenases toward oxygen.
• Rapid and ability to utilize a variety of organic substrates and higher purity of hydrogen with potentially higher theoretical yield.	• Both processes can have low hydrogen yield and high hydrogen cost.

6.7 FACTORS AFFECTING THE PERFORMANCE OF BIOELECTRICITY GENERATION

Having established that $BioH_2$ is a source of bioelectricity through fuel cells, Table 6.5 summarizes the specific process parameters affecting the performance of MFCs during bioelectricity production.

6.8 LIMITATIONS OF BIOHYDROGEN AND BIOELECTRICITY PRODUCTION

Despite the promising potential of the production and technologies for both $BioH_2$ and bioelectricity, Table 6.6 presents the various limitations affecting the performance of these processes.

TABLE 6.5
Factors Affecting Microbial Fuel Cells in Relation to Bioelectricity Production

Factor	Notes	References
pH	- Near neutral pH environment. - Alkaline pH lowers power production in the cathode chamber. - Acidic pH leads to around 232–284 to 311–339 mV and power density of about 95–116 to 182–237 mV/m^3.	Zhang et al. (2011); Jadhav & Ghangrekar (2009)
	Recommendation: Always avoid low pH.	
Substrate concentration	- COD removal efficiency of >90% results in OLR of 435–720 mg COD/L/d. - Decline of COD removal efficiency to 70% prompts a lower rate (870 mg COD/L/d).	Ye et al. (2019)
Temperature	- Room temperature higher than 20°C–35°C. - Low temperature (4°C–30°C) requires a long time to provide consistent power performance. - No electricity generation at temperatures below 15°C.	Pandit et al. (2021); Choi et al. (2004)
	Recommendation: Thermophilic temperatures.	
Salinity	- About 20 g/L of NaCl enhances the efficiency of MFCs. - Decreases internal resistance by 33% and improves maximum output by 30%.	Lefebvre et al. (2012)

TABLE 6.6
Limitations of Biohydrogen and Bioelectricity Production

Biohydrogen	Bioelectricity
• Despite the attractive performance of direct biophotolysis, it suffers from intractable barriers of oxygen sensitivity and tedious economics.	• The technology has low power generation compared to the conventional one. • Limited oxygen diffusion rate with high internal resistance.
• Low conversion efficiencies of solar energy and the use of nitrogenases with a high demand for energy are regarded as insurmountable factors to the production.	• Require significant improvement in the system efficiency and substrate management to ensure maximal energy output and cost-effectiveness.
• Feedstock processing is costly, hence the need to develop low-cost methods for the growing, harvesting, transporting, and pretreatment of biomass waste.	• High cost on a large scale is a limitation that needs to be resolved.
• The efficiency of microbial fuel cells is decreased by the sensitivity of hydrogenases to oxygen and hydrogen partial pressure.	• Pilot-scale affected by high operational cost and setup, low power density, and temperature fluctuation.
• Insufficient and lack of knowledge on the improvement of economics by combining hydrogen production with other processes.	• Biofueling and energy losses due to resistance of the conducting material poses some limitations.
• Appropriate bioreactor design, scale-up processes, hydrogen recovery, and difficulty in sustaining steady continuous hydrogen production rate need to be addressed.	

Sources: Pandu & Joseph (2012); Hassan et al. (2024); Apollon (2023).

6.9 FURTHER RESEARCH AND DIRECTIONS

- Based on the limitations mentioned above, numerous improvements are needed in future research, such as advancing scientific research (e.g. the development of bioreactor designs, the engineering of hydrogen enzymes, and genetic modification of microorganisms to improve production yields).
- Studies on the economic, technological, and environmental constraints of $BioH_2$ generation are needed, with an emphasis on making $BioH_2$ a viable alternative to fossil fuels.
- The most suitable method, as well as a promising alternative to hydrogen production, is fermentative technology (Thulasisingh et al. 2023). Therefore, a future study focusing on wastewater sources for $BioH_2$ production could be a useful and interesting area of research, given its sustainable, economic, and commercial feasibility.
- Investigate genetic modification of microorganisms, develop integrated systems capable of producing both hydrogen and electricity, and explore coupling with renewable electricity to create resilient hybrid energy grids.

6.10 CONCLUSION AND RECOMMENDATIONS

6.10.1 Conclusion

$BioH_2$ and bioelectricity production are among the most promising frontiers in renewable bioenergy systems. Advances in microbial physiology, genetic engineering, and process optimization have improved the yield, stability, and economic feasibility of $BioH_2$. Similarly, MFCs have evolved as multifunctional systems capable of generating electricity, while addressing environmental challenges, such as wastewater treatment and pollutant degradation. This chapter demonstrates that diverse feedstocks, from agricultural residues and lignocellulosic biomass to agro-industrial effluents and domestic wastewater, can be successfully valorized for bioenergy generation. $BioH_2$ offers high energy density and clean combustion properties, while bioelectricity provides decentralized energy solutions through low-cost, small-scale MFC systems. However, challenges remain, including the low conversion efficiency of biological systems, scale-up limitations, high costs of electrode materials, and the need for pretreatment processes to improve biomass biodegradability. Furthermore, both technologies must overcome barriers to commercialization, particularly in achieving consistent performance at an industrial scale and ensuring economic competitiveness with conventional fossil-based energy systems. Despite these challenges, the convergence of biotechnology, nanotechnology, and process engineering continues to advance both $BioH_2$ and bioelectricity technologies toward practical application. If adequately supported by policy frameworks, financial investment, and further research, these bioenergy solutions could play a pivotal role in the global energy transition.

6.10.2 General Recommendation

Having concluded and provided references for future research, the authors offered the following recommendations based on the review's findings. The recommendations broadly focus on various areas of the study related to $BioH_2$ and bioelectricity production.

1. Process optimization and integration: Combine dark and photo-fermentation processes for enhanced $BioH_2$ yield. Explore hybrid bioenergy systems integrating MFCs with anaerobic digestion or hydrogen fermentation.
2. Feedstock diversification: Prioritize the use of low-cost and abundant substrates, such as wastewater, agricultural residues, and lignocellulosic biomass. Develop energy-efficient and scalable pretreatment technologies.
3. Material and design improvements: Invest in cost-effective, durable electrode materials and innovate reactor configurations for decentralized applications.
4. Environmental and economic assessment: Conduct comprehensive life cycle assessments and explore circular economy models that couple bioenergy production with waste management.
5. Policy and commercialization support: Provide incentives for pilot-scale demonstration projects and foster public-private partnerships for industrial adoption.

ACKNOWLEDGMENTS

The authors would like to express their appreciation to the Department of Science and Innovation (DSI), Technology Innovation Agency (TIA), National Research Foundation (NRF), ESKOM and the Renewable Energy (Wind–Biogas) Research Niche Area of the Department of Research and Innovation at the University of Fort Hare for their academic support.

REFERENCES

Ahmed, S.F., M. Mofijur, N. Islam, T.A. Parisa, N. Rafa, A. Bokhari, J.J. Klemeš, & T.M. Mahlia. 2022. Insights into the development of microbial fuel cells for generating biohydrogen, bioelectricity, and treating wastewater. *Energy* 254: 124163.

Albuquerque, M.M., W.J. Martinez-Burgos, G. De Bona Sartor, L.A. Letti, J.C. De Carvalho, C.R. Soccol, & A.B. Medeiros. 2024. Advances and perspectives in biohydrogen production from palm oil mill effluent. *Fermentation* 10: 141.

Angelaalincy, M.J., R.N. Krishnaraj, G. Shakambari, B. Ashokkumar, S. Kathiresan, & P. Varalakshmi. 2018. Biofilm engineering approaches for improving the performance of microbial fuel cells and bioelectrochemical systems. *Front. Energy Res.* 6: 63.

Apollon, W. 2023. An overview of microbial fuel cell technology for sustainable electricity production. *Membranes* 13: 884.

Arimi, M.M., J. Knodel, A. Kiprop, S.S. Namango, Y. Zhang, & S.U. Geißen. 2015. Strategies for improvement of biohydrogen production from organic-rich wastewater: A review. *Biomass Bioenergy.* 75: 101–118.

Baranitharan, E., M.R. Khan, A. Yousuf, W.F. Teo, G.Y. Tan, & C.K. Cheng. 2015. Enhanced power generation using controlled inoculum from palm oil mill effluent fed microbial fuel cell. *Fuel* 143: 72–79.

Barghash, H., K.E. Okedu, & A. Al Balushi. 2021. Biohydrogen production using landfill leachate considering different photofermentation processes. *Frontiers Bioeng. Biotechnol.* 9: 644065.

Chakrawarti, A., R. Tripathi, R. Singh, A. Singh, M. Ahmad, P. Gupta, B. Singh, K.M. Siddiqui, & S. Mishra. 2024. Studies on bioelectricity production from industrial and domestic waste: Current trends and future perspective. *J. Electr. Electron. Eng.* 19: 1–5.

Chandrasekhar, K., L. Yong-Jik, & L. Dong-Woo. 2015. Biohydrogen production: Strategies to improve process efficiency through microbial routes. *Int. J. Mol. Sci.* 16: 8266–8293.

Cheng, J., X. Zhu, J. Ni, & A. Borthwick. 2010. Palm oil mill effluent treatment using a two-stage microbial fuel cells system integrated with immobilized biological aerated filters. *Bioresour. Technol.* 101: 2729–2734.

Choi, Y., E. Jung, H. Park, S.R. Paik, S. Jung, & S. Kim. 2004. Construction of microbial fuel cells using thermophilic microorganisms, *Bacillus licheniformis* and *Bacillus thermoglucosidasius*. *Bull. Korean Chem. Soc.* 25: 813–818.

Das, S.R., & N. Basak. 2024. Optimization of process parameters for enhanced biohydrogen production using potato waste as substrate by combined dark and photo fermentation. *Biomass Convers. Biorefin.* 14: 4791–4811.

Dessie, Y., & S. Tadesse. 2022. Advancements in bioelectricity generation through nanomaterial-modified anode electrodes in microbial fuel cells. *Front. Nanotechnol.* 4: 876014.

Dessie, Y., S. Tadesse, & R. Eswaramoorthy. 2021. Surface roughness and electrochemical performance properties of biosynthesized α-MnO_2/NiO-based polyaniline ternary composites as efficient catalysts in microbial fuel cells. *J. Nanomater.* 2021: 7475902.

dos Santos, M.C., W.C. Nadaleti, G. Przybyla, & J.B. Machado. 2025. Biogas and biohydrogen production using the main agricultural biomasses generated in Brazil: An integrative analysis to contribute to the implementation of energy decarbonization. *Int. J. Hydrog. Energy* 137: 190–201.

ElMekawy, A., H.M. Hegab, K. Vanbroekhoven, & D. Pant. 2014. Techno-productive potential of photosynthetic microbial fuel cells through different configurations. *Renew. Sustain. Energy Rev.* 39: 617–627.

Gazali, T.A., & M.A. Moqsud. 2017. The effectiveness of animal dungs and leaf mold for bioelectricity generation using microbial fuel cell with soils. *J. Sustain. Bioenergy Syst.* 7: 165–181.

Ghazali, N.F., N.A. Mahmood, K.A. Ibrahim, S.A. Muhammad, & N.S. Amalina. 2017. Electricity generation from palm oil tree empty fruit bunch (EFB) using dual chamber microbial fuel cell (MFC). *IOP Conf. Ser. Mater. Sci. Eng.* 206: 012025.

Guadarrama-Pérez, O., J. Hernández-Romano, L. García-Sánchez, Gutierrez-Macias, T., & E.B. Estrada-Arriaga. 2019. Simultaneous bioelectricity and biohydrogen production in a continuous flow single microbial electrochemical reactor. *Environ. Prog. Sustain. Energy* 38: 297–304.

Guellout. Z., E. Francois-Lopez, Y. Benguerba, C. Dumas, K.K. Yadav, A.M. Fallatah, A. Pugazhendhi, & B. Ernst. 2022. Dark fermentative biohydrogen production from vinicultural biomass without exogenous inoculum in a semi-batch reactor: A kinetic study. *J. Environ. Manag.* 305: 114393.

Guo, F., G. Fu, Z. Zhang, & C. Zhang. 2013. Mustard tuber wastewater treatment and simultaneous electricity generation using microbial fuel cells. *Bioresour. Technol.* 136: 425–430.

Gurung, A., & S.E. Oh. 2015. Rice straw as a potential biomass for generation of bioelectrical energy using microbial fuel cells (MFCs). *Energy Sources A: Recovery Util. Environ. Eff.* 37: 2625–2631.

Han, H., L. Wei, B. Liu, H. Yang, & J. Shen. 2012. Optimization of biohydrogen production from soybean straw using anaerobic mixed bacteria. *Int. J. Hydrog. Energy.* 37: 13200–13208.

Hassan, M., S. Kanwal, R.S. Singh, S.A.M. Ali, M. Anwar, & C. Zhao. 2024. Current challenges and future perspectives associated with configuration of microbial fuel cell for simultaneous energy generation and wastewater treatment. *Int. J. Hydrog. Energy.* 50: 323–350.

Hassan, S.H., A.Z. Abd el Nasser, & R.M. Kassim. 2019. Electricity generation from sugarcane molasses using microbial fuel cell technologies. *Energy* 178: 538–543.

He, J., X. Xin, Z. Pei, L. Chen, Z. Chu, M. Zhao, X. Wu, B. Li, X. Tang, & X. Xiao. 2021. Microbial profiles associated improving bioelectricity generation from sludge fermentation liquid via microbial fuel cells with adding fruit waste extracts. *Bioresour. Technol.* 337: 125452.

Hong, C., L. Yong-feng, Z. Yong-juan, & J. Ying. 2010. *Biohydrogen production from molasses at continuous hydrogen production system by Biohydrogenbacterium R3 sp.* Paper presented at the *International Conference on Mechanic Automation and Control Engineering*, 26–28 June 2010, Wuhan, China.

IEA. 2022. *World Energy Outlook 2022*. Paris: IEA.

In-Seop, C., B. Orianna, J. Jae-Kyung, & H.N. Kenneth. 2006. Electrochemically active bacteria (EAB) and mediator-less microbial fuel cells. *J. Microbiol. Biotechnol.* 16: 163–177.

Ioannou, L.A., G.L. Puma, & D. Fatta-Kassinos. 2015. Treatment of winery wastewater by physicochemical, biological and advanced processes: A review. *J. Hazard. Mater.* 286: 343–368.

Jadhav, G.S., & M.M. Ghangrekar. 2009. Performance of microbial fuel cell subjected to variation in pH, temperature, external load and substrate concentration. *Bioresour. Technol.* 100: 717–723.

Javed, M.M., M.A. Nisar, B. Muneer, & M.U. Ahmad. 2017. Production of bioelectricity from vegetable waste extract by designing a U-shaped microbial fuel cell. *Pak. J. Zool.* 49: 711–716.

Jung, S., & J.M. Regan. 2007. Comparison of anode bacterial communities and performance in microbial fuel cells with different electron donors. *Appl. Microbiol. Biotechnol.* 77: 393–402.

Kaewkannetra, P., W. Chiwes, & T.Y. Chiu. 2011. Treatment of cassava mill wastewater and production of electricity through microbial fuel cell technology. *Fuel* 90: 2746–2750.

Kafaei, R., A.R. Yazdanbakhsh, M. Sadani, & N. Alavi. 2023. Bioelectricity and biohydrogen production using high solid content of oily-kitchen wastes in air cathode microbial fuel cells. *Iran. J. Chem. Chem. Eng.* 42: 2895.

Kim, D., W. Kim, C. Yun, D. Son, D. Chang, H. Bae, Y. Lee, Y. Sunwoo, & K. Hong. 2013. Agro-industrial wastewater treatment by electrolysis technology. *Int. J. Electrochem. Sci.* 8: 9835–9850.

Kotay, S.M., & D. Das. 2008. Biohydrogen as a renewable energy resource – Prospects and potentials. *Int. J. Hydrog. Energy* 33: 258–263.

Kundu, P., S.V. Vineetha, A. Mohan, & A. Ravikumar. 2025. Biohydrogen production from various waste resources through circular economy: Current technologies and future perspective. *J. Mater. Cycles Waste Manag.* 27: 1263–1282.

Lefebvre, O., Z. Tan, S. Kharkwal, & H.Y. Ng. 2012. Effect of increasing anodic NaCl concentration on microbial fuel cell performance. *Bioresour. Technol.* 112: 336–340.

Li, W.W., H.Q. Yu, & Z. He. 2014. Towards sustainable wastewater treatment by using microbial fuel cells-centered technologies. *Energy Environ. Sci.* 7: 911–924.

Li, Y., Z. Zhang, S. Zhu, H. Zhang, Y. Zhang, T. Zhang, & Q. Zhang. 2018. Comparison of bio-hydrogen production yield capacity between asynchronous and simultaneous saccharification and fermentation processes from agricultural residue by mixed anaerobic cultures. *Bioresour. Technol.* 247: 1210–1214.

Logan, B.E., & K. Rabaey. 2012. Conversion of wastes into bioelectricity and chemicals by using microbial electrochemical technologies. *Science* 10: 686–690.

Logan, B.E., & J.M. Regan. 2006. Electricity-producing bacterial communities in microbial fuel cells. *Trends Microbiol.* 14: 512–518.

Logan, B.E., M.J. Wallack, K.Y. Kim, W. He, Y. Feng, & P.E. Saikaly. 2015. Assessment of microbial fuel cell configurations and power densities. *Environ. Sci. Technol. Lett.* 2: 206–214.

Lu, H., M. Peng, G. Zhang, B. Li, & Y. Li. 2019. Brewery wastewater treatment and resource recovery through long term continuous-mode operation in pilot photosynthetic bacteria-membrane bioreactor. *Sci. Total Environ.* 646: 196–205.

Lubbe, F., J. Rongé, T. Bosserez, & J.A. Martens. 2023. Golden hydrogen. *Curr. Opin. Green Sustain. Chem.* 39: 100732.

Mamimin, C., A. Singkhala, P. Kongjan, B. Suraraksa, P. Prasertsan, T. Imai, & S. O-Thong. 2015. Two-stage thermophilic fermentation and mesophilic methanogen process for biohythane production from palm oil mill effluent. *Int. J. Hydrog. Energy.* 40: 6319–6328.

Mari, A.G., C.L. Andreani, T.U. Tonello, L.C. Leite, J.R. Fernandes, D.D. Lopes, J.A. Rodrigues, & S.D. Gomes. 2020. Biohydrogen and biomethane production from cassava wastewater in a two-stage anaerobic sequencing batch biofilm reactor. *Int. J. Hydrog. Energy* 45: 5165–5174.

Mejia-Saucedo, C., G. Buitron, M.F. León-Galván, & J. Carrillo-Reyes. 2022. Biomass purge strategies to control the bacterial community and reactor stability for biohydrogen production from winery wastewater. *Int. J. Hydrog. Energy.* 47: 5891–5900.

Miran, W., M. Nawaz, J. Jang, & D.S. Lee. 2016. Sustainable electricity generation by biodegradation of low-cost lemon peel biomass in a dual chamber microbial fuel cell. *Int. Biodeterior. Biodegradation* 106: 75–79.

Mohan, Y., S.M. Kumar, & D. Das. 2008. Electricity generation using microbial fuel cells. *Int. J. Hydrog. Energy* 33: 423–426.

Moqsud, M.A. 2021. Bioelectricity from organic solid waste. In *Strategies of Sustainable Solid Waste Management*, ed. H.M. Saleh, 1–10. IntechOpen. doi:10.5772/intechopen.95297

Moqsud, M.A., K. Omine, N. Yasufuku, M. Hyodo, & Y. Nakata. 2013. Microbial fuel cell (MFC) for bioelectricity generation from organic wastes. *Waste Manage.* 33: 2465–2469.

Moradian, J.M., A. Ali, K. Yang, M. Nazar, H. Chen, X. Yan, S.N. Haider, Y. Zhang, & Y.C. Yong. 2024. Direct transformation of rice straw to electricity and hydrogen by a single yeast strain: Performance and mechanism. *Fuel* 376: 132697.

Moradian, J.M., Z. Fang, & Y.C. Yong. 2021. Recent advances on biomass-fueled microbial fuel cell. *Bioresour. Bioprocess.* 8: 14.

Moreno-Andrade, I., M.J. Berrocal-Bravo, & I. Valdez-Vazquez. 2023. Biohydrogen production from food waste and waste activated sludge in codigestion: Influence of organic loading rate and changes in microbial community. *J. Chem. Technol. Biotechnol.* 98: 230–237.

Muddasar, M., R. Liaquat, A. Aslam, M.Z. Ur Rahman, A. Abdullah, A.H. Khoja, K. Latif, & A. Bahadar. 2022. Performance efficiency comparison of microbial electrolysis cells for sustainable production of biohydrogen - A comprehensive review. *Int. J. Energy Res.* 46: 5625–5645.

Nasr, M., A. Tawfik, S. Ookawara, M. Suzuki, S. Kumari, & F. Bux. 2015. Continuous biohydrogen production from starch wastewater via sequential dark-photo fermentation with emphasize on maghemite nanoparticles. *J. Ind. Eng. Chem.* 21: 500–506.

Nevin, K.P., H. Richter, S.F. Covalla, J.P. Johnson, T.L. Woodard, A.L. Orloff, H. Jia, M. Zhang, & D.R. Lovley. 2008. Power output and columbic efficiencies from biofilms of *Geobacter sulfurreducens* comparable to mixed community microbial fuel cells. *Environ. Microbiol.* 10: 2505–2514.

Nguyen, T.A., K.R. Kim, M.S. Kim, & S.J. Sim. 2010. Thermophilic hydrogen fermentation from Korean rice straw by *Thermotoga neapolitana. Int. J. Hydrog. Energy* 35: 13392–13398.

Obileke, K., H. Onyeaka, E.L. Meyer, & N. Nwokolo. 2021. Microbial fuel cells, a renewable energy technology for bioelectricity generation: A mini-review. *Electrochem. Commun.* 125: 107003.

Palanisamy, G., H.Y. Jung, T. Sadhasivam, M.D. Kurkuri, S.C. Kim, & S.H. Roh. 2019. A comprehensive review on microbial fuel cell technologies: Processes, utilization, and advanced developments in electrodes and membranes. *J. Clean. Prod.* 221: 598–621.

Pandit, S., N. Savla, J.M. Sonawane, A.M. Sani, P.K. Gupta, A.S. Mathuriya, A.K. Rai, D.A. Jadhav, S.P. Jung, & R. Prasad. 2021. Agricultural waste and wastewater as feedstock for bioelectricity generation using microbial fuel cells: Recent advances. *Fermentation* 7: 169.

Pandu, K., & S. Joseph. 2012. Comparisons and limitations of biohydrogen production processes: A review. *Int. J. Adv. Eng. Technol.* 2: 342.

Plakar, G.P.Z., S.K. Abdulsalami, L.O. Kanayo, & E.O. Emeka. 2025. Valorization of mixed blackwater/agricultural wastes for bioelectricity and biohydrogen production: A microbial treatment pathway. *Heliyon* 11: e41126.

Policastro, G., F. Carraturo, M. Compagnone, M. Guida, & M. Fabbricino. 2022. Enhancing hydrogen production from winery wastewater through fermentative microbial culture selection. *Bioresour. Technol. Rep.* 19: 101196.

Prado, E.R., & R.C. Rial. 2025. Biohydrogen production from residual biomass: The potential of wheat, corn, rice, and barley straw – Recent advances. *Bioresour. Technol.* 432: 132638.

Rahimnejad, M., N. Mokhtarian, G. Najafpour, W. Daud, & A. Ghoreyshi. 2009. Low voltage power generation in a biofuel cell using anaerobic cultures. *World Appl. Sci. J.* 6: 1585–1588.

Rahman, A., M.S. Borhan, & S. Rahman. 2018. Evaluation of microbial fuel cell (MFC) for bioelectricity generation and pollutants removal from sugar beet processing wastewater (SBPW). *Water Sci. Technol.* 77: 387–397.

Rao, R., & N. Basak. 2021. Fermentative molecular biohydrogen production from cheese whey: Present prospects and future strategy. *Appl. Biochem. Biotechnol.* 193: 2297–2330.

Ren, N., J. Li, B. Li, Y. Wang, & S. Liu. 2006. Biohydrogen production from molasses by anaerobic fermentation with a pilot-scale bioreactor system. *Int. J. Hydrog. Energy* 31: 2147–2157.

Rojas-Villacorta, W., S. Rojas-Flores, S.M. Benites, R. Nazario-Naveda, C.V. Romero, M. Gallozzo-Cardenas, D. Delfín-Narciso, F. Díaz, & E. Murga-Torres. 2023. Preliminary study of bioelectricity generation using lettuce waste as substrate by microbial fuel cells. *Sustainability* 15: 10339.

Roy, S., & S. Pandit. 2019. Microbial electrochemical system: Principles and application. In *Microbial Electrochemical Technology: Biomass, Biofuels and Biochemicals*, ed. S.V. Mohan, S. Varjani, & A. Pandey, 19–48. Elsevier.

Sai, K., Y.K. Ravi, G. Ma, I. Chattopadhyay, S. Palani, V. Kumar, G. Kumar, & R.B. Jeyakumar. 2023. Development of an integrated biorefinery system for bioconversion of lignocellulosic biomass to polyhydroxyalkanoates and biohydrogen. *ACS Sustainable Chem. Eng.* 11: 4606–4622.

Sani, A.D., N. Savla, S. Pandit, A.S. Mathuriya, P.K. Gupta, N. Khanna, R.P. Babu, & S. Kumar. 2021. Recent advances in bioelectricity generation through the simultaneous valorization of lignocellulosic biomass and wastewater treatment in microbial fuel cell. *Sustain. Energy Technol. Assess.* 48: 101572.

Sarma, M.K., M.G. Quadir, R. Bhaduri, S. Kaushik, & P. Goswami. 2018. Composite polymer coated magnetic nanoparticles-based anode enhances dye degradation and power production in microbial fuel cells. *Biosens. Bioelectron.* 119: 94–102.

Seifert, K., M. Waligórska, & M. Laniecki. 2010. Brewery wastewaters in photobiological hydrogen generation in presence of *Rhodobacter sphaeroides* OU 001. *Int. J. Hydrog. Energy.* 35: 4085–4091.

Sevda, S., P.J. Sarma, K. Mohanty, T.R. Sreekrishnan, & D. Pant. 2018. Microbial fuel cell technology for bioelectricity generation from wastewaters. In *Waste to Wealth. Energy, Environment, and Sustainability*, ed. R. Singhania, R. Agarwal, R. Kumar, & R. Sukumaran, 237–258. Singapore: Springer.

Sganzerla, W.G., L. Sillero, T. Forster-Carneiro, R. Solera, & M. Perez. 2023. Determination of anaerobic cofermentation of brewery wastewater and brewer's spent grains for biohydrogen production. *BioEnergy Res.* 16: 1073–1083.

Shaikh, R., A. Rizvi, M. Quraishi, S. Pandit, A.S. Mathuriya, P.K. Gupta, J. Singh, & R. Prasad. 2021. Bioelectricity production using plant-microbial fuel cell: Present state of art. *S. Afr. J. Bot.* 140: 393–408.

Soares, J.F., T.C. Confortin, I. Todero, L. Luft, G.A. Ugalde, L.P. Tovar, F.D. Mayer, & M.A. Mazutti. 2022. Estimation of bioethanol, biohydrogen, and chemicals production from biomass wastes in Brazil. *Clean Soil Air Water* 50: 2200155.

Sołowski, G. 2018. Biohydrogen production – Sources and methods: A review. *Int. J. Bioprocess. Biotech.* 101.

Thulasisingh, A., K. Ananthakrishnan, J. Nandakumar, & S. Kannaiyan. 2023. Fermentative biohydrogen fuel production utilizing wastewater: A review. *Clean Soil Air Water* 51: 2300112.

Wang, D., S.C. Pillai, S.H. Ho, J. Zeng, Y. Li, & D.D. Dionysiou. 2018. Plasmonic-based nanomaterials for environmental remediation. *Appl. Catal. B: Environ.* 237: 721–741.

Wang, M., Z. Yan, B. Huang, J. Zhao, & R. Liu. 2013. Electricity generation by microbial fuel cells fueled with *Enteromorpha prolifera* hydrolysis. *Int. J. Electrochem. Sci.* 8: 2104–2111.

Wu, J., S. Upreti, & F. Ein-Mozaffari. 2013. Ozone pretreatment of wheat straw for enhanced biohydrogen production. *Int. J. Hydrog. Energy.* 38: 10270–10276.

Xie, H., Q. Yu, Z. Zuo, Z. Han, X. Yao, & Q. Qin. 2016. Hydrogen production via sorption-enhanced catalytic steam reforming of bio-oil. *Int. J. Hydrog. Energy* 41: 2345–2353.

Xu, X., Q. Zhao, M. Wu, J. Ding, & W. Zhang. 2017. Biodegradation of organic matter and anodic microbial communities' analysis in sediment microbial fuel cells with/without Fe (III) oxide addition. *Bioresour. Technol.* 225: 402–408.

Ye, Y., H.H. Ngo, W. Guo, S.W. Chang, D.D. Nguyen, Y. Liu, L.D. Nghiem, X. Zhang, & J. Wang. 2019. Effect of organic loading rate on the recovery of nutrients and energy in a dual-chamber microbial fuel cell. *Bioresour. Technol.* 281: 367–373.

Zang, G.L., G.P. Sheng, Z.H. Tong, X.W. Liu, S.X. Teng, W.W. Li, & H.Q. Yu. 2010. Direct electricity recovery from *Canna indica* by an air-cathode microbial fuel cell inoculated with rumen microorganisms. *Environ. Sci. Technol.* 44: 2715–2720.

Zhang, L., C. Li, L. Ding, K. Xu, & H. Ren. 2011. Influences of initial pH on performance and anodic microbes of fed-batch microbial fuel cells. *J. Chem. Technol. Biotechnol.* 86: 1226–1232.

Zhang, T., D. Jiang, H. Zhang, D.J. Lee, Z. Zhang, Q. Zhang, Y. Jing, Y. Zhang, & C. Xia. 2020. Effects of different pretreatment methods on the structural characteristics, enzymatic saccharification and photo fermentative biohydrogen production performance of corn straw. *Bioresour. Technol.* 304: 122999.

Zhang, Y.J., C.Y. Sun, X.Y. Liu, W. Han, Y.X. Dong, & Y.F. Li. 2013. Electricity production from molasses wastewater in two-chamber microbial fuel cell. *Water Sci. Technol.* 68: 494–498.

7 Pyrolytic Products from Biomass Feedstocks

Denzel C. Makepa, Paul Zvawanda, and Chido H. Chihobo

7.1 INTRODUCTION

7.1.1 Background on Biomass and Pyrolysis

The global transition toward a sustainable energy system necessitates the rapid deployment of conversion technologies capable of utilizing abundant, non-fossil carbon sources. Modern bioenergy is currently the single largest contributor to renewable energy worldwide. In recent years, it has supplied approximately 55% of global renewable energy (excluding traditional biomass use), representing over 6% of total global energy demand (IEA 2025). This dominant role arises from biomass being a near-zero net-emissions energy source, and CO_2 released during its conversion is effectively offset by the CO_2 absorbed during plant growth via photosynthesis (Kashif et al. 2025). However, the successful and sustained deployment of bioenergy depends critically on strict sustainability criteria. Global reports indicate that rapidly increasing demand for bioresources could potentially outstrip sustainable supply, threatening climate mitigation efforts and biodiversity if not carefully managed (Iyiola & Ogwu 2025). Consequently, biomass use must be limited primarily to waste streams, residues, or dedicated energy crops grown on degraded or marginal lands, ensuring it does not compete with food production or trigger adverse land-use change. This stringent resource limitation elevates the importance of developing conversion technologies that achieve maximum resource efficiency, yielding high-value products from every component of the sustainably sourced feedstock.

Pyrolysis stands out as a versatile and promising thermochemical technology that fulfils this requirement. It involves the thermal decomposition of organic matter such as lignocellulosic biomass in an inert or oxygen-free environment. This rapid conversion process generates three distinct product fractions: solid biochar, condensable liquid bio-oil, and non-condensable pyrolysis gas (or syngas). Beyond generating renewable energy, pyrolysis offers significant environmental advantages by providing a robust method for waste management, converting agricultural and forestry residues into useful commodities (Makepa et al. 2025). Furthermore, the resulting

DOI: 10.1201/9781003650751-7

biochar offers a proven mechanism for long-term carbon sequestration and soil enhancement, providing a dual pathway for reducing fossil fuel reliance and mitigating climate change.

7.1.2 Types of Pyrolysis and Process Control

Pyrolysis is not a uniform or rigid process; rather, its product distribution is governed by complex reaction kinetics and is highly responsive to operational parameters. Through the adjustment of key conditions such as temperature, heating rate, and vapor residence time, engineers can strategically steer the process toward the preferential formation of solid, liquid, or gaseous products (Mutsengerere et al. 2019).

Three major modes of pyrolysis are recognized based on these kinetic factors:

1. **Slow pyrolysis:** This mode typically operates at lower temperatures (300°C–550°C) with very low heating rates (<0.1°C/s) and long residence times ranging from minutes to several hours. This lengthy confinement of vapors allows for extensive secondary reactions between the solid and volatile phases, maximizing the yield of the solid biochar product (Mutsengerere et al. 2019).
2. **Fast pyrolysis:** This technique operates at moderate temperatures (450°C–650°C) but is characterized by extremely high heating rates (10–1,000°C/s) and very short vapor residence times (typically less than 2 seconds). These precise conditions facilitate rapid devolatilization of the biomass and immediate quenching of the resulting vapors, effectively suppressing secondary cracking reactions that would otherwise convert liquid precursors into permanent gases (Mutsengerere et al. 2019).
3. **Flash pyrolysis:** Representing an extreme refinement of fast pyrolysis, this mode utilizes even higher heating rates (exceeding 1,000°C/s) and minimal residence times (<0.5 seconds), often at higher temperatures, further pushing the maximization of liquid products (Mutsengerere et al. 2019).

Maximizing liquid yield in fast and flash pyrolysis presents notable engineering challenges, primarily due to the poor thermal diffusivity of raw biomass. To achieve the high heating rates required for efficient conversion, feedstock must be finely ground to ensure uniform heat transfer and consistent thermal decomposition (Luo & Zhou 2025). This constraint highlights that commercial viability depends not only on chemical pathway control but also on optimized reactor design and feedstock logistics. Technologies such as fluidized bed reactors are particularly effective as they facilitate rapid heat transfer and support scalable operation (Raza et al. 2021). Moreover, process conditions (especially temperatures above 600°C and extended vapor residence times) can trigger secondary cracking reactions, shifting product distribution toward gaseous outputs like syngas at the expense of bio-oil and biochar yields (Mutsengerere et al. 2019).

7.1.3 The Integrated Valorization Strategy

While the existing body of literature includes comprehensive reviews on biomass pyrolysis, many focus primarily on either bio-oil upgrading for bulk fuels or biochar applications in soil science, often treating the gaseous stream merely as an internal energy source (Liu et al. 2023; Zhang et al. 2023). This chapter differentiates itself by adopting an integrated biorefinery perspective, providing equal analytical depth to the valorization pathways of all three co-products (solid, liquid, and gas) within a unified framework of sustainable resource management.

The economic feasibility of large-scale pyrolysis facilities is usually unfavorable when competing solely on the basis of replacing crude fossil fuels. The technological strategy presented here shifts the focus from commodity production to high-margin specialty chemicals and advanced functional materials. Through the comprehensive detailing of the conversion of primary pyrolysis products into high-value intermediates, the chapter establishes a framework wherein pyrolysis transforms from a bulk energy producer into a specialty chemical and material manufacturer, significantly enhancing the economic proposition.

7.2 GASEOUS PRODUCTS

7.2.1 Composition and Characteristics

The gaseous product of biomass pyrolysis, often referred to as pyrolysis gas or syngas, is the non-condensable fraction generated during the thermal decomposition process (Mutsengerere et al. 2019). This product stream is primarily composed of light, permanent gases including H_2, CO, CO_2, CH_4, and other light hydrocarbons (He et al. 2025).

The yield and composition of syngas are highly sensitive to operational parameters that promote secondary cracking reactions (Talwar et al. 2025). Among them, temperature is the most statistically significant lever. High pyrolysis temperatures, typically exceeding 600°C, accelerate the cracking of volatile components, which significantly increases the total gas yield at the expense of liquid bio-oil and solid biochar (Talwar et al. 2025). Studies have demonstrated that syngas efficiency can increase substantially with temperature, rising from a yield of 38.92% at 400°C to 56.64% at 600°C (Altıkat et al. 2024). Similarly, residence time plays a crucial role. A longer residence time promotes extensive secondary reactions between the solid and volatile phases, diminishing bio-oil yield while concurrently increasing the synthesis gas yield (Altıkat et al. 2024).

The specific chemical structures within the biomass feedstock also fundamentally influence the composition of the gaseous output.

- **Lignin** pyrolysis favors the production of CH_4, primarily due to the cracking of its stable aromatic rings and methoxy groups (Chen et al. 2022).
- **Hemicellulose** decomposition is the main driver for the generation of CO_2 (Chen et al. 2022).
- **Cellulose** cracking, particularly at high temperatures (above 550°C), results in the maximum yield of CO (Chen et al. 2022).

Furthermore, the inorganic mineral content (ash) within the biomass can act as an in situ catalyst promoting specific decomposition pathways that alter both the quantity and chemical composition of the final gas stream (Parvari et al. 2025).

7.2.2 Applications and Chemical Synthesis Pathways

Pyrolysis of biomass yields a gaseous mixture (syngas) rich in CO, CO_2, H_2, CH_4, and light hydrocarbons (Robazza et al. 2024). This raw gas contains condensable tars and entrained char, which must be removed before utilization. Effective gas cleaning and conditioning is the first upgrading step. Thermal or catalytic tar cracking/reforming (e.g. Ni-based catalysts at 600°C–900°C) can convert heavy organics into additional H_2 and CO (Koritár & Haydary 2025). Cyclones or hot-gas filters remove particulates (char). Acid gases (H_2S, HCl, and NH_3) are scrubbed out (e.g. alkali solutions, amine units) to prevent downstream catalyst poisoning (Hrbek et al. 2025). After cleaning, the syngas stream is often shifted to optimize the H_2/CO ratio. For instance, the water-gas shift (WGS) reaction ($CO + H_2O \rightarrow CO_2 + H_2$) over Fe/Cr or Cu/Zn catalysts raises the H_2 content for fuel-cell or chemical needs. As noted in recent reviews, tailoring syngas composition is critical for later synthesis steps. Figure 7.1 provides a comprehensive overview of the main upgrading pathways and end-use applications of pyrolysis syngas.

7.2.2.1 Catalytic Reforming

A common upgrading step is catalytic reforming of the pyrolysis vapors in a second reactor. Here, volatile hydrocarbons and tars are steam- or dry (CO_2)-reformed to H_2 and CO over metal catalysts (often Ni on oxides or char) (Koritár & Haydary

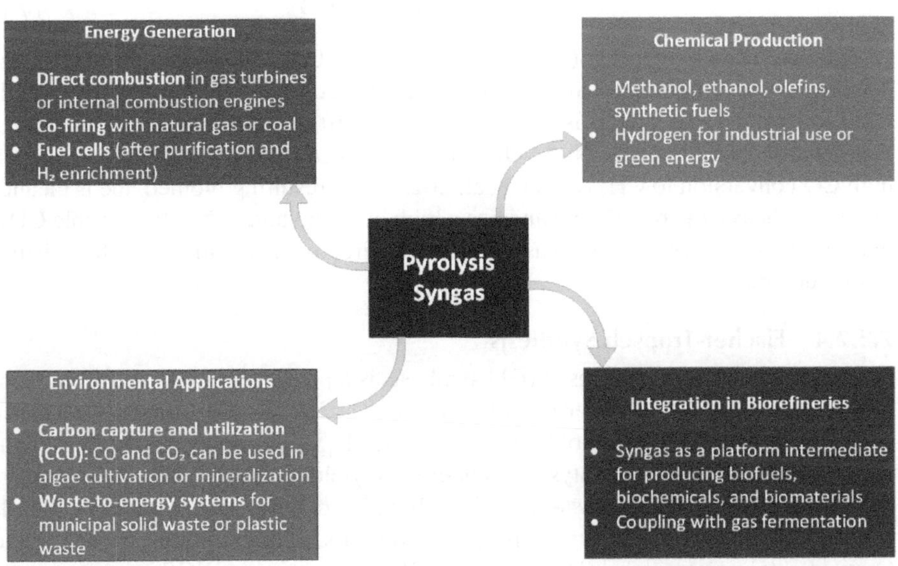

FIGURE 7.1 Schematic overview of pyrolysis syngas upgrading and applications.

2025). Two-stage pyrolysis + reforming setups have been widely studied for H_2-rich syngas. Reported results show dramatic H_2 gains. For example, Ni–Al_2O_3 catalysts raised H_2 in the product from ~18 (no catalyst) to ~44 vol% (Kabeh et al. 2025). Overall, two-stage reforming can double or triple the H_2 yield compared to non-catalytic pyrolysis. Besides boosting H_2, reforming eliminates tars at moderate temperatures (500°C–800°C) (Koritár & Haydary 2025). In practice, Ni-based catalysts (Ni/ZrO_2, Ni/char, limonite, etc.) are most effective, often combined with steam addition. Process variables like higher reforming temperature and steam-to-biomass ratio further improve conversion, though they add energy cost (Kabeh et al. 2025). Thus, catalytic reforming furnishes a cleaner, H_2-rich syngas suitable for synthesis.

7.2.2.2 Water–Gas Shift Reaction

After reforming, the syngas may undergo the WGS reaction to fine-tune the H_2/CO ratio. This equilibrium step ($CO + H_2O \rightleftarrows CO_2 + H_2$) is used industrially with Fe–Cr (350°C) or Cu–Zn (200°C) catalysts. Through the consumption of CO and steam, WGS increases H_2 yield, a key step for fuel cells or any H_2-demanding application. In practice, shift reactors convert much of the remaining CO into H_2 and CO_2. Although specific pyrolysis studies are few, the principle is well-known. High-temperature WGS (350°C–450°C) combined with low-temperature polishing maximizes H_2 production. The extra H_2 is then available for clean energy uses or as feedstock in downstream processes (Lee et al. 2023).

7.2.2.3 Methanation (Synthetic Natural Gas)

When the goal is gaseous fuel, methanation (the Sabatier reaction) converts CO and H_2 into CH_4:

$$CO + 3H_2 \rightarrow CH_4 + H_2O \tag{7.1}$$

This exothermic reaction (Ni catalyst, ~300°C–400°C, moderate pressure) produces synthetic natural gas (SNG) that can be injected into pipelines or used in boilers. Biomass-derived syngas is well-suited, after shift, the H_2/CO ratio (~3) matches methanation stoichiometry. Ni/Al_2O_3 and Ru/activated carbon catalysts have shown high CO conversion to CH_4 (Hrbek et al. 2025). The resulting biomethane is cleaner (no tars or heavy hydrocarbons) and has a high heating value. SNG (renewable CH_4) thus provides a renewable gas substitute directly compatible with existing infrastructure.

7.2.2.4 Fischer-Tropsch Synthesis

For liquid fuels, Fischer-Tropsch (FT) synthesis is a proven pathway. In FT reactors, clean syngas (H_2/CO ≈ 2) over Fe or Co catalysts at 200°C–300°C and 20–40 bar is polymerized into a spectrum of hydrocarbons (C_nH_{2n+2}) and water. The carbon chain length can be tailored by catalyst and conditions, yielding diesel-range fuels, waxes, or jet fuel precursors. Biomass-derived FT liquids are essentially renewable diesel/kerosene. Critical to FT is syngas purity. Sulfur, tars, and alkali must be negligible to avoid catalyst poisoning (Hrbek et al. 2025). In practice, a WGS stage may precede FT to reach the ideal H_2/CO, and treating units (amine wash, etc.) remove trace poisons. Numerous studies have demonstrated biomass-syngas-FT to produce paraffinic fuels, often coprocessing with a FT oxygenate integration (Adhikari et al. 2018; Mahmoudi et al. 2020).

7.2.2.5 Alcohol Synthesis

Catalytic routes also allow production of alcohols from syngas. The classic example is methanol synthesis ($CO + 2H_2 \rightarrow CH_3OH$) over $Cu–ZnO/Al_2O_3$ catalysts at ~50–100 bar and 200°C–300°C. Pyrolysis syngas, after adjustment of H_2/CO and purification, can be fed to methanol plants. Research has demonstrated direct methanol yields from "bio-syngas" with only minor feed modifications. Beyond methanol, mixed-alcohol or higher-alcohol catalysts (e.g. modified FT, Rh- or Co-based systems) can convert syngas into ethanol or butanol. In parallel, biological alcohol routes exist. Acetogenic bacteria (*Clostridium* spp.) ferment CO/H_2 into ethanol or butanol under mild conditions. Such syngas fermentation is especially attractive. It tolerates CO_2 and some residual gas impurities and has reached industrial scale for ethanol (e.g. steel mill gas to ethanol). In fact, syngas fermentation by *C. autoethanogenum* is known to produce acetate and ethanol from biomass-derived gas streams (Robazza et al. 2024). These biochemical routes complement the chemical catalysis approaches in a biorefinery context.

7.2.2.6 Ammonia Synthesis

The H_2-rich syngas can be used in NH_3 production via the Haber-Bosch process. In this route, the cleaned syngas is first shifted and stripped to maximize H_2, then blended with nitrogen (from air) and fed to an Fe (or Ru) catalyst at high pressure (150–300 bar) and 350°C–550°C. The result is renewable NH_3, an essential fertilizer and chemical feedstock. Converting pyrolysis syngas to NH_3 (often called "green ammonia") closes the loop on biomass energy, effectively storing biomass energy in a stable nitrogen carrier (Farsi 2023). While few pilot projects exist yet, the pathway is well-established in concept. Any H_2 source (even from biomass) can substitute for natural gas in NH_3 making. Table 7.1 summarizes the key syngas upgrading pathways and products.

7.2.2.7 Energy Generation

Cleaned pyrolysis syngas is a versatile energy source. It can be directly combusted in boilers, gas turbines, or internal-combustion engines for heat and electricity. For example, bio-oil plants often burn the non-condensable gas to provide process heat. Advanced systems cofire syngas with coal or natural gas in power plants. Syngas combustion produces lower emissions of particulates and sulfur (when cleaned) compared to raw biomass. In recent work, it was shown that firing biomass-derived syngas in a turbine can yield stable power output, though the lower heating value (typically 5–15 MJ/Nm^3) requires design adjustments (Hrbek et al. 2025). At smaller scales, fuel cells (especially solid oxide fuel cell (SOFC)) can convert syngas to electricity with high efficiency. After WGS enrichment, the mainly H_2 syngas can feed a Proton Exchange Membrane fuel cell or SOFC directly, yielding electricity and useful process heat without combustion. Such applications are still emerging but have been demonstrated in pilot biorefineries.

TABLE 7.1
Key Syngas Upgrading Pathways and Products

Route	Main Reactions	Typical Conditions/ Catalyst	Products/Uses
Gas cleaning	Tar cracking/ reforming, particulate filtration	Ni-based catalyst (700°C–1,000°C); cyclones/bag filters; wet scrubbers (NaOH, amines)	Clean H_2–CO gas, free of tars, H_2S, and HCl
Catalytic reforming	Steam reforming/ CO_2 reforming of hydrocarbons	Ni catalyst, steam (~2–10 atm), 600°C–800°C	Increased H_2, CO (high H_2/CO syngas)
Water-gas shift	$CO + H_2O \rightarrow CO_2 + H_2$	Fe–Cr (350°C) or Cu–Zn (200°C) catalysts	Higher H_2/CO ratio for fuel cells or synthesis
Methanation	$CO + 3H_2 \rightarrow CH_4 + H_2O$	Ni/Al_2O_3 catalyst, 300°C–400°C, 20–30 bar	Syngas natural gas (renewable CH_4)
Fischer-Tropsch	$xCO + (2x+1)H_2 \rightarrow C_xH_{2x+2} + xH_2O$	Fe or Co catalyst, 200°C–300°C, 20–40 bar	Liquid fuels (diesel/jet), waxes (H_2/CO ~ 2)
Methanol synthesis	$CO + 2H_2 \rightarrow CH_3OH$	$Cu–ZnO/Al_2O_3$ catalyst, 50–100 bar, 200°C–300°C	Methanol (fuel, H_2 carrier, chemical feedstock)
Ammonia synthesis	$N_2 + 3H_2 \rightarrow 2NH_3$	Fe-based (Haber-Bosch), 150–300 bar, 350°C–550°C	Ammonia (fertilizer, industrial)

7.2.2.8 Chemical Production

Beyond fuels, pyrolysis syngas is a feedstock for chemical manufacturing. The H_2 fraction alone is valuable: it can support refining or be sold as a fuel gas or H_2 carrier (Al-Qahtani 2023). H_2 can also be used as a reducing agent in the steel-making industry, thus facilitating the manufacture of green steels. CO can be used in carbonylations or FT. Crucially, syngas (CO/H_2) can be fermented by specialized microbes to make biobased chemicals. Acetogenic fermenters can convert CO/H_2 to acetic acid, ethanol, butyrate, or butanol. For instance, *C. autoethanogenum* fermentations yield mainly acetate and ethanol from syngas (Robazza et al. 2024). Gas fermentation is already commercial (e.g. steel mill syngas to ethanol) and can be applied to biomass syngas. In parallel, catalytic chemistry can yield light olefins. Methanol made from syngas can undergo "methanol-to-olefins" or "methanol-to-gasoline" processing to form ethylene, propylene or gasoline-range chemicals. Thus, pyrolysis syngas serves as a platform intermediate in the biorefinery (Figueirêdo et al. 2022), a feedstock for wide-ranging petrochemical-equivalent products, but from renewable carbon.

7.2.2.9 Environmental and Waste-to-Energy Applications

Pyrolysis syngas can also play a role in environmental strategies. The CO_2 coproduced, especially after WGS, can be captured and utilized. For example, CO_2 can be fed to microalgae photobioreactors to produce algal biomass or biocrude, effectively recycling carbon. Mineral carbonation (reacting CO_2 with alkaline wastes) or even carbon capture and storage could be coupled to syngas plants to make bioenergy with carbon capture. Moreover, pyrolysis itself is often applied to wastes and residues like municipal solid waste, agricultural residues, or plastics. Gasifying or pyrolyzing these wastes generates a syngas that is substantially cleaner than landfilling or incineration. For instance, refuse-derived fuel pellets can yield ~50% of the input as combustible gas (lower heating value (LHV) ~11–12 MJ/Nm^3) (Koritár & Haydary 2025), turning waste into energy and chemicals. Clean syngas from waste pyrolysis has been demonstrated in lab and pilot units, showing efficient removal of H_2S, HCl, and tars (Koritár & Haydary 2025). In this way, pyrolysis syngas contributes to waste-to-energy solutions, reducing landfill and creating renewable energy carriers.

7.2.2.10 Integration in Biorefineries

In a modern biorefinery, pyrolysis syngas is not used in isolation but integrated with other processes. It complements liquid bio-oil and biochar production by valorizing the gaseous byproduct. For example, a biorefinery might use pyrolysis to fractionate biomass. In this process, bio-oil goes to a hydrotreater, biochar to soil amendment, and syngas to fuels. Synergies arise when syngas is coprocessed: fermenters can detoxify bio-oil aqueous phases while consuming syngas to produce carboxylates or alcohols (Robazza et al. 2024). Blending biomass with carbonaceous waste (plastics, natural gas) before pyrolysis can boost H_2 yield, making downstream H_2 or synthetic fuel production more efficient (Kabeh et al. 2025). As a platform intermediate, syngas thus links thermal conversion to catalytic and biological conversion steps. Its versatile downstream options (fuels, power, chemicals, and fertilizers) make it a key nexus in sustainable biorefineries, allowing full utilization of biomass carbon and energy.

7.3 LIQUID PRODUCTS (BIO-OIL/PYROLYSIS OIL)

7.3.1 Production and Yield (Fast Pyrolysis)

Fast pyrolysis is the thermal decomposition of biomass under oxygen-limited conditions with very rapid heating (typically 400°C–600°C), short vapor residence times (0.5–5.0 seconds), and rapid quenching of the volatiles to maximize liquid product formation (Makepa et al. 2025). The principal mechanism begins with the thermal depolymerization of cellulose, hemicellulose, and lignin into a distribution of reactive primary vapors (sugars, furans, light oxygenates, and phenolic fragments). Many of these primary vapors undergo secondary reactions in the vapor phase (recombination, fragmentation, and dehydration) that ultimately determine the molecular complexity of the condensed liquid called bio-oil (Makepa et al. 2025). The interplay of primary cleavage (fragmentation of biopolymers) and secondary vapor phase chemistry explains why reactor design (conduits and quench location), heating rate, and residence time strongly influence yield and product composition (Oasmaa et al. 2021).

Typical liquid yields from optimally operated bench- and pilot-scale fast pyrolysis units are reported in the literature as around 50–75 wt% on dry feedstock, with many commonly cited values clustering in ~60–75 wt% for wood and many agricultural residues under optimized conditions. Yield depends on feedstock (lignin-rich feedstocks generally give lower liquid yield), reactor type (fluidized bed, bubbling versus circulating, and ablative), and condensation/collection strategy; char and non-condensable gas typically make up the remainder (char ≈ 10–25 wt%, gas ≈ 10–20 wt%). Published compilations and recent reviews report the 50%–75% range as an industry benchmark for conventional fast pyrolysis (Jerzak et al. 2024; Arbelaez et al. 2015).

7.3.2 Physico-Chemical Properties

7.3.2.1 Challenges

Crude bio-oil produced by fast pyrolysis has a combination of properties that make it chemically and physically distinct from fossil liquid fuels and that create barriers to direct use as a transport fuel:

- High oxygen content (commonly 15–40 wt% O_2 depending on feedstock and process), which lowers heating value and causes reactive functionality (Rowland et al. 2024).
- **Acidity/corrosivity:** Measured pH values approximately 2–4, due to organic acids (acetic, formic, and levoglucosan fragments) and phenolic acids. This causes corrosion of storage and handling equipment if unmitigated (Rowland et al. 2024).
- **High water content:** Water content can be 15–30 wt% (or higher for certain feedstocks/condensation regimes), reducing LHV and degrading combustion behavior (Han et al. 2019).
- **Thermal/chemical instability:** Bio-oil ages during storage due to polymerization/condensation of reactive oxygenates, leading to increases in viscosity and formation of insoluble phases (gum and sediments) (Rowland et al. 2024).
- **High viscosity:** Viscosity increases with oxygen content and molecular weight distribution. Aged oils can be gel-like, hindering pumping and atomization (Rowland et al. 2024).

7.3.2.2 Chemical Composition

Bio-oil is a complex, multi-phase mixture composed of hundreds to thousands of compounds. According to Cesari et al. (2019) and Sewsynker-Sukai et al. (2023), major families of compounds include:

- Phenolic compounds (e.g. monomeric lignin pyrolysis products such as guaiacol, syringol, and alkylated phenols),
- Carboxylic acids (e.g. acetic acid, formic acid, and levulinic-type species),
- Aldehydes and ketones (e.g. furfural, hydroxyacetaldehyde, and acetol),
- Anhydrosugars and sugar fragments (e.g. levoglucosan from cellulose),

- Polymeric oligomers (repolymerized fragments that form the heavy "tar" fractions),
- Nitrogen and sulfur heteroatoms (from proteinaceous or contaminant feedstock, usually minor but problematic for some downstream catalysts).

The precise speciation depends strongly on feedstock and pyrolysis conditions. Lignin-rich streams show higher phenolic content, while hemicellulose-rich feeds produce more furans and small aldehydes. Analytical studies and fractionation experiments show that phenolic fractions are particularly valuable as chemical feedstocks if they can be separated economically (Cesari et al. 2019; Sewsynker-Sukai et al. 2023).

7.3.3 Upgrading and Refining

7.3.3.1 Why Upgrading Is Necessary

Unmodified bio-oil generally fails transportation fuel specifications because of low energy density, high oxygen content, poor ignition properties, and corrosivity. Many industrial applications therefore require upgrading to (a) reduce oxygen, acidity, and water; (b) raise heating value; (c) stabilize the liquid against storage-ageing; and (d) provide compatibility with existing refinery processes if "coprocessing" (integration with petroleum refinery units) is the route chosen. Upgrading is a critical step if the bio-oil is to become a "drop-in" fuel or a refinery feed (Han et al. 2019).

7.3.3.2 Principal Upgrading Techniques

7.3.3.2.1 Hydrodeoxygenation

Hydrodeoxygenation (HDO) represents a cornerstone technology for upgrading bio-oils derived from fast pyrolysis of biomass. Through a cascade of hydrogenolysis, hydrogenation-dehydration, decarbonylation, and decarboxylation reactions, HDO systematically removes oxygenated functional groups from bio-oil molecules. The overarching stoichiometry can be summarized as: R–O → RH + $H_2O/CO/CO_2$, where R signifies the organic moiety. The result is a significant reduction in oxygen content and transformation of reactive oxygenates into hydrocarbon-analogous fractions, thereby improving energy density, reducing acidity, and enhancing stability (Díaz et al. 2024).

Industrial implementations of HDO often employ sulfided Ni–Mo or Co–Mo catalysts supported on γ-Al_2O_3, drawing from well-established hydrotreating practices in petroleum refining. These catalysts operate in their sulfided state and are effective in hydrogenolytic deoxygenation, primarily via removal of oxygen as water, CO, or CO_2. In parallel, research into noble metal catalysts (e.g. Ru, Pt, and Rh) and supported base-metal systems (such as Ni, and Ni–Mo) has accelerated, particularly for applications requiring sulfur-free processing of oxygenates. These catalysts may function alone or in bifunctional architectures, combining metallic hydrogenation sites with acidic supports to concurrently catalyze deoxygenation and selective cracking (Díaz et al. 2024; Qu et al. 2021).

Recent studies reinforce the prevalence of these catalysts: Ni–Mo/Al_2O_3 and Ru/C remain widely recognized as workhorse HDO materials. For instance, Seo et al.

(2025) demonstrated that Ni–Mo/Al_2O_3 can effectively deoxygenates pyrolysis oils at ~300°C, shifting carbon ranges toward higher boiling fractions and elevating hydrocarbon yields. Similarly, Aqsha et al. (2014) reported that Ni–Mo/TiO_2 catalysts can convert about 98% of guaiacol (a typical bio-oil model compound) into phenolic and hydrocarbon products at ~350°C using a 2:1 H_2-to-feed molar ratio.

Typical operational conditions span 200°C–400°C under elevated H_2 pressures. Reported H_2 pressures range from ~10 to 200 bar, with 50–150 bar being common in continuous pilot systems. To ensure complete conversion and to avoid operational constraints, experiments most commonly employ low liquid space velocities of ~0.05–2/hour, facilitating extended fluid-catalyst contact. Reactor configurations such as fixed-bed trickle-phase and ebullated-bed reactors are widely leveraged. The latter is particularly advantageous for managing solids and thermal regulation during intensive deoxygenation (Qu et al. 2021).

7.3.3.2.2 Catalytic Cracking/Zeolite Upgrading (In Situ Vapor Upgrading)

Catalytic cracking, particularly via zeolite-based upgrading, is a widely studied and industrially relevant technique for enhancing the quality of pyrolysis vapors and bio-oils. This process aims to convert oxygen-rich vapors or liquids into lighter hydrocarbons, primarily through acid-catalyzed cracking, deoxygenation, and aromatization reactions. When applied in the vapor phase (commonly referred to as in situ catalytic pyrolysis), zeolites such as ZSM-5, HY, and fluid catalytic cracking (FCC) catalysts facilitate the transformation of oxygenates into gasoline-range compounds, including monoaromatics like benzene, toluene, and xylenes (Reza et al. 2023). These reactions occur rapidly at elevated temperatures and are driven by the strong Brønsted acidity and shape-selective properties of the zeolite framework.

In contrast, when catalytic cracking is applied to liquid bio-oils, typically in FCC coprocessing configurations, the goal shifts toward converting heavier fractions into distillates and gaseous hydrocarbons. This approach allows bio-oils to be integrated into existing petroleum refining infrastructure, leveraging FCC units to coprocess renewable and fossil feedstocks. The cracking reactions in this context are more complex due to high viscosity, oxygen content, and instability of bio-oils, which require pretreatment or blending strategies to ensure compatibility with FCC operations (Reza et al. 2023).

The most commonly employed catalysts in both vapor phase and liquid phase upgrading include HZSM-5, metal-promoted ZSM-5 variants (e.g. Ni-ZSM-5 and Co-ZSM-5), and hybrid formulations combining FCC catalysts with zeolitic components. These catalysts operate effectively within a temperature range of 400°C–600°C, where the thermal energy supports cracking and deoxygenation while preserving aromatic selectivity (Chen et al. 2025). Metal promotion enhances deoxygenation activity and can improve resistance to coke formation, although it may also alter product distribution and reduce selectivity toward desired aromatics.

7.3.3.2.3 Emulsification and Blending

In contrast to intensive upgrading techniques such as HDO, emulsification and blending offer a more accessible and cost-effective strategy for the near-term utilization of pyrolysis bio-oils. These approaches are particularly relevant for static heat applications, decentralized energy systems, and early-stage deployment scenarios where full refinery integration or deep catalytic treatment may not be feasible. The central idea is to incorporate bio-oils into existing fossil fuel streams, either through physical blending or coprocessing, thereby introducing biogenic carbon into conventional fuel products without extensive infrastructure modifications (Leng et al. 2018).

Emulsification involves the formation of stable mixtures between bio-oils and petroleum-derived fuels such as diesel or vacuum gas oil (Han et al. 2021). This is typically achieved using surfactants and high-shear mixing, which help overcome the immiscibility and polarity differences between the aqueous, oxygen-rich bio-oil and hydrophobic fossil fuels. The resulting emulsions can be directly combusted in boilers or engines or used as feedstocks for further upgrading. Stability of these emulsions depends on several factors, including the type and concentration of surfactants, temperature, and pretreatment of the bio-oil (e.g. filtration, pH adjustment, or mild catalytic stabilization) (Leng et al. 2018). Studies have shown that with appropriate surfactant systems, emulsions can remain stable for several weeks and exhibit combustion characteristics comparable to diesel blends.

7.3.4 Applications

7.3.4.1 Direct Combustion/Industrial Heat and Power

Crude or mildly upgraded bio-oil can be used as a boiler fuel for industrial heat and steam generation and for combined heat and power systems. For stationary combustion applications (furnaces and boilers), issues such as atomization, corrosion, and deposit formation have been addressed with modified burners or with pretreatment (neutralization and solids removal). Cofiring or co-combustion with heavy fuel oil is a near-term application for supplying renewable heat. Pilot and demonstration systems (multi-kW to MW scale) have validated stable combustion when appropriate pretreatment/materials choices are applied (Hou et al. 2016).

7.3.4.2 Precursor for Advanced Biofuels (Drop-in Fuels)

Upgraded bio-oil is a feedstock for second-generation liquid fuels. Pathways include deep HDO followed by hydrocracking or integration into refinery hydrotreating/FCC units to produce gasoline, diesel, and jet fractions. Catalytic fast pyrolysis coupled with in situ upgrading (vapor upgrading with zeolites) also produces more hydrocarbon-like liquids directly. Coprocessing in refinery units is an attractive industrial route because it leverages existing refinery infrastructure. Recent technoeconomic and demonstration studies indicate modest coprocessing fractions (few percent to low tens of percent) can be introduced without major plant modifications. For aviation fuels (bio-jet), HDO followed by fractionation (or FT synthesis from upgraded intermediates) are commonly discussed industrial routes (Valle et al. 2019; van Dyk & Saddler 2024).

7.3.4.3 Source of Specialty Chemicals (Phenols, Vanillin, and Platform Molecules)

Beyond fuels, bio-oil contains value-added chemical precursors: phenolic compounds (from lignin), low-molecular-weight acids, furans, and other oxygenated building blocks that can be separated and upgraded into chemicals such as phenol derivatives, adhesives, and fine chemicals. Vanillin is routinely discussed as a high-value target obtained from lignin valorization. Fractionation (liquid–liquid extraction, solvent fractionation, and membrane separations) followed by targeted catalytic or biological conversion (oxidation, hydrogenation, and biotransformations) can produce specialty chemicals at a higher margin than fuel. Several recent reviews and studies emphasize separation strategies for phenolic fractions and biocatalytic/chemical routes to vanillin and other aromatics from lignin-rich pyrolysis streams (Cesari et al. 2019; D'Arrigo et al. 2024).

7.4 SOLID PRODUCTS (BIOCHAR)

7.4.1 Production and Yield (Slow Pyrolysis)

Slow pyrolysis is conventionally defined as the thermal decomposition of biomass in an oxygen-limited environment with relatively low heating rates and long solid residence times (seconds to hours). The process is designed to favor solid char formation rather than liquids or syngas. In slow pyrolysis, biomass particles are heated more slowly and held at moderate temperatures so that extensive devolatilization is followed by secondary reactions that consolidate a carbonaceous, porous residue (biochar). Mechanistically, biochar forms by (a) thermal cleavage of biopolymer linkages in cellulose, hemicellulose, and lignin, producing reactive radical fragments and oligomers; (b) cross-linking, polymerization, and aromatization of reactive fragments on the particle surface; and (c) progressive loss of volatiles and oxygenated functional groups, leaving a carbon-rich, condensed aromatic network. Secondary tar condensation and polymerization on the solid surface can also increase char mass and influence porosity. These mechanistic pathways are summarized in recent studies (Awad et al. 2024; Yaashikaa et al. 2020).

Under well-controlled slow pyrolysis conditions, biochar yields are substantially higher than in fast pyrolysis. Published surveys and technoeconomic studies report typical solid yields in the range of 25–50 wt% on a dry biomass basis for slow pyrolysis (actual yields vary with reactor design, residence time, heating rate, and feedstock type). Some studies show yields outside this band for particular feedstocks or very low heating rates, but the 25%–50% range is a reliable working benchmark for slow pyrolysis technology assessments (Safarian 2023).

7.4.2 Characteristics and Surface Chemistry

7.4.2.1 Bulk Composition and Carbon Stability

Biochars are characterized by high fixed carbon fractions (aromatic carbon content) and low volatile matter relative to the parent biomass. Increasing pyrolysis temperature generally raises the fraction of fixed carbon and lowers hydrogen/carbon and oxygen/carbon atomic ratios, a signature of increasing aromaticity and chemical stability (i.e. longer persistence in soils and improved recalcitrance to microbial decomposition). High-temperature chars (≥600°C–700°C) are generally more graphitic/aromatic and thus more stable over multi-decadal timescales. Lower-temperature chars (<400°C–500°C) retain more oxygenated functionalities and labile carbon (Awad et al. 2024; Ghorbani et al. 2022).

7.4.2.2 Porosity, Surface Area, and Microstructure

One of biochar's most important functional characteristics is its porous structure and specific surface area (SSA). As pyrolysis temperature increases, devolatilization opens internal pores and removes volatile matter, which usually increases SSA and microporosity up to a feedstock-dependent limit. Feedstock morphology strongly modulates absolute surface areas: lignocellulosic woods typically produce different pore networks from agricultural residues, manures or biosolids (the latter often produce higher ash content and different pore development). The combination of feedstock and temperature therefore controls pore size distribution (micro/meso/macropores), SSA (ranging typically from a few m^2/g to several hundred m^2/g for highly activated chars), and accessible surface for sorption or catalytic applications (Awad et al. 2024; Ghorbani et al. 2022).

7.4.2.3 Surface Chemistry and Functional Groups

Biochar surfaces present a range of oxygen-containing functional groups (carboxyl, phenolic, carbonyl, and lactone) whose abundance decreases with increasing pyrolysis temperature. Lower-temperature biochars (300°C–450°C) typically have more surface acidity (and cation exchange capacity, CEC) and a richer distribution of polar groups that enhance nutrient retention and metal complexation in soils. High-temperature chars are more alkaline (due to ash and inorganic concentrates), have higher pH, and fewer polar functionalities but enhanced aromaticity and electrical conductivity, traits valuable for sorbents and electrode materials. The choice of temperature/feedstock therefore permits tuning of surface chemistry for desired applications (Awad et al. 2024; Ghorbani et al. 2022).

7.4.3 Applications

7.4.3.1 Soil Amendment and Carbon Sequestration

Biochar is widely studied and increasingly applied as a soil amendment because it can (a) increase soil organic carbon and so contribute to long-term carbon sequestration, (b) improve soil physical properties (bulk density, porosity, and water retention), and (c) enhance nutrient retention and microbial habitat, often leading to yield benefits depending on soil type and crop. Low-temperature biochars with higher surface polarity and CEC are commonly effective for nutrient retention, whereas higher-temperature chars provide more persistent carbon pools. Studies show that carefully matched biochar applications can sequester a portion of feedstock carbon for decades to centuries and can reduce emissions of N_2O in some soil systems (Premalatha et al. 2023; Rodrigues et al. 2023). The sequestration efficiency and agronomic outcome depend heavily on application rate, soil texture, and biochar properties.

7.4.3.2 Adsorbent for Water and Wastewater Treatment (Heavy Metals and Organics)

Biochar's porous network, SSA, and surface functional groups make it an effective, low-cost sorbent for removing heavy metals (Pb^{2+}, Cd^{2+}, Cr^{6+}, Cu^{2+}, etc.), nutrients (NH_4^+, PO_4^{3-}), and organic contaminants (phenols, dyes, and pharmaceuticals) from aqueous media. Surface complexation, electrostatic attraction, ion exchange, and pore filling are the dominant sorption mechanisms. The relative importance of each depends on pH, ionic strength, and char surface chemistry. Activation (physical or chemical) and surface modification (e.g. oxidation, magnetization, and impregnation with metal oxides) are widely used to enhance sorption capacity and selectivity. Recent reviews compile numerous batch and column studies demonstrating very large removal capacities for modified biochars and highlight the potential for cost-effective wastewater treatment and resource recovery (Ngaba et al. 2025; Yaashikaa et al. 2020).

7.4.3.3 Catalyst Support and Materials for Energy Storage (Supercapacitors and Batteries)

Because high-temperature biochars can display high electrical conductivity, hierarchical porosity and tunable surface chemistry, they are being intensively investigated as low-cost carbonaceous electrode materials for supercapacitors, lithium-ion batteries, sodium-ion batteries, and as supports for heterogeneous catalysts. Activation, templating, and heteroatom (N, P, S) doping are common strategies to increase capacitance, electrical conductivity, and catalytic activity. Lab-scale studies show that properly engineered biochar-derived carbons can approach the performance of conventional activated carbons in supercapacitors and in some cases deliver promising specific capacitance and rate capability. Biochar also provides a robust, low-cost support for metal nanoparticles for catalysis (e.g. for HDO and oxygen reduction reaction) with the porous network facilitating dispersion and mass transport.

Reviews surveying these material science applications emphasize that scale-up, reproducibility and cost-effective activation remain active challenges (Altıkat et al. 2024; Barszcz et al. 2024).

7.5 INFLUENCE OF FEEDSTOCK AND PROCESS PARAMETERS

Feedstock composition and process settings jointly determine whether carbon from biomass exits a pyrolysis unit as gas, liquid, or solid. Different feedstocks carry distinct macromolecular signatures (cellulose, hemicellulose, and lignin) and non-structural components (lipids and proteins) plus minerals and ash that bias thermal chemistry. In broad terms, high-cellulose materials tend to yield large fractions of volatile, liquid precursors (favoring high bio-oil yields), hemicelluloses decompose at lower temperatures to light oxygenates and gases, and lignin produces phenolic fragments and a greater tendency toward char formation. Agricultural residues rich in alkali metals often catalyze cracking reactions that increase gas and char at the expense of oil quality, while algae and sludges (because of high lipid, protein, or mineral content) give higher proportions of N-containing volatiles, inorganic contaminants, and altered oil composition unless pretreated (e.g. lipid extraction, dewatering, or demineralization) or blended with lignocellulosic feedstocks (Jerzak et al. 2024; Kan et al. 2016).

Superimposed on feedstock effects are the primary operating levers (temperature, heating rate, and residence time), which jointly steer the product slate. Temperature is the dominant control: moderate temperatures (roughly

450°C–500°C in fast pyrolysis regimes) favor maximal liquid production, whereas higher temperatures (≳600°C) accelerate bond scission and secondary cracking, shifting yield toward permanent gases (CO, CO_2, H_2 and light hydrocarbons), producing more condensed, chemically stable char. Heating rate controls the mode of devolatilization. Very fast heating with rapid vapor quench preserves primary vapors and maximizes bio-oil while slow heating encourages progressive devolatilization, secondary polymerization, and higher char yields. Vapor residence time is particularly critical for oil quality. Short vapor residence coupled with rapid quench limits secondary cracking and polymerization and thus preserves liquid yield, whereas long vapor residence promotes gas formation and tar deposition. Finally, intrinsic catalysts in the feedstock (K, Na, Ca, Fe, etc.) can markedly alter outcomes by promoting cracking, reforming, or char formation, so washing, blending, or pretreatment are common process responses (Montoya et al. 2017; Riesco-Avila et al. 2022).

For process design, this means selecting feedstock and pretreatment to match the target product and then choosing reactor and quench strategies to enforce the desired thermal history: high-cellulose, low-ash feeds combined with very fast heating and minimal vapor residence to maximize liquids; slow heating, long solid residence, and lignin/ash-rich feeds to maximize char; and high temperatures with catalytic reforming or prolonged vapor residence when syngas production is the objective. In practice, the optimal solution is a co-optimization of feedstock sourcing, pretreatment, reactor design, and downstream upgrading to meet technical and economic goals while managing the consequences of ash, N, and other contaminants.

7.6 INTEGRATION OF PYROLYTIC COPRODUCTS WITHIN A CIRCULAR AND SUSTAINABLE BIOREFINERY FRAMEWORK

Figure 7.2 provides an integrated scheme that encapsulates the holistic valorization of biomass-derived pyrolytic products within a biorefinery framework. As shown, the solid (biochar), liquid (bio-oil), and gaseous (syngas) fractions are not treated as isolated outputs but as interconnected streams within a circular and sustainable resource management system. Syngas can be recycled to supply process heat for the pyrolysis reactor, thereby reducing external energy demands and improving overall energy efficiency. Biochar, beyond its direct uses as a soil amendment or catalyst support, contributes to long-term carbon sequestration and nutrient retention, closing the loop between energy production and agricultural productivity. Similarly, upgraded bio-oil and extracted platform chemicals provide renewable feedstocks for fuels and materials, displacing fossil-based alternatives and enhancing the economic viability of the system. Together, these pathways exemplify the principles of circularity (where energy, carbon, and material flows are continuously reused and valorized), minimizing waste generation, reducing greenhouse gas emissions, and advancing the transition toward a low-carbon, sustainable bioeconomy.

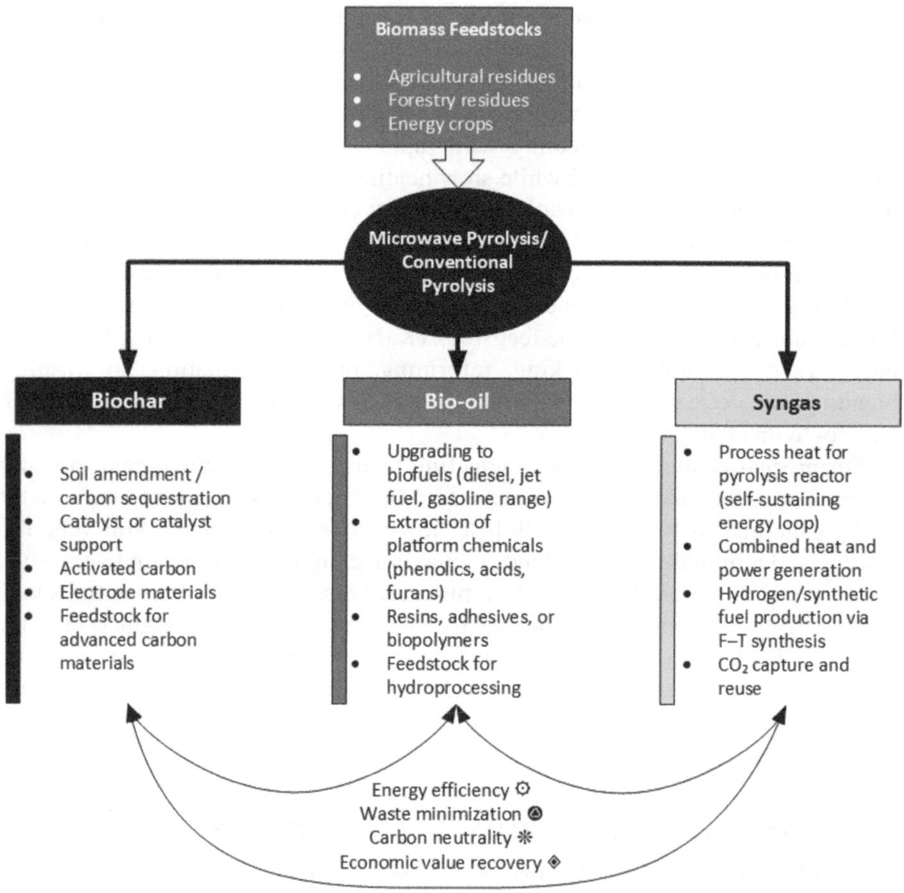

FIGURE 7.2 Integrated schematic of biomass pyrolysis within a biorefinery framework.

7.7 CONCLUSIONS

Pyrolysis converts diverse biomass into syngas, bio-oil, and biochar, with product distributions governed by feedstock composition. Fast pyrolysis favors liquid production, slow pyrolysis yields more char, and higher temperatures shift carbon to permanent gases. Each product stream requires targeted upgrading. For example, syngas demands robust cleaning and catalytic conditioning for the synthesis of fuels or chemicals, bio-oil needs stabilization and deoxygenation (e.g. HDO or catalytic cracking) to meet fuel specifications, and biochar offers agronomic, sorbent, and materials uses when tailored by feedstock and temperature. Integrated biorefineries that couple thermal conversion with catalytic and biological downstream routes, along with appropriate pretreatment and fractionation, enhance value recovery and reduce waste. Research priorities include low-severity upgrading methods, tar-resistant catalysts, scalable demineralization, and rigorous life cycle assessment to verify climate benefits. With coordinated feedstock selection, reactor design, and upgrading pathways, pyrolysis can provide renewable fuels, chemical intermediates, and stable carbon products that support decarbonization and circularity.

REFERENCES

Adhikari, S., H. Nam, & A. Kulkarni. 2018. Biomass gasification and effect of physical properties on products. In *Biomass Preprocessing and Pretreatments for Production of Biofuels. Mechanical, Chemical and Thermal Methods*, ed. J.A. Tumuluru, 25–37. Boca Raton, FL: CRC Press.

Al-Qahtani, A.M.. 2023. A comprehensive review in microwave pyrolysis of biomass, syngas production and utilization. *Energies* 16: 6876.

Altıkat, A., M.H. Alma, A. Altıkat, M.E. Bilgili, & S. Altıkat. 2024. A comprehensive study of biochar yield and quality concerning pyrolysis conditions: A multifaceted approach. *Sustainability* 16: 937.

Aqsha, A., N. Mahinpey, L. Katta, L. Gras, & C. J. Lim. 2014. Synthesis of novel catalysts for hydrodeoxygenation of bio-oil: Guaiacol as a model component. *WIT Trans. Ecol. Environ.* 186: 489–498.

Arbelaez, J.I.M., F.C. Janna, & M. Garcia-Pérez. 2015. Fast pyrolysis of biomass: A review of relevant aspects. Part I: Parametric study. *DYNA* 82: 239–248.

Awad, M.I., Y. Makkawi, & N.M. Hassan. 2024. Yield and energy modeling for biochar and bio-oil using pyrolysis temperature and biomass constituents. *ACS Omega* 9: 18654–18667.

Barszcz, W., M. Łożyńska, & J. Molenda. 2024. Impact of pyrolysis process conditions on the structure of biochar obtained from apple waste. *Sci. Rep.* 14: 10501.

Cesari, L., L. Canabady-Rochelle, & F. Mutelet. 2019. Separation of phenols from lignin pyrolysis oil using ionic liquid. *Sep. Purif. Technol.* 209: 528–534.

Chen, D., K. Cen, X. Zhuang, Z. Gan, J. Zhou, Y. Zhang, & H. Zhang. 2022. Insight into biomass pyrolysis mechanism based on cellulose, hemicellulose, and lignin: Evolution of volatiles and kinetics, elucidation of reaction pathways, and characterization of gas, biochar and bio-oil. *Combust. Flame* 242: 112142.

Chen, X., A. Shafizadeh, H. Shahbeik, M.H. Nadian, M. Golvirdizadeh, W. Peng, S.S. Lam, M. Tabatabaei, & M. Aghbashlo. 2025. Enhanced bio-oil production from biomass catalytic pyrolysis using machine learning. *Renew. Sustain. Energy Rev.* 209: 115099.

D'Arrigo, P., L.A.M. Rossato, A. Strini, & S. Serra. 2024. From waste to value: Recent insights into producing vanillin from lignin. *Molecules* 29: 442.

Díaz, A.C., L. Abdelouahed, N. Brodu, V. Montes-Jiménez, & B. Taouk. 2024. Upgrading of pyrolysis bio-oil by catalytic hydrodeoxygenation, a review focused on catalysts, model molecules, deactivation, and reaction routes. *Molecules* 29: 4325.

Farsi, M. 2023. Ammonia production from syngas: Plant design and simulation. In *Advances in Synthesis Gas: Methods, Technologies and Applications*, ed. M.R. Rahimpour, M.A. Makarem, & M. Meshksar, 381–399. Amsterdam: Elsevier.

Figueirêdo, M.B., I. Hita, P.J. Deuss, R.H. Venderbosch, & H.J. Heeres. 2022. Pyrolytic lignin: A promising biorefinery feedstock for the production of fuels and valuable chemicals. *Green Chem.* 24: 4680–4702.

Ghorbani, M., E. Amirahmadi, R.W. Neugschwandtner, P. Konvalina, M. Kopecký, J. Moudrý, K. Perná, & Y.T. Murindangabo. 2022. The impact of pyrolysis temperature on biochar properties and its effects on soil hydrological properties. *Sustainability* 14: 14722.

Han, X., H. Wang, Y. Zeng, & J. Liu. 2021. Advancing the application of bio-oils by coprocessing with petroleum intermediates: A review. *Energy Convers. Manag. X* 10: 100069.

Han, Y., M. Gholizadeh, C.-C. Tran, S. Kaliaguine, C.-Z. Li, M. Olarte, & M. Garcia-Perez. 2019. Hydrotreatment of pyrolysis bio-oil: A review. *Fuel Process. Technol.* 195: 106140.

He, Q., Q. Wang, & J. Luan. 2025. A review on biomass pyrolysis and pyrolysis mechanisms. *Biofuels*. DOI: 10.1080/17597269.2025.2537515

Hou, S.-S., W.-C. Huang, F.M. Rizal, & T.-H. Lin. 2016. Cofiring of fast pyrolysis bio-oil and heavy fuel oil in a 300-kW_{th} furnace. *Appl. Sci.* 6: 326.

Hrbek, J., C. Pfeifer, P. Blanco-Sanchez, R. Baldwin, M. Swanson, & J. Strege. 2025. *Gas Cleaning from Gasification for Production of Biofuels and Biochemicals*. Paris: IEA Bioenergy.

IEA. 2025. Bioenergy – IEA. https://www.iea.org/energy-system/renewables/bioenergy

Iyiola, A.Ovie, & M.C. Ogwu. 2025. Environmental management using renewable biological resources in the global south. In *Sustainable Bioeconomy Development in the Global South: Volume II Bioeconomy Techniques*, ed. M.C. Ogwu, S.C. Izah, J.G. Vazquez-Arenas, S.T. Feleke, & X. Wei, 299–323. Singapore: Springer.

Jerzak, W., E. Acha, & B. Li. 2024. Comprehensive review of biomass pyrolysis: Conventional and advanced technologies, reactor designs, product compositions and yields, and techno-economic analysis. *Energies* 17: 5082.

Kabeh, K.Z., M. Prussi, & D. Chiaramonti. 2025. Advances in bio-hydrogen production: A critical review of pyrolysis gas reforming. *Appl. Sci.* 15: 3995.

Kan, T., V. Strezov, & T.J. Evans. 2016. Lignocellulosic biomass pyrolysis: A review of product properties and effects of pyrolysis parameters. *Renew. Sustain. Energy Rev.* 57: 1126–1140.

Kashif, M., M.A. Sabri, M. Aresta, A. Dibenedetto, & F. Dumeignil. 2025. Sustainable synergy: Unleashing the potential of biomass in integrated biorefineries. *Sustain. Energy Fuels* 9: 338–400.

Koritár, M., & J. Haydary. 2025. Reduction of tar, sulfur, chlorine and CO_2 in syngas produced by gasification of refuse-derived fuel pellets. *Sci. Rep.* 15: 18446.

Lee, R.-R., I.-J. Jeon, W.-J. Jang, H.-S. Roh, & J.-O. Shim. 2023. Advances in catalysts for water-gas shift reaction using waste-derived synthesis gas. *Catalysts* 13: 710.

Leng, L., H. Li, X. Yuan, W. Zhou, & H. Huang. 2018. Bio-oil upgrading by emulsification/microemulsification: A review. *Energy* 161: 214–232.

Liu, J., X. Chen, W. Chen, M. Xia, Y. Chen, H. Chen, K. Zeng, & H. Yang. 2023. Biomass pyrolysis mechanism for carbon-based high-value products. *Proc. Combust. Inst.* 39: 3157–3181.

Luo, Z., & J. Zhou. 2025. Thermal conversion of biomass. In *Handbook of Climate Change Mitigation and Adaptation*, ed. M. Lackner, B. Sajjadi, & W.Y. Chen, 1233–1289. Cham: Springer.

Mahmoudi, H., H. Jahangiri, O. Doustdar, N. Akbari, J. Wood, A. Tsolakis, M.L. Wyszynski. 2020. Maximizing paraffin to olefin ratio employing simulated nitrogen-rich syngas via Fischer-Tropsch process over Co_3O_4/SiO_2 catalysts. *Fuel Process. Technol.* 208: 106477.

Makepa, D.C., C.H. Chihobo, & D. Musademba. 2025. Technological advancements in biofuel, bioproducts, and bioenergy production from fast pyrolysis of lignocellulosic biomass. In *Biofuels and Bioenergy. Biorefinery and Circular Bioeconomy Approaches*, ed. B. Gurunathan, & G. Venkatesh, 85–119. Cambridge, MA: Woodhead Publishing.

Montoya, J., B. Pecha, D. Roman, F.C. Janna, & M. Garcia-Perez. 2017. Effect of temperature and heating rate on product distribution from the pyrolysis of sugarcane bagasse in a hot plate reactor. *J. Anal. Appl. Pyrolysis* 123: 347–363.

Mutsengerere, S., C.H. Chihobo, D. Musademba, & I. Nhapi. 2019. A review of operating parameters affecting bio-oil yield in microwave pyrolysis of lignocellulosic biomass. *Renew. Sustain. Energy Rev.* 104: 328–336.

Ngaba, M.J.Y., O.M. Yemele, B. Hu, & H. Rennenberg. 2025. Biochar application as a green clean-up method: Bibliometric analysis of current trends and future perspectives. *Biochar* 7: 83.

Oasmaa, A., J. Lehto, Y. Solantausta, & S. Kallio. 2021. Historical review on VTT fast pyrolysis bio-oil production and upgrading. *Energy Fuels* 35: 5683–5695.

Parvari, E., D. Mahajan, & E.L. Hewitt. 2025. A Review of biomass pyrolysis for production of fuels: Chemistry, processing, and technoeconomic analysis. *Biomass* 5: 54.

Premalatha, R.P., J.P. Bindu, E. Nivetha, P. Malarvizhi, K. Manorama, E. Parameswari, & V. Davamani. 2023. A review on biochar's effect on soil properties and crop growth. *Front. Energy Res.* 11: 1092637.

Qu, L., X. Jiang, Z. Zhang, X. Zhang, G. Song, H. Wang, Y. Yuan, & Y. Chang. 2021. A review of hydrodeoxygenation of bio-oil: Model compounds, catalysts, and equipment. *Green Chem.* 23: 9348–9376.

Raza, M., A. Inayat, A. Ahmed, F. Jamil, C. Ghenai, S.R. Nagvi, A. Shanableh, M. Ayoub, A. Waris, & Y.-K. Park. 2021. Progress of the pyrolyzer reactors and advanced technologies for biomass pyrolysis processing. *Sustainability* 13: 11061.

Reza, Md.S., Z.B. Iskakova, S. Afroze, K. Kuterbekov, A. Kabyshev, K.Z. Bekmyrza, M.M. Kubenova, M.S.A. Bakar, M.S.A. Azad, H. Roy, & Md.S. Islam. 2023. Influence of catalyst on the yield and quality of bio-oil for the catalytic pyrolysis of biomass: A comprehensive review. *Energies* 16: 5547.

Riesco-Avila, J.M., J.R. Vera-Rozo, D.A. Rodríguez-Valderrama, D.M. Pardo-Cely, & B. Ramón-Valencia. 2022. Effects of heating rate and temperature on the yield of thermal pyrolysis of a random waste plastic mixture. *Sustainability* 14: 9026.

Robazza, A., F.C.F. Baleeiro, S. Kleinsteuber, & A. Neumann. 2024. Two-stage conversion of syngas and pyrolysis aqueous condensate into L-malate. *Biotechnol. Biofuels Bioprod.* 17: 85.

Rodrigues, L., A. Budai, L. Elsgaard, B. Hardy, S.G. Keel, C. Mondini, C. Plaza, & J. Leifeld. 2023. The importance of biochar quality and pyrolysis yield for soil carbon sequestration in practice. *Eur. J. Soil Sci.* 74 (4): e13396.

Rowland, S.M., R. Martinez, C.J. Wrasman, K. Iisa, M.R. Nimlos, & M.B. Griffin. 2024. Thermal reactivity of bio-oil produced from catalytic fast pyrolysis of biomass. *Energy Fuels* 38: 19626–19638.

Safarian, S. 2023. Performance analysis of sustainable technologies for biochar production: A comprehensive review. *Energy Rep.* 9: 4574–4593.

Seo, D.-J., J.B. Lee, Y.-J. Kim, H.-R. Cho, S.-Y. Kim, G.-E. Kim, Y.-D. Park, G.-H. Kim, J.-C. An, K. Oh, & J.-I. Park. 2025. The characteristics of hydrodeoxygenation of biomass pyrolysis oil over alumina-supported Ni–Mo catalysts. *Catalysts* 15: 6.

Sewsynker-Sukai, Y., A.N. David, D.C.S. Rorke, I.A. Sanusi, G.S. Aruwajoye, E.L. Meyer, & E.B.G. Kana. 2023. Integrated biorefineries: The path forward. In *Applied Biotechnology Reviews. Advances in Lignocellulosic Biofuel Production Systems*, ed. P. Moodley, R.C. Ray, & E.B.G. Kana, 267–304. Oxford: Woodhead Publishing.

Talwar, P., M.A. Agudelo, & S. Nanda. 2025. Pyrolysis process, reactors, products, and applications: A review. *Energies* 18: 2979.

Valle, B., A. Remiro, N. García-Gómez, A.G. Gayubo, & J. Bilbao. 2019. Recent research progress on bio-oil conversion into bio-fuels and raw chemicals: A review. *J. Chem. Technol. Biotechnol.* 94: 670–689.

van Dyk, S., & J. Saddler. 2024. *Progress in Commercialization of Biojet/Sustainable Aviation Fuels (SAF): Technologies and Policies, IEA Bioenergy Task 39*. Paris: IEA Bioenergy.

Yaashikaa, P.R., P.S. Kumar, S. Varjani, & A. Saravanan. 2020. A critical review on the biochar production techniques, characterization, stability and applications for circular bioeconomy. *Biotechnol. Rep.* 28: e00570.

Zhang, Y., Y. Liang, S. Li, Y. Yuan, D. Zhang, Y. Wu, H. Xie, K. Brindhadevi, A. Pugazhendhi, & C. Xia. 2023. A review of biomass pyrolysis gas: Forming mechanisms, influencing parameters, and product application upgrades. *Fuel* 347: 128461.

8 Advancements in Liquid Biofuels from Biomass Biorefining

Ifeanyichukwu Edeh and Ifreke Akpan

8.1 INTRODUCTION

The continuous quest for alternative and sustainable sources of energy keeps on pushing research boundaries for new energy sources that would not only replace conventional energy sources (fossil fuels) but also mitigate carbon footprint (Sharma et al. 2025; Sundaram & Raj Muniyandi 2025). Over the years, different biomass, such as algae, cereals, oil plants, have been tested to check their potential in generating energy (biofuel) suitable to be used as an alternative to fossil fuel (Malode et al. 2021). The use of this biomass for energy production is accompanied by enormous challenges, especially in terms of optimal production technique(s), which limit its ability to compete with conventional fuel. Nevertheless, researchers are developing novel strategies to enhance production, reduce production costs, and meet regulatory specifications.

Some of the areas considered for biofuel production advancement include feedstock, pretreatment method, fermentation techniques, production techniques, recovery method, and the catalyst used to aid in the production. Several methods have been developed over the years to maximize biofuel production. Thus, this chapter focuses on the advancement in biofuel production with a major concern on biodiesel, biobutanol, and bioethanol.

8.2 BIODIESEL

Biodiesel is a liquid fuel synthesized from both plant and animal lipids. In addition to being a sustainable energy source, it has the capability of replacing conventional fuel. Biodiesel is a renewable fuel that is ecologically benign, efficient, and a replacement fuel. Thus, biodiesel has the potential to lower the levels of environmental pollution caused by the use of conventional diesel (Malode et al. 2021).

The main raw material for biodiesel production is triglycerides (lipid) derived from various sources of biomass such as palm oil, castor oil, olive oil, and animal fat (Elgharbawy et al. 2025). Biodiesel has several desirable properties, which include non-toxicity, non-corrosive, and environmental friendliness. It has a high

DOI: 10.1201/9781003650751-8

TABLE 8.1
Specification for Biodiesel Based on ASTM D6751 and EN14214 (B100) Standards

Property	ASTM D6751	EN 14214
Density at 15°C (kg/m³)	880	860–900
Viscosity at 40°C (mm²/s)	1.9–6.0	3.5–5
Cetane number	Min. 47	Min 51.0
Acid value (mg KOH/g)	≤0.50	≤0.5
Pour point (°C)	−15 to −16	—
Flash point (°C)	≥130	≥101
Cloud point (°C)	−3 to −12	—
Water content (mg/kg)	≤500	≤500
Iodine number (g I2/100 g)	—	≤120

Source: Ramos et al. (2019).

flash point, efficient combustion, low sulfur content, high cetane number, etc. Biodiesel is non-explosive and non-flammable, making it safer to store, handle, and transport than fossil diesel. For example, it has a flash point of 423°C, which is higher than that of the conventional diesel 337°C (Ling et al. 2014). Table 8.1 presents the standard specifications of biodiesel as stipulated by ASTM D6751 and EN14214 for B100.

8.2.1 Feedstocks for Biodiesel Production

Because the cost of the raw materials accounts for up to 80% of the overall cost of producing biodiesel, these materials are essential. The ease of getting a high-quality feedstock in large quantities and the comparative manufacturing costs are mostly considered in selecting a feedstock to use. There are basically two primary categories of feedstocks for biodiesel production, namely, edible oil and non-edible oil. Previously, edible oils accounted for more than 95% of the feedstock for the production of biodiesel. This has caused food crises in the world. The current trend in raw materials is geared toward the use of non-edible oil as a feedstock for the production of biodiesel. Edible oils include, but are not limited to, soybean oil, peanut oil, sunflower oil, palm oil, coconut oil, and non-edible oils include rapeseed oil, Jatropha oil, Jojoba oil, and waste-cooking oil (Banković-Ilić et al. 2012).

8.2.1.1 Algae Oil

Algal biomass is a promising source of oil for biodiesel production. It consists of about 20%–80% oil content, depending on the species. This oil is essential for the production of kerosene or biodiesel. Some identified species that are rich in oils are

Euglena, *Tribonema*, etc. The oil composition differs from one species to another. Some are very rich in saturated fatty acids, while some have low fatty acid content (Usha & Nwaigwe 2016).

8.2.1.2 Municipal Sewage Sludge

This is another promising feedstock for biodiesel production obtained from a wastewater treatment plant. It contains several amounts of organic and inorganic substances, including lipids, proteins, lignin, synthetic detergents, soap, carbohydrates, as well as various synthetic chemicals from industries. The disposal of this sludge to the environment causes pollution. Different strategies have been adopted for sewage sludge management, and one of which is 'reuse'. The lipid content in the range of 10%–25% makes sewage sludge an ideal feedstock for biodiesel. The major challenge of using this sludge as a feedstock is in lipid extraction due to the large water content (Selvam et al. 2025).

8.2.2 Pretreatment Methods for Oils Used in Biodiesel Production

Oils used for the production of biodiesel often contain contaminants, and as such, the percentage yield of the biodiesel is reduced during synthesis. Therefore, before employing it in biodiesel synthesis, pretreatment is necessary to reduce the level of contaminants. Some pretreatment steps for some of the oils used as biodiesel feedstock are discussed in this section.

8.2.2.1 Free Fatty Acid Reduction Using Acid

It is imperative to reduce the free fatty acids of the oil whenever the concentration is high to prevent side reactions like saponification. Commonly used acids include H_2SO_4, H_2PO_3, and HNO_3. The pretreatment reaction is called esterification. The use of H_2SO_4 was studied by Hayyan et al. (2011) for reducing free fatty acids in palm oil sludge. The experiment used a temperature range of 40°C–80°C and a methanol to oil ratio ranging from 6:1 to 14:1, for a duration of 33–120 min. Results demonstrated that under optimal conditions of 8:1 methanol to oil ratio, temperature of 60°C, and treatment time of 60 min, a concentration of 0.75% H_2SO_4 decreased the free fatty-acid concentration from 24% to 2%.

8.2.2.2 Free Fatty Acid Reduction Using Glycerolysis

Due to corrosion of equipment and excessive alcohol consumption caused by the use of acids, it is critical to develop alternative techniques of free fatty acid reduction. Glycerolysis is one of these ways. The process of glycerolysis converts free fatty acids to glycerides, which then react with alcohol to produce monoglycerides, diglycerides, and triglycerides (Mamtani et al. 2021).

8.2.3 Biodiesel Production Techniques

There are several methods used for the production of biodiesel, depending on the conditions of the raw materials, available resources, and technology. Most commonly used methods include microemulsion, transesterification, and pyrolysis.

8.2.3.1 Microemulsion

Microemulsion is an equilibrium state process, involving the distribution of an optically isotropic microstructure with a dimension between 1 and 50 nm, which forms as a result of combining two immiscible liquids and one or more active substance(s) in the right proportion. Once the microdroplets are evenly distributed, the system becomes thermodynamically stable with low viscosity. In biodiesel production, the components include vegetable oil, alcohol, water, and a catalyst. The alcohol is mostly methanol or ethanol and acts as a surfactant/emulsifier to allow miscibility of the other components. Biodiesel is produced by blending the different components in the right proportion. The major problem associated with biodiesel produced via this method includes incomplete combustion and accumulation of carbon in the engine, which may lead to nozzle failure (Elgharbawy et al. 2025).

8.2.3.2 Pyrolysis

It involves heating the complex structural hydrocarbons in an oxygen-free environment to break the polymer and generate simpler molecules for biodiesel production. In the production, the feedstock (oil or fat) is first converted to bio-oil, and the bio-oil is then upgraded via hydrotreating or hydrocracking to obtain biodiesel. In hydrotreating, hydrogen reacts with oil to remove traces of sulfur and oxygen, while in hydrocracking, hydrogen is added to the bio-oil to create smaller chain hydrocarbons to meet biodiesel specifications (Brahma et al. 2022). Figure 8.1 shows the thermal cracking route for biodiesel production.

This method requires minimal application of technology for the production process. Catalysts like alumina or zeolite are used for this method. The pyrolysis for biodiesel is consistently carried out at temperatures between 250°C and 350°C. Biodiesel is produced by heating oil in a reactor, which then condenses the resulting

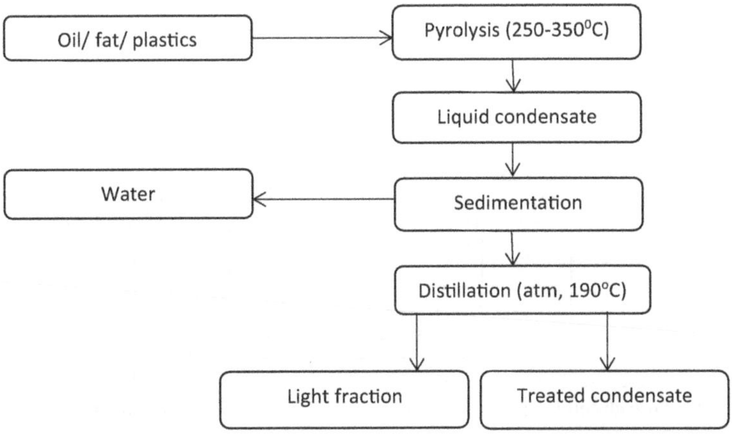

FIGURE 8.1 Schematic diagram for thermal cracking route for biodiesel production.

vapor. This method is useful for lowering the density and viscosity of the biodiesel. These two factors influence biodiesel in internal combustion in diesel engines (Elgharbawy et al. 2025).

8.2.3.3 Transesterification Reaction

In transesterification or alcoholysis, glycerol and fatty acid alkyl esters are produced when triglycerides of fatty acids combine with alcohol in a nucleophilic way. It is a three-step reaction. Firstly, the triglyceride is reduced to diglyceride. This is followed by the reduction of diglyceride to monoglyceride. Lastly, the monoglyceride is converted to glycerol. Three ester molecules are produced in the overall process since an ester is created at each stage. A catalyst, which may be a base or an acid, an enzyme, a bifunctional catalyst, or any combination thereof, is used to speed up the process. The amount of free fatty acids in the oil determines the choice of the catalyst. The process involves mixing alcohol and catalyst and stirring while heating. Thereafter, the mixture is poured into the oil in the reacting vessel, where the reaction occurs. The products obtained are subjected to separation for biodiesel recovery. Figure 8.2 shows a schematic diagram of different stages involved in the production of biodiesel via transesterification (Brahma et al. 2022).

During biodiesel production via transesterification, other side reactions take place depending on the composition of the fatty acid or the type of catalyst. If the triglyceride of high free fatty acid content is used with a base catalyst, the saponification

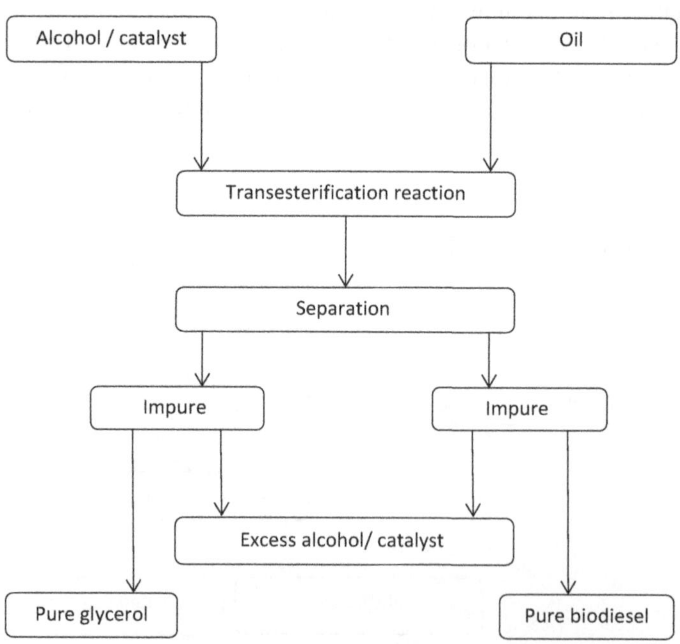

FIGURE 8.2 A schematic diagram of the transesterification reaction.

reaction would occur, or when an acid catalyst is used, the esterification reaction would take place before transesterification (Elgharbawy et al. 2025).

8.2.4 Technological Advancements in Biodiesel Production

The traditional transesterification reaction has some limitations, such as high cost of production, low efficiency, etc. To tackle these challenges, some advanced methods have been developed, including supercritical fluid transesterification, microwave-assisted transesterification, and ultrasonic-assisted transesterification (Elgharbawy et al. 2025).

8.2.4.1 Ultrasonic-Assisted Transesterification

In this method, ultrasonic waves are used to cause cavitation formation and the collapsing of the microbubbles within the reacting medium. This cavitation increases the rate of mixing of the reactants, causing mass transfer, which subsequently increases the rate of reaction. It is a very effective method, when high-viscosity feedstock is used (Elgharbawy et al. 2025).

8.2.4.2 Microwave-Assisted Transesterification

In this method, microwave radiation is used to maintain uniform and rapid heating of the reacting vessel. This reduces energy consumption compared to the traditional transesterification reaction. The overall yield of the reaction increases, with less cost of production. It can easily be integrated into existing reactors with promising scale-up potential (Elgharbawy et al. 2025).

8.2.4.3 Supercritical Fluid Transesterification

In supercritical fluid transesterification, alcohol is used at its supercritical state (temperature of 200°C–400°C and pressure of about 200 bar). At this state, it exhibits both liquid and gaseous behavior, making it hyperreactive. This method does not require the use of a catalyst. The use of this method has several advantages, which include fast reaction time, no separation of catalyst, high yield, and the capacity to produce biodiesel from feedstock with high free fatty acid (Rathnam et al. 2020).

8.2.5 Reagents Used for Biodiesel Production

8.2.5.1 Alcohol

One of the key reactants in biodiesel production is alcohol. Based on the number of functional groups connected to the main chain, alcohols are classified into three basic categories: primary, secondary, and tertiary. Primary alcohols like methanol and ethanol are the most commonly used in biodiesel production, while other alcohols have also been studied, such as branched-chain alcohols, and other lower-chain alcohols, such as propanol, butanol, and isopropanol (Brahma et al. 2022).

8.2.5.2 Triglycerides

Triglycerides used in the production of biodiesel are obtained from different sources, such as edible and non-edible feedstocks. Examples of primary sources include olive, palm fruit, babassu, linseed, lard, tallow from beef, grease, etc. These oils contain fatty acids, either the same chain length or a different chain length. Examples of triglycerides include palmitic acid, oleic acid, linolenic acid, and stearic acid. Proper choice of the type of oil to be used for biodiesel production is very important as it influences the energy content of the biodiesel and the economic viability (Brahma et al. 2022).

8.2.6 Factors Affecting Biodiesel Production via Transesterification

8.2.6.1 Alcohol-to-Oil Ratio

In biodiesel production, alcohol to oil ratio is essential as it determines the yield. For example, the alcohol-to-oil ratio is set at 6:1, when using a base catalyst. Because transesterification is a reversible process, a significant quantity of methanol is required to maintain product-side balance. The equilibrium shift toward the product would be enhanced by increasing the methanol-to-oil ratio. Reaction rate and percentage yield are unaffected by increasing the ratio above a certain critical threshold (Yaqoob et al. 2020). When working with oil feedstock that has a high free fatty acid content, it is necessary to use more alcohol than catalyst when using an alkali catalyst for neutralization. However, using too much alcohol can raise production costs due to purification and alcohol removal.

8.2.6.2 Temperature

Among other factors influencing biodiesel production via transesterification reaction is reaction temperature. Depending on the oil properties and catalyst type, the transesterification reaction often occurs at a range of temperatures. It may happen at room temperature or close to the alcohol's boiling point, depending on the oil source. As the temperature rises, the oil's viscosity decreases, which leads to an increase in the reaction rate and a decrease in reaction time. Therefore, a higher reaction temperature would facilitate oil-alcohol miscibility. However, the temperature must be carefully controlled so as not to exceed the reaction's maximum; otherwise, parallel reactions would occur (Nabaterega et al. 2026).

8.2.6.3 Reaction Time

During the transesterification reaction, the percentage yield varies with reaction time. At the beginning, the reaction rate is relatively slow because of less solvent dispersion and incomplete mixing of the reactants. During this stage, the reaction occurs at the outer surface of the oil and triglyceride, but with time, agitation increases, causing more reactant interaction, which brings about the subsequent increase in yield (Nabaterega et al. 2026). But beyond a certain threshold time, reaction starts to decrease, and reverse reaction may occur.

8.2.6.4 Type and Amount of Catalyst

The catalyst used in the transesterification reaction has a significant impact on biodiesel production. Using a large quantity of catalyst increases the available active sites, where the reaction will occur; thus, the rate of reaction and yield are both increased. The produced biodiesel may have a high viscosity, if a large amount of catalyst is used. Therefore, to avoid the high cost of purification, it is essential to maintain a controlled amount of catalyst. Generally, the biodiesel yield is increased, when the catalytic loading is increased, up to a point where it begins to decrease (Nabaterega et al. 2026).

8.2.6.5 Water and Free Fatty Acid Content

When choosing the reaction pathway to use during transesterification, the amount of water and free fatty acids in the reactants must be considered. In order for the reaction to take place, the feedstock must adhere to certain limits for both water and free fatty content. To prevent the saponification reaction, for instance, base-catalysts may only be employed in oils with a free fatty content of less than 1%. Similarly, oils that contain a lot of water may result in ester production when an acid catalyst is used, but water or free fatty acid removal is not necessary for the transesterification reaction at supercritical conditions.

8.2.6.6 Degree of Agitation

Oil and alcohol do not completely mix with one another, when they are combined. Thus, the reaction takes place at the boundary between the two liquids. The transesterification reaction is therefore slow because of this scenario. The process is accelerated by allowing the oil and solvent to contact, which is made possible through agitation. According to Tabatabaei et al. (2019), using a moderate agitation speed of 400 rpm yields the best result. It has been shown that reaction rates are reduced with slow agitation speeds, but soap formation takes place with high agitation speeds due to irreversibility.

8.2.7 Reactors Used for Biodiesel Production

Reactors are equipment that provides enabling conditions for chemical reactions to take place. In a reactor, reactants are transformed into products after being subjected to the required conditions. When designing a reactor, many factors are considered. These include, but are not limited to, the nature of the reactant(s), method by which the reactant(s) are to be charged in, and the product(s) removed, materials of construction, type of stirring system, hydrodynamic, intake and discharge of heat, and size of the reactor. This is to ensure safety, productivity, and the economy of the process.

In biodiesel production, the reactor design used is determined by whether the production process is considered to be batch or continuous. Some reactors used for biodiesel production include batch reactors, fixed bed reactors, continuous stirrer tank reactors, bubble column reactors, hybrid catalytic plasma reactors, membrane reactors, microchannel reactors, and reactive distillation (Brahma et al. 2022).

8.2.8 Techno-Economic Analysis of Biodiesel

Techno-economic analysis is an effective tool used in assessing the economic feasibility of different processes relating to biodiesel production. A comprehensive techno-economic analysis encompasses the cost of different feedstock, the intended production method to be adopted, and the scalability of such a process for industrial purposes. The technical aspect of the analysis involves the development of the process flow diagram, comprehensive material, and energy balance calculations using simulation software, such as Aspen Plus, while the economic aspect involves the development of the capital and project cost estimation, discounted cash flow, and determination of the minimum biodiesel selling price.

In biodiesel production, the major costs are those of feedstock and technology. The choice of feedstock and technology impacts the entire process cost. Using non-edible feedstock with a base-catalyst in the transesterification reaction is more effective in cost reduction. Advanced technologies like microwave, ultrasonic, and superfluid transesterification, though producing high yield, are capital-intensive (Akpan et al. 2023).

8.2.9 Life Cycle Analysis of Biodiesel

Life cycle analysis (LCA) is an important tool for assessing the environmental impact of products and processes involved in the production process throughout the entire life of the product. The LCA tool helps in the identification of potential impacts during a process design and for decision-making to improve the process before scaling up for commercialization. Biodiesel LCA involves the evaluation of material and energy input, output, and the subsequent environmental impact of cultivating the feedstock to the emission of gases in an internal combustion engine. Biodiesel LCA involves four stages: feedstock production, feedstock transportation and processing, production of biodiesel, and utilization of biodiesel in internal combustion engines. The analysis must also take into consideration input and output at each stage of the process. Biodiesel LCA indicates that it has the capacity to reduce greenhouse gas emissions and global warming substantially and hence help in the protection of the ozone layer from depletion due to low carbon footprint (Nogales-Delgado 2025).

8.3 BIOETHANOL

8.3.1 Bioethanol Production

Bioethanol is a liquid biofuel obtained from biomass (plantain plant waste, sugar cane bagasse, forestry residues, paper, and pulp waste, etc.) through different processes, which include pretreatment, hydrolysis, fermentation, and purification. These feedstocks are classified under different generations, which include first-generation (edible feedstock), second-generation (lignocellulose), third-generation (algae), and fourth-generation (modified microbes like engineered *Escherichia coli* strains) (Kamusoko et al. 2021). Many countries around the globe are currently producing and using bioethanol for different purposes, with the US leading in 2021 with 1,435.8 PJ of production (OECD 2022). Other countries like Brazil, Canada, the EU, and China are also major producers of bioethanol. In 2023, world production capacity stood at 106.76 billion liters, and it is estimated to reach 137.27 billion in 2028 (Research & Markets 2023).

Bioethanol is a potential alternative to conventional gasoline and can be used either directly or blended with conventional gasoline, which helps in the reduction of greenhouse gas emissions and overdependence on conventional gasoline. The direct form is classified as (E100), while the blended forms are E85 and E10. If E100 is to be used in an existing internal combustion engine, the timing system has to be modified. However, the use of E100 is challenging, with difficulty in starting the engine at a low temperature (poor cold-starting property due to the high heat of vaporization of bioethanol). The blended forms (E85 and E10) do not require modification of the engine. Rather, they help in improving the ignition and engine performance. Advantages of bioethanol include a high-octane rating resulting in increased engine efficiency and performance, broad flammability, higher compression ratio, and comparable energy content (Carrillo-Nieves et al. 2019). The major disadvantage of bioethanol is the high cost of production, which results from the use of some feedstock, enzymes, detoxification, and ethanol recovery. Nevertheless, efforts have been made to produce bioethanol from low-cost lignocellulosic feedstock through the use of modified processes.

8.3.2 Bioethanol Production from Lignocellulose

Lignocellulose biomass is a second-generation feedstock for bioethanol production. It is non-edible and not expensive and is readily available around the globe. The lignocellulose energy consumption globally stands at about 40% and the annual output of about 10–50 billion dry tons. Its common sources include barley straw, cotton stalk, banana, plantain plant waste, sugar cane bagasse, forestry residues, paper and pulp waste, etc., (Khaliq et al. 2020). Figure 8.3 illustrates the structural components of lignocellulose biomass. Lignocellulose biomass comprises 25%–30% hemicellulose, 40%–50% cellulose, 15%–20% lignin, and some traces of pectin, nitrogenous compounds, and inorganic compounds (Mori et al. 2015).

Cellulose is a linear syndiotactic polymer made up of glucose linked together by β-1,4-glycosidic bonds. Its peculiar polymer chain arrangement makes it a stable compound. Lignin is a 3-D heterogeneous polycrystalline mesh polymer that belongs to polyphenol compounds. It inhibits hydrolysis due to its complex hydrophobic polymer structure. Hemicellulose is a mixture that comprises polysaccharides with straight and branched chains (Kamusoko et al. 2022). The polysaccharide in hemicellulose has a low degree of polymerization, which makes it easy to degrade to monosaccharides such as dextrose, fructose, xylose, mannose, galactose, etc.

The lignocellulose structure poses challenges in its development for utilization, especially for bioethanol production, due to the presence of lignin, which surrounds the entire structure of the cellulose and hemicellulose. Thus, this necessitates proper pretreatment of the biomass to make it susceptible to microorganisms, enzymes, etc., to act on it. The following steps are involved in the conversion of the biomass to bioethanol: pretreatment, saccharization/hydrolysis, fermentation, and separation/purification (Kamusoko et al. 2021).

8.3.3 Pretreatment

Pretreatment of lignocellulose is the most expensive stage in bioethanol production. It accounts for about US$ 0.3/gallon of ethanol produced (Edeh 2020). Various

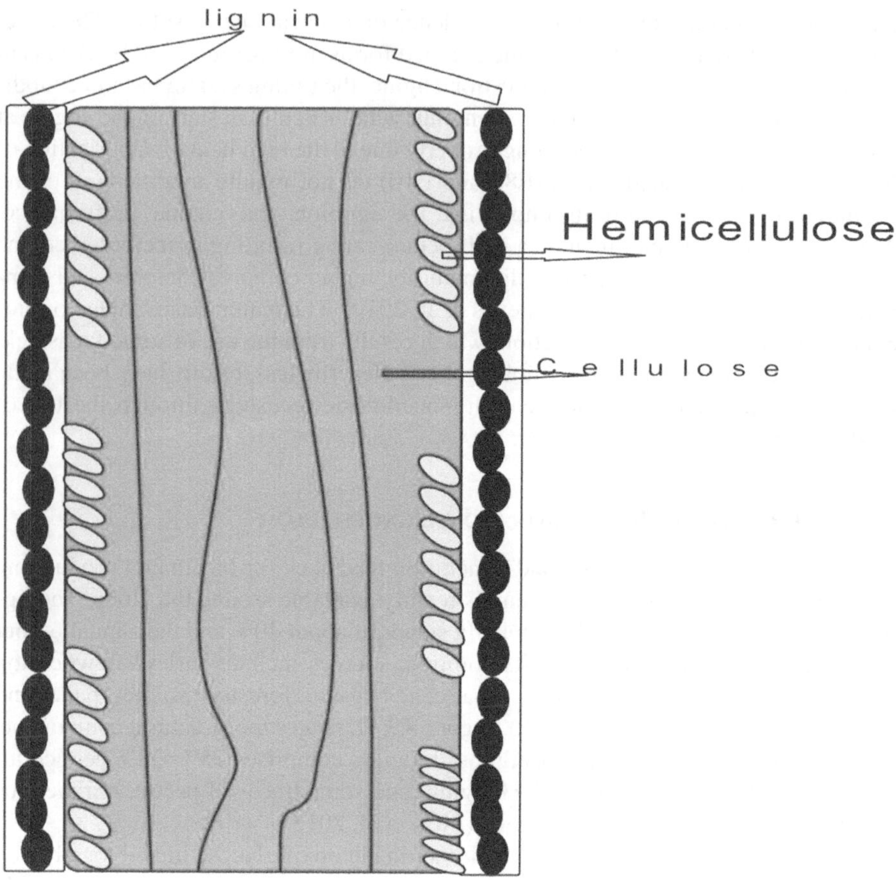

FIGURE 8.3 Lignocellulose biomass and its structural components.

pretreatment methods have been developed in the last 80 years for efficient conversion of lignocellulose to bioethanol in view to minimize the cost and mitigating environmental impact. A good pretreatment method maximizes the efficiency of the hydrolysis process, which will in turn increase the yield of monomeric sugar, minimize carbohydrate degradation, and reduce inhibitor formation in the byproduct (Procentese et al. 2017). Pretreatment methods include biological, chemical, physicochemical, physical, ionic liquid, and deep eutectic solvent pretreatments. It is an essential step because it helps to unwind the complex polymer matrix, which consists of lignin, hemicellulose, and cellulose. This exposes and increases the accessibility of enzymes to work on the cellulose during the hydrolysis stage.

8.3.3.1 Physical Pretreatment

Physical pretreatment involves the reduction of the size of the biomass after dehydration. Methods comprises extrusion, irradiation via microwave, and size reduction.

8.3.3.1.1 Size Reduction

There are different methods of size reduction, including milling, grinding, coarse size reduction, chipping, and shredding. These are used to improve the digestibility of the lignocellulose during hydrolysis, which decreases the degree of polymerization and crystallization, while increasing the surface area of the biomass for a faster rate of hydrolysis.

8.3.3.1.2 Extrusion

This method is designed to expose the biomass active sites for accessibility of the enzyme during hydrolysis. It involves using operating conditions like high shear, rapid mixing, less resident time, no furfural, no conditioning, and moderate barrel temperature. It is very important because it can be applied on an industrial scale and it is environmentally friendly.

8.3.3.1.3 Microwave (Irradiation)

This pretreatment makes use of an electromagnetic field to disrupt the lignocellulose polar bond by vibrating the structures until it is energized. As a result of the vibration, the complex polymer structures are broken down for easy attack by the enzyme. This pretreatment method is essential because of low energy demand, less inhibitor formation, ease of operation, high heating value, and degradation of the structural organization of cellulose fraction.

8.3.3.2 Chemical Pretreatment

Chemical pretreatments are used to disrupt and remove lignin or hemicellulose from the polymer mixture. It includes the use of acid, alkali, organic acid, and oxidative delignification. The use of this method is highly selective, depending on the type of biomass processing. It is very effective, though it involves harsh operating conditions, which may affect the end-product, and it is not environmentally friendly.

8.3.3.2.1 Acid Pretreatment

There are two types of acid pretreatment: dilute acid and concentrated acid. The use of dilute acid leads to the formation of inhibiting byproducts, though very useful in bioethanol production on an industrial scale. Depending on the end-product needed, two types of operating conditions are used for dilute acid pretreatment: temperature less than 120°C for a long duration between 30 and 90 min and high temperature more than 180°C for a short duration between 1 and 5 min, respectively. Before further treatment of the biomass, the inhibitors produced by the acid treatment must be removed to make the process economically viable. Some of the acids that are used include H_3PO_4, HCl, HNO_3, dilute H_2SO_4, lactic acid, acetic acid, maleic acid, and peracetic acid. One major disadvantage of this method is corrosion of the reactor by acid (Lee & Jeffries 2011).

8.3.3.2.2 Alkaline Pretreatment

Unlike acid pretreatment, alkali pretreatment operates at ambient temperature and pressure to disrupt the lignin complex structure into fractions (Refaat 2012). Some alkali used for pretreatment includes hydroxyl derivatives of sodium, calcium, ammonium, and potassium salts. The solubility of hemicellulose and cellulose in alkali solution is less than that of acid solution. Comparatively, alkali pretreatment operates in less harsh conditions than acid, but when a longer operating time is required, a higher temperature is recommended.

8.3.3.2.3 Organosolv Pretreatment

The use of organic solvents proved to be effective in lignin deconstruction for enzyme accessibility during hydrolysis. Examples of organic solvents used are acetone, ethanol, methanol, glycol, ethylene, etc. The application of an organic solvent requires a catalyst for the process to be effective. Examples of such catalysts include lime, H_2SO_4, NaOH, ammonia, etc. Aside from lignin, which is a valuable byproduct, syrup of hemicellulose C_5 and C_6 and cellulose fraction are also produced (Agbor et al. 2011). Due to the low boiling points of organic solvents, the process requires high pressure, which is a major drawback for the use of organic solvents. However, bioethanol produced from this process is of high quality.

8.3.3.3 Biological Pretreatment

This is the application of microorganisms to degrade lignin in biomass. These microbes are capable of disrupting the cell wall of the biomass and promote accessibility of the enzymes during hydrolysis. Some microbes used for this process include white rot, brown rot, soft rot fungi, and bacteria. Biological pretreatment does not produce unwanted products like chemical and physical pretreatment. It does not require the application of acid or base or harsh reaction conditions like high temperature or pressure. Thus, it requires ambient conditions. Some lignin-degrading enzymes present in microbes include laccases, polyphenol oxidases, lignin peroxidase, etc. The main disadvantage of this treatment method is a very low process rate, which results in a long residence time of the reaction.

8.3.3.4 Physico-Chemical Method

This is the combination of both chemical and physical pretreatment methods to deconstruct the complex lignocellulose structure. It includes methods like steam explosion, CO_2 explosion, liquid hot water, ammonia fiber explosion, and wet oxidation.

8.3.3.4.1 Wet Oxidation

Pretreatment using the wet oxidation requires subjecting the biomass to high temperatures and pressures in the ranges of 170°C–200°C and 10–12 bar O_2, respectively. The process occurs for 10–15 min (Merklein et al. 2016). This method is very suitable for biomass with high lignin residue. Wet oxidation degrades the lignocellulose and produces fewer inhibitors, disrupts the lignin matrix, and eliminates lower-grade cellulose from the polymer matrix. This provides an enabling environment for the enzyme to work on cellulose during hydrolysis. Nevertheless, the common drawback of this method is the high energy requirement, which increases the overall production cost. The overall efficiency of this method depends on using appropriate temperature, oxygen pressure, and suitable residence time for the reaction (Merklein et al. 2016). In wet oxidation, chemical agents like Na_2CO_3 and alkaline H_2O_2 help to promote the rate of reaction by decreasing reaction temperature and improving hemicellulose degradation.

8.3.3.4.2 CO_2 Explosion

In this method, CO_2 is subjected to supercritical conditions, which causes the gas to act as a solvent to disrupt the lignocellulose polymer matrix. The gas at supercritical conditions is made to pass through the vessel containing the biomass. The vessel is maintained at high temperature for several minutes. At this high temperature, the gas enters the biomass to form carbonic acid, which dissolves hemicellulose. High pressure of the gas disrupts the biomass structure, which promotes enzyme accessibility (Hendriks & Zeeman 2009). The use of CO_2 explosion depends on the type of biomass treated. It can only function well with biomass that has high moisture content. This method is very effective because it produces fewer inhibitors and eliminates lignin in non-harsh conditions (non-acidic and without corrosion of the vessel).

8.3.3.4.3 Ammonia Pretreatment

Ammonia pretreatment includes ammonia fire explosion, soaking aqueous ammonia, and ammonia recycled percolation. This method uses ammonia to pretreat lignocellulose. In ammonia fire explosion, lignocellulose is heated with liquid ammonia at a temperature range of 60°C–100°C and pressure between 250 and 300 psi for a few minutes. The treatment causes dilation and phase change of the biomass cellulose crystalline, which promotes reactivity of leftover carbohydrate after the pretreatment. Some operating parameters required for ammonia fiber explosion include appropriate ammonia loading, good water loading, and temperature. For the soaking aqueous ammonia method, the biomass is soaked in ammonia in a batch reactor at 30°C–60°C for some minutes, depending on the nature of the biomass (Agbor et al. 2011).

In ammonia recycle percolation, liquid ammonia is percolated through the biomass to disrupt the complex polymer matrix. Compared to other pretreatment methods, ammonia recycle percolation produces no inhibitor. The general drawback of this method is high energy consumption.

8.3.3.4.4 Steam Explosion

In this method, biomass is exposed to high-temperature steam between 160°C and 260°C under high pressure (Agbor et al. 2011). The degradation of hemicellulose into xylose and glucose monomer is caused by acetic acid produced from hemicellulose acetyl groups during the treatment. This pretreatment is sometimes called auto-hydrolysis.

Catalysts like H_2SO_4, CO_2, or SO_2 are used to improve the efficiency of this process. Steam explosion has been found to minimize the formation of inhibitors, improve the recovery of hemicellulose sugar, and facilitate efficient hydrolysis. The major drawback of this treatment is the formation of fermentation inhibitors and compounds like lignin-carbohydrate matrix and the need to wash the hydrolysate (Agbor et al. 2011).

8.3.3.4.5 Liquid Hot Water

This method uses liquid water instead of steam at a very high pressure of 5 MPa and a temperature varying between 170°C and 230°C. It is also called hydro-thermolysis.

It removes lignin for easy accessibility of the enzyme into hemicellulose and cellulose. It is very effective in treating biomass like softwood (Rabemanolontsoa & Saka 2016). For the effectiveness of this method, a pH of 4–7 is required to avoid sugar degradation and the formation of inhibitors. The general low cost of this treatment makes it attractive.

8.3.3.5 Advanced Pretreatment Methods

With the evolution of the concept of green engineering, which is developed to mitigate negative impacts caused by the use of conventional chemicals on the environment, more robust and environmentally friendly methods are developed. These include ionic liquid (IL) and deep eutectic solvents (DES)-based pretreatment methods.

8.3.3.5.1 Ionic Liquid

IL is considered to be environmentally benign, due to its high thermal and chemical stability, ability to exist as a liquid at relatively low temperatures and vapor pressure, non-flammable, etc. It is made up of a significant quantity of organic cations and a small amount of inorganic ammonia, which combine to form a salt solution with strong electrostatic bonding. The characteristics of the solution can be designed for a specific product by adjusting the combination ratio of the components.

Many studies revealed that lignocellulose biomass treated with IL shows an increased surface area, reduced percentage of lignin, and cellulose depolymerization (Kassaye et al. 2017). Examples of ILs include 1-ethyl-3-methylimidazolium acetate, 1-ethyl-3-methylimidazolium diethyl phosphate, 1-butyl-3-methylimidazolium chloride, tetrabutylammonium hydroxide, etc. The major drawback of this pretreatment method is the high cost of chemicals, which influences the overall cost of the process.

8.3.3.5.2 Deep Eutectic Solvent Pretreatment

DES is a tunable or designer solvent prepared from a mixture of organic compounds with a hydrogen bond acceptor, for example, potassium carbonates, glycerol, choline chloride, and lactic acid (Lugani et al. 2020). Over the years, DES has been used as an electrolyte, reaction medium, and solvent for extraction and recently as a solvent for pretreatment. A report by Li et al. (2023) indicates that lignocellulose treated with DES prevents loss of sugars and exposes the cellulose structure to enzymatic hydrolysis. The use of DES was also found to reduce energy utilization by 28% as compared to acid or alkaline pretreatment (Procentese et al. 2017).

8.3.4 Saccharization

After pretreatment, saccharization becomes imperative since the enzymes responsible for bioethanol production can digest only monomers. Saccharization can be catalyzed, either by an acid or by enzymes. The use of an acid catalyst for the hydrolysis of sugar monomers is to break the hydrogen bond of the cellulose chain and convert the crystallized structure to an amorphous form. Both concentrated and dilute acids are used for this process. Concentrated acid requires low temperature condition at short period of time, while dilute acid requires high temperature conditions.

Examples of acids predominantly used are H_2SO_4 and HCl. However, the disadvantages of using acid catalysts include the formation of an inhibition agent, the corrosive nature of the acid, degradation of the produced sugar, recovery of the acid, disposal, energy requirement of the process, etc. (Azhar et al. 2017).

Enzyme hydrolysis requires mild operating conditions, and it is environmentally friendly method, without the formation of inhibiting agents. The efficiency of enzyme hydrolysis is affected by pH, temperature, enzyme loading, reaction time, substrate concentration, and the presence of impurities in the reacting vessel. Enzymes used for hydrolysis include modular and non-modular glycosyl hydrolases, which are cellulases and hemicellulases (Patel et al. 2019), carbohydrate esterases (Imran et al. 2016), and auxiliary activity protein (Kawaguchi et al. 2017). Lignocellulolytic enzymes are produced from bacterial (e.g. *Clostridium*, *Piromyces*, etc.) and fungal (*Trichoderma reesei*, *Aspergillus* sp., etc.) strains, although fungi are the most widely used (Patel et al. 2019).

Over the years, strain improvement strategies have been adopted to increase the efficiency of these microbes, while reducing the overall cost of production. Some of these strategies include random mutagenesis, site-directed mutagenesis, heterogeneous expression of protein and genome, and metabolic engineering (Wen et al. 2020). All these methods have been found to increase the efficiency of the targeted microorganisms.

8.3.5 Fermentation

This is a vital step in the ethanol production process. It involves the biological conversion of the monomeric sugars obtained from hydrolysis to ethanol, acids, and gases. Microbial strains include *Saccharomyces cerevisiae*, *E. coli*, *Candida shehatae*, *Scheffersomyces stipites*, *Zymomonas mobilis*, etc., (Lugani et al. 2020). Yeast, especially *S. cerevisiae* is the most widely reported strain for ethanolic fermentation. Fermentation is controlled by factors like pH, temperature, aeration rate, and salt, sugar and ethanol concentration. The efficiency of the process depends on microbial strength and the ability of the microbes to withstand high concentrations of substrate, sugar, ethanol, and other byproducts. The initial challenges of natural microbes withstanding harsh operating conditions for optimum yield have been overcome by the development of genetically modified microbes (Kamusoko et al. 2021). The fermentation process can be done in batch, fed-batch, or continuous mode, depending on the capacity of the production process.

8.3.5.1 Methods Adopted for Fermentation

Many approaches have been adopted for the ethanolic fermentation process. Some of the reported methods include separate hydrolysis and fermentation, simultaneous saccharification and fermentation, and simultaneous saccharification and cofermentation.

8.3.5.1.1 Separate Hydrolysis and Fermentation

In this method, enzymatic hydrolysis and monomer fermentation are performed in a separate system, which allows for optimum reaction conditions for both processes.

They are separated and subjected to different optimum reaction conditions. This helps maximizing the yield, though a costly and longer reaction time is needed (Damayanti et al. 2021).

8.3.5.1.2 Simultaneous Saccharification and Fermentation

In this method, two different processes, that is, enzymatic hydrolysis and fermentation, are performed simultaneously in a single reactor unit. The process can easily convert cellulosic materials to monomeric sugars, which gives way to ethanol production. The process is very efficient, lowers initial costs, and minimizes environmental impacts (Damayanti et al. 2021). However, the reaction requires more enzymes for the saccharification process.

8.3.5.1.3 Simultaneous Saccharification and Cofermentation

This method is suitable for bioethanol production under the biorefinery approach. It involves the concurrent enzymatic hydrolysis of pretreated biomaterials with cofermentation of hexose and pentose sugars in a single system. A modified strain of *S. cerevisiae* is used for this process because of its ability to ferment pentose sugar, with decreased cost and inhibition. The only challenge with this method is controlling process parameters for optimum yield since both processes are performed in a single system (Singh et al. 2022).

8.3.6 Ethanol Recovery

This is one of the most important stages in ethanol production. It involves the separation of the ethanol from the broth after the fermentation process. The water content of the broth is reduced to about 0.5%, which raises the concentration of the anhydrous ethanol to about 99.5%. The reduction is profound using processes like simple distillation and azeotrophic distillation. Dehydration, membrane separation, and extractive distillation, among these processes. The most used method is extractive distillation (Hajinezhad et al. 2021).

8.3.7 Life Cycle Analysis

Life cycle analysis (LCA) of bioethanol is performed to assess the environmental impact of producing and using bioethanol from different feedstock, right from the start of production till disposal of the final waste. It assesses all environmental impacts like gas emissions, resource depletion, and land used for cultivation. The basic methodology used for LCA includes goal and scope definition, life cycle inventory (LCI), life cycle impact assessment (LCA), and result interpretation. The usefulness of LCA cannot be overemphasized because it helps in the identification of the potential impact of bioethanol production from different biomass before optimization of the process.

Tools used for LCA include LUCAS, Lune, TrACI, ReCiPe, Ecoindicator 99, and CML 2022 (Rosenbaum et al. 2017). Several LCA performed on bioethanol production and utilization showed that it has substantial potential to replace fossil fuels, while mitigating carbon footprint, which in turn protects the ozone layer from depletion (Edeh 2021).

8.4 BIOBUTANOL

Biobutanol is an exceptional alcohol, which has a unique property that makes it a promising substitute for conventional gasoline. Such properties include but not limited to high combustion heat, low volatility, low hygroscopicity, elevated energy density, and less corrosive nature. In addition, it can be mixed with gasoline at a ratio up to 85% without modification of the conventional Otto engine cycle. These exceptional characteristics make biobutanol a more valid source of energy than even the conventional gasolines. Globally, the estimated market value of biobutanol in 2025 stood at US$ 1.18 billion, and it is projected to reach US$ 3.54 billion by 2034. The projected rapid increase is attributed to novel production methods, which minimize the gross production cost (Ganeshan & Tülbek 2025).

8.4.1 Biobutanol Production

8.4.1.1 Microbial Species for Biobutanol Production

The isolation of the genus *Clostridium* by Weizmann in 1912 marks the beginning of research into various microbes that can ferment biomass for the production of bioethanol. The genus *Clostridium* comprises over 40 species of Gram-positive anaerobic bacteria, which are solventogenic in nature. The microbes are rod-shaped in shape, with peritrichous flagella that enable them to be mobile, although some environmental factors may influence the shape of the microorganism. *Clostridium* species undergoes fermentation through two destructive processes of acidogenesis and solventogenesis. The fermentation process of this kind is sometimes referred to as biphase fermentation. The main products of this fermentation are acetone, biobutanol, and ethanol (ABE). Nevertheless, other products like acetic and butyne acids, gases (CO_2 and H_2) are also obtained (Sánchez-Ramírez et al. 2023). Acidogenesis involves the utilization of glucose to generate ATP in the presence of NADH. The acids produced in this process are reabsorbed as substrates during solventogenesis for the production of ABE (Díaz et al. 2023). The production of these trio solvents at >13 g/L concentration affects bacterial cell metabolism. As a result, the metabolic activities in the cell will cease.

Clostridium also produces biobutanol from two other pathways using different sugars. The pathways are the isopropanol-butanol ethanol (IBE) and the hexanol-butanol-ethanol (HBE). The major distinction of these pathways is in substrates and the species of Clostridium used for the fermentation process (Liberato et al. 2019). Aside from the solventogenic bacteria, non-solventogenic bacteria also produce butanol following a different pathway.

8.4.1.2 Feedstock for Biobutanol Production

Just like the most biofuels, the biobutanol feedstock is classified into four generations, which are the first-generation, second-generation, third-generation, and fourth-generation. Syngas, a mixture of CO_2, CO, and H_2, generated from biomass gasification, has also been used as a source of carbon for biobutanol synthesis (Phillips et al. 2015).

8.4.1.2.1 First-Generation Biobutanol Feedstock

The first-generation biobutanol feedstock is made up of edible biomass, such as corn, millet, cassava, molasses, sugarcane, rice, etc. Substrates, such as lactose, sucrose, mannose, fructose, glucose, melezitose, xylose, dextrin, trehalose, and galactose are extracted from these feedstocks for solventogenesis to produce biobutanol. Butanol production from these materials involves soaking the substrate in a mild solution of sodium metasulfite, as a concentrated solution would affect microbial performance. The microorganism widely used in the first-generation is *C. acetobutylicum* DSM 792. The main drawbacks of this type of feedstock include food crisis and the high cost of the raw material (Dehhaghi et al. 2020).

8.4.1.2.2 Second-Generation Feedstock

Lignocellulose and hemicellulose biomass are the major sources of second-generation biobutanol production. Second-generation feedstocks come from municipal residues, forestry waste, agro-raw material, dedicated crops, etc. These materials can eradicate the food crisis witnessed in the first-generation. However, the production of sugar from cellulosic materials is the major challenge. It involves physical and chemical processes to break down the complex recalcitrant structure of the biomass. Hence, the conversion of lignocellulose and hemicellulose to butanol requires pretreatment to expose the sugars for easy access of the fermentation microbes and reduction or removal of toxic compounds (Chen et al. 2024).

After the pretreatment, hydrolysis of the biomass is performed to obtain fermentable sugars from the feedstocks. Hydrolysis products are subjected to fermentation using different *Clostridium* sp., to obtain either ABE, IBE, or HBE. The pretreatment and hydrolysis processes for biobutanol are similar to those reported earlier on, for bioethanol production (Chen et al. 2024). Figure 8.4 shows the ABE fermentation pathway for biobutanol synthesis. In this diagram, acidogenesis and solventogenesis are indicated on the left- and right-hand side, respectively.

Separation and purification of biobutanol from the fermentation broth is more expensive compared to bioethanol. This can be attributed to the complexity and low concentration of biobutanol (2% w/w). The other challenge is the close boiling point of butanol (93°C) and water (100°C), which mitigates the use of ordinary distillation for butanol purification. These factors increase the overall cost of production of biobutanol from the second-generation feedstock (Chadni et al. 2023).

8.4.1.2.3 Third-Generation Feedstock

The third-generation feedstock makes use of marine and aquatic biomass, particularly microalgae and macroalgae (seaweed) as a source of biobutanol. Macroalgae are highly nutritious and are often used for a meal rather than as a source of energy. Microalgae are photosynthetic microorganisms that can grow in fresh water, wastewater, seawater, ponds, and so on. It does not require special land like energy crops; thus, it poses no threat to food security. More so, microalgae do not require harsh pretreatment conditions due to the absence of lignin in its cellular structure. However, it releases trace amounts of furfural and 5-hydroxymethyl furfural, when exposed to 3% H_2SO_4 hydrolysis. Approximately 1.83 kg of CO_2 is produced per kilogram of microalgal biomass (Jambo et al. 2016).

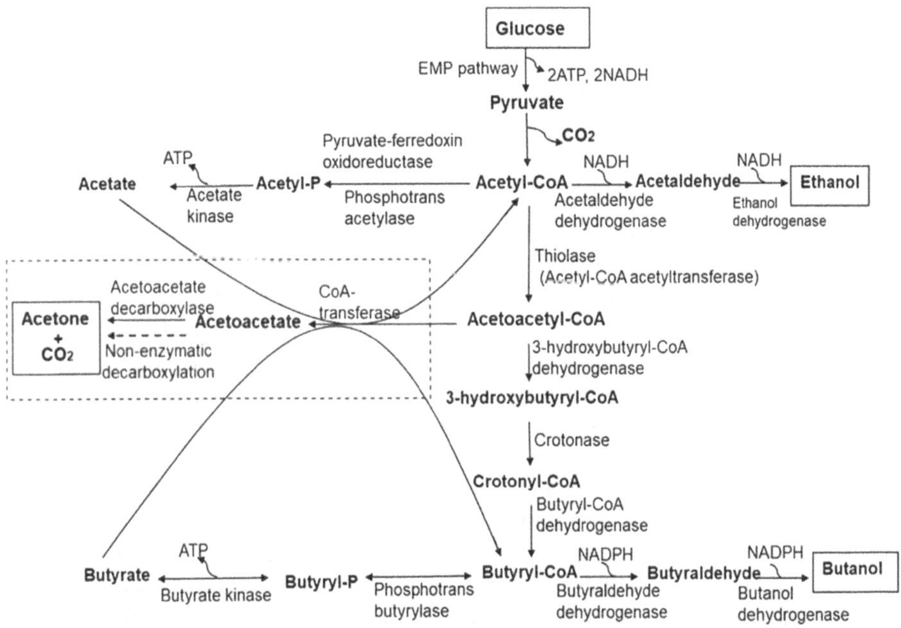

FIGURE 8.4 Acetone-butanol-ethanol metabolic pathway using *C. beijerinckii* and *C. acetobutylicum*.

(Adapted from Han et al. 2011).

Microalgae are useful microorganisms due to their utility as animal feed, remediating oil-spilt polluted water, and fertilizer. There are basically two groups of microalgae: the lipid-rich group (e.g. *Chlorella, Nannochloropsis, Scenedesmus, Butryococcus brauiri*) and carbohydrate-rich (*Dunaliella tertiolecta, Phaeodactylum tricornutium, Nitzschia closterium, Chlamydomaonas reinharatii*, etc.). Carbohydrate-rich microalgae contain more than 40% starch, which makes them an ideal feedstock for biobutanol. These species are processed through fermentation to generate biobutanol (Jambo et al. 2016).

8.4.1.2.4 Fourth-Generation Feedstock

Unlike other biofuels, little research has been done on the production of biobutanol from fourth-generation feedstocks. These are genetically engineered strains of microalgae and macroalgae. As an example, algae species like *Chlorella, Neochloris, and Nannochloropsis* have been modified, and the results showed high lipid content with potential for optimization to improve biobutanol yield (Shokravi et al. 2022).

One of the ways to optimize biobutanol yield is by the addition of reducing agents like ascorbic acid, L-cysteine, etc., to the algal strain. The reducing agent acts as an electron donor, influencing specific genetic expression, which in turn improves metabolite flux in the direction of the biosynthesis route (Chandgude et al. 2021). At present, only 3,000 algal species out of over 40,000 species have the potential of being used as biofuel feedstock. Among them, just about 30 species have been genetically constructed for biobutanol production. The challenge is the construction of some of the algal genes for biobutanol production due to their complexity.

8.4.1.3 Biobutanol Fermentation

Fermentation converts the monomer sugar units into biobutanol using *Clostridium* bacteria. It occurs within specific operating conditions of temperature, pH, agitation speed, and time. The ABE fermentation is achieved via batch, fed-batch, and continuous cultures. Batch fermentation is the most studied because it is simple and easy to operate. It takes place at around 35°C–37°C and pH 5, for a period of 32–72 h, depending on the type of microbial strain. During this process, the substrate, nutrients, and microbes are charged into the fermenter at the beginning of the process. Thereafter, incubation at specified conditions is performed for optimum growth of the microbes and subsequent yield of butanol.

The fermentation process may be either anaerobic or aerobic, depending on the targeted product(s) (Veza et al. 2021).

However, aerobic conditions boost the metabolic processes in the cell, producing more ATP. Thus, anaerobic conditions are preferred for bio-butanol production. At the beginning of anaerobic fermentation, nitrogen or CO_2 gas is used to maintain the osmotic pressure of the system until the microorganism begins to generate gases to stabilize the pressure of the system. When fermentation is completed, processes like centrifugation and filtration, among others, are used to separate the cell biomass from the broth. Pure butanol is then obtained via distillation. Typically, the ABE production ratio of 3:6:1 (w/v) and about 11–13 g/L of butanol are achieved (Maiti et al. 2015).

Continuous culture requires the fermentation process to be continuous for a long time without any downtime using one inoculum. It is essential because it reduces the cost and time of operation and produces a high percentage yield, if properly managed. Chang et al. (2016) reported ABE solvents in the range of 18–22 g/L from continuous fermentation of corn stalk. Whereas fed-batch culture has been developed to surpass challenges of limiting initial substrate concentration, targeting higher butanol titer (Ibrahim et al. 2018).

8.4.1.4 Biobutanol Purification and Recovery

The separation and purification of the ABE solvents is a daunting task in butanol production from biomass. The traditional distillation method has been practiced over the past years. However, the boiling point (118°C) and the energy cost of butanol are much higher, which in turn increases the overall cost of production. Several improved integrated methods have been proposed to overcome these challenges. These downstream technologies include pervaporation, gas stripping, liquid-liquid extraction, perstraction, and gas stripping (Veza et al. 2021).

8.4.1.4.1 Perstraction

This is an advanced liquid-liquid extraction, which entails the use of a selective membranes to extract butanol. The extraction solution and fermentation broth are on either side of the membrane. This prevents direct contact of the two solutions. This method prevents the harsh or toxic nature of the extractant from contaminating the fermentation medium. Examples of such extractants include oleyl alcohol,

1-dodecanol, 1-octanol, tributyrin, etc. The efficiency of this method depends largely on the type and thickness of the membrane and the type of extraction used. Its drawbacks are the high cost of the membrane and fouling (Solis et al. 2024).

8.4.1.4.2 Pervaporation

This technology involves partial vaporization of a multicomponent mixture through a permeable membrane. In this method, only the target liquid is allowed to pass through the membrane. It is an efficient method of separating liquids of the same boiling points. The principle is to evaporate the effluent from the fermenting medium, which contains butanol, contacting it with one side of the membrane, and transferring it as vapor with the aid of sweep gas. The method is based on diffusivity rather than volatility (Wang et al. 2017). The permeated vapor is condensed to recover the butanol. Frequently used membranes include silicon and aluminum-free zeolite.

8.4.1.4.3 Gas Stripping

In this method, fermentation effluent is made to pass through the rotating shaft of the bioreactor by pumping oxygen-free nitrogen through the fermenter. This method helps to separate volatile components, mainly ABE, from the effluent. Volatile components are condensed, while stripping gases are recycled into the fermenter. Gas stripping neither negatively affects the physiology of the microbes nor the fermenting medium. The major constraint of this method is the hyperfoaming at the initial stages of fermentation in the bioreactor. This can be overcome by the use of an anti-foaming agent like polypropylene glycol. To improve the efficiency of this process, appropriate conditions in the fermenter, an appropriate dosage of anti-foaming agent, and gas flow rate must be carefully selected (Veza et al. 2021).

8.4.1.4.4 Liquid–Liquid Extraction

In liquid-liquid extraction, two or more immiscible liquids are mixed with the purpose of reducing the toxicity of the fermenting medium and removing the targeted solvent. Effluents from the broth are moved to the extractant and separated by simple distillation (Veza et al. 2021). The efficiency of the process depends on the choice of the extractant and other operating conditions. Though a very successful integration process, it is very expensive, and if not properly managed, it can result in environmental toxicity, corrosion of the equipment, emulsion formation, and loss of essential byproducts.

8.4.2 Technologies for Advancement of Biobutanol Fermentation

Due to challenges encountered during the biobutanol fermentation, some novel techniques have been developed to improve the process, and subsequently, the quality of the biobutanol. Process advancements comprise strain improvement (mutation and genetic engineering), process optimization, research on novel substrates and pretreatment methods, modifying different fermentation processes, and many more. Among these, developing the fermentation process through innovation and modern fermentation knowledge is fundamental to boosting biobutanol production (Ibrahim et al. 2018).

One of the most important approaches in the advancement of butanol fermentation is the coculturing technique. Coculture fermentation of aerobic and anaerobic microorganisms has been proposed to increase the resultant butanol yield. It involves the application of a second strain into the bioreactor using the general idea that bacteria live in a consortium and complement each other. The second microbe introduced has a specific task to perform in the fermentation process. This synergistic interaction reduces the burden on one bacterium, thereby increasing the efficiency of the process. For example, mixed cultures of *Bacillus subtilis* and *C. butylicum* produced 5.4–6.5 times more butanol than monoculture experiments (Ethiraj et al. 2023). The use of cellulolytic strains, which are responsible for producing monomeric sugars from lignocellulosic biomass, is also key to developing efficient cocultures.

Additional efficiency of the fermentation process can be obtained by the use of two different strains of solventogenic bacteria. This boasts the strength of the process with a resulting high yield. Some microbes that secrete amino acids can be introduced into the bioreactor, because studies revealed that some solventogenic bacteria use amino acids for energy production, which in turn would boost the production process. Examples of such microorganisms include the yeast *S. cerevisiae* (Wu et al. 2019).

The use of electron acceptors is a new approach utilized in optimizing biobutanol yield during fermentation. In this approach, a reducing agent is introduced into the broth to enhance the redox potential of the broth toward the target metabolite. It influences the generation of a specific gene that improves the yield of biobutanol (Ding et al. 2018). The application of this method was adopted by Ding et al. (2018), when sodium sulfate was added as an electron receptor to corn meal substrate using *C. acetobutylicum* as microbes. The result revealed a 34.8% increase in biobutanol titer.

8.5 CONCLUSION

Over the decades, the development of biofuel production technologies has evolved with the advancement of new methods that focus on optimization of yield and mitigation of the high cost of production. To achieve this, genetically constructed microbes, integrated fermentation-separation processes, and integrated separation-separation processes are researched for best practices that will facilitate high yield of biodiesel, butanol, and bioethanol from biomass. As researchers continue to push boundaries, it is believed that in the near future, the utilization of liquid biofuels as an alternative energy source will become established and help in solving global energy crises and reducing carbon footprint.

REFERENCES

Agbor, V.B., N. Cicek, R. Sparling, A. Berlin, & D.B. Levin. 2011. Biomass pretreatment: Fundamentals toward application. *Biotechnol. Adv.* 29: 675–685.

Akpan, I.O., I. Edeh, & L. Uyigue. 2023. A review on biodiesel production. *Pet. Chem. Ind. Int.* 6: 131–141.

Azhar, S.H.M., R. Abdulla, S.A. Jambo, H. Marbawi, J.A. Gansau, A.A.M. Faik, & K.F. Rodrigues. 2017. Yeasts in sustainable bioethanol production: A review. *Biochem. Biophys. Rep.* 10: 52–61.

Banković-Ilić, I.B., O.S. Stamenković, & V.B. Veljković. 2012. Biodiesel production from non-edible plant oils. *Renew. Sustain. Energy Rev.* 16: 3621–3647.

Brahma, S., B. Nath, B. Basumatary, B. Das, P. Saikia, K. Patir, & S. Basumatary. 2022. Biodiesel production from mixed oils: A sustainable approach towards industrial biofuel production. *Chem. Eng. J. Adv.* 10: 100284.

Carrillo-Nieves, D., M.J.R. Alanís, R. de la Cruz Quiroz, H.A. Ruiz, H.M.N. Iqbal, & R. Parra-Saldívar. 2019. Current status and future trends of bioethanol production from agro-industrial wastes in Mexico. *Renew. Sustain. Energy Rev.* 102: 63–74.

Chadni, M., M. Moussa, V. Athès, F. Allais, & I. Ioannou. 2023. Membrane contactors-assisted liquid-liquid extraction of biomolecules from biorefinery liquid streams: A case study on organic acids. *Sep. Purif. Technol.* 317: 123927.

Chandgude, V., T. Välisalmi, J. Linnekoski, T. Granström, B. Pratto, T. Eerikäinen, G. Jurgens, & S. Bankar. 2021. Reducing agents assisted fed-batch fermentation to enhance ABE yields. *Energy Convers. Manag.* 227: 113627.

Chang, Z., D. Cai, Y. Wang, C. Chen, C. Fu, G. Wang, P. Qin, Z. Wang, & T. Tan. 2016. Effective multiple stages continuous acetone-butanol-ethanol fermentation by immobilized bioreactors: Making full use of fresh corn stalk. *Bioresour. Technol.* 205: 82–89.

Chen, X., Y. Li, X. Li, J. Shi, & L. Liu. 2024. Exploring the potential of multiple lignocellulosic biomass as a feedstock for biobutanol production. *Fuel* 357: 129697.

Damayanti, D., D. Supriyadi, D. Amelia, D.R. Saputri, Y.L.L. Devi, W.A. Auriyani, & H.S. Wu. 2021. Conversion of lignocellulose for bioethanol production, applied in biopolyethylene terephthalate. *Polymers* 13: 2886.

Dehhaghi, M., H.K.S. Panahi, G.S. Jouzani, S. Nallusamy, V.K. Gupta, M. Aghbashlo, & M. Tabatabaei. 2020. Anaerobic rumen fungi for biofuel production. In *Fungi in Fuel Biotechnology*, ed. G.S. Jouzani, M. Tabatabaei, & M. Aghbashlo, 149–175. Springer International Publishing.

Díaz, M.J., P.M. Ferrero, & M. Moya. 2023. Sequential acid/alkali pretreatment for an olive tree pruning biorefinery. *Agronomy* 13: 2682.

Ding, J., H. Luo, F. Xie, H. Wang, M. Xu, Z. Shi. 2018. Electron receptor addition enhances butanol synthesis in ABE fermentation by *Clostridium acetobutylicum*. *Bioresour. Technol.* 247: 1201–1205.

Edeh, I. 2020. Biodiesel production as a renewable resource for the potential displacement of the petroleum diesel. In *Biorefinery Concepts, Energy and Products*, ed. V. Beschkov, 1–36. IntechOpen. doi:10.5772/intechopen.93013

Edeh, I. 2021. Bioethanol production: An overview. *Bioethanol Technologies*, ed. F. Inambao, 1–22. IntechOpen. doi:10.5772/intechopen.94895

Elgharbawy, A.S., M. Farghali, A.I. Osman, M.A. Hanafy, & A.H. Al-Muhtaseb. 2025. Innovative biodiesel production for sustainable energy: Advances in feedstocks, transesterification, and cost efficiency. *Biomass Bioenergy.* 201: 108114.

Ethiraj, B., T. Raihan, S. Sarmin, A. Karim, M.M.R. Khan, K.S. Muthuvelu, S. Muthusamy, A. Yousuf, & M.A. Islam. 2023. Strategies to enhance biobutanol production from lignocellulosic biomass. In *Microbiology of Green Fuels*, ed. A. Yousuf, & E. Tomás-Pejó, 21. CRC Press.

Ganeshan, S., & M.Ç. Tülbek. 2025. Fermentative butanol production – Perspectives and scale-up challenges. *Encyclopedia* 5: 50.

Hajinezhad, A., M. Lak, & H. Alimoradiyan. 2021. Optimum energy analysis of separation and purification units for bioethanol production by the use of lignocellulose feedstock. *Int. J. Energy Res.* 45: 15452–15462.

Han, B., V. Gopalan, & T.C. Ezeji. 2011. Acetone production in solventogenic *Clostridium* species: New insights from non-enzymatic decarboxylation of acetoacetate. *Appl. Microbiol. Biotechnol.* 91: 565–576.

Hayyan, A., Md Z. Alam, M.E.S. Mirghani, N.A. Kabbashi, N.I.N.M. Hakimi, Y.M. Siran, & S. Tahiruddin. 2011. Reduction of high content of free fatty acid in sludge palm oil via acid catalyst for biodiesel production. *Fuel Process. Technol.* 92: 920–924.

Hendriks, A.T.W.M., & G. Zeeman. 2009. Pretreatments to enhance the digestibility of lignocellulosic biomass. *Bioresour. Technol.* 100: 10–18.

Ibrahim, M.F., S.W. Kim, & S. Abd-Aziz. 2018. Advanced bioprocessing strategies for biobutanol production from biomass. *Renew. Sustain. Energy Rev.* 91: 1192–1204.

Imran, M., Z. Anwar, M. Irshad, M.J. Asad, & H. Ashfaq. 2016. Cellulase production from species of fungi and bacteria from agricultural wastes and its utilization in industry: A review. *Adv. Enzyme Res.* 4: 44–55.

Jambo, S.A., R. Abdulla, S.H.M. Azhar, H. Marbawi, J.A. Gansau, & P. Ravindra. 2016. A review on third-generation bioethanol feedstock. *Renew. Sustain. Energy Rev.* 65: 756–769.

Kamusoko, R., R.M. Jingura, Z. Chikwambi, & W. Parawira. 2021. Strategies for valorization of crop residues into biofuels and other value-added products. *Biofuels Bioprod. Bioref.* 15: 1950–1964.

Kamusoko, R., R.M. Jingura, W. Parawira, & Z. Chikwambi. 2022. Characterization of lignocellulosic crop residues for potential biogas production in Zimbabwe. *Biofuels Bioprod. Bioref.* 16: 1165–1171.

Kassaye, S., K.K. Pant, & S. Jain. 2017. Hydrolysis of cellulosic bamboo biomass into reducing sugars via a combined alkaline solution and ionic liquid pretreatment steps. *Renew. Energy* 104: 177–184.

Kawaguchi, H., Y. Katsuyama, D. Danyao, P. Kahar, S. Nakamura-Tsuruta, H. Teramura, K. Wakai, K. Yoshihara, H. Minami, C. Ogino, Y. Ohnishi, & A. Kondo. 2017. Caffeic acid production by simultaneous saccharification and fermentation of kraft pulp using recombinant *Escherichia coli*. *Appl. Microbiol. Biotechnol.* 101: 5279–5290.

Khaliq, Ab D., A. Chafidz, M.A. Lukman, & I. Kholil. 2020, Making of bioethanol banana weevil as renewable energy. *IOP Conf. Ser. Mater. Sci. Eng.* 722: 012080.

Lee, J.-W., & T.W. Jeffries. 2011. Efficiencies of acid catalysts in the hydrolysis of lignocellulosic biomass over a range of combined severity factors. *Bioresour. Technol.* 102: 5884–5890.

Li, P., C. Yang, Z. Jiang, Y. Jin, & W. Wu. 2023. Lignocellulose pretreatment by deep eutectic solvents and related technologies: A review. *J. Bioresour. Bioprod.* 8: 33–44.

Liberato, V., C. Benevenuti, F. Coelho, A. Botelho, P. Amaral, N. Pereira Jr, & T. Ferreira. 2019. *Clostridium* sp. as biocatalyst for fuels and chemicals production in a biorefinery context. *Catalysts* 9: 962.

Ling, H., W. Teo, B. Chen, S.S.J. Leong, & M.W. Chang. 2014. Microbial tolerance engineering toward biochemical production: From lignocellulose to products. *Curr. Opin. Biotechnol.* 29: 99–106.

Lugani, Y., R. Rai, A.A. Prabhu, P. Maan, M. Hans, V. Kumar, S. Kumar, A.K. Chandel, & R.S. Sengar. 2020. Recent advances in bioethanol production from lignocelluloses: A comprehensive review with a focus on enzyme engineering and designer biocatalysts. *Biofuel Res. J.* 7: 1276–1295.

Maiti, S., S.J. Sarma, S.K. Brar, Y.L. Bihan, P. Drogui, G. Buelna, M. Verma, & C.R. Soccol. 2015. Novel spectrophotometric method for detection and estimation of butanol in acetone–butanol–ethanol fermenter. *Talanta* 141: 116–121.

Malode, S.J., K.K. Prabhu, R.J. Mascarenhas, N.P. Shetti, & T.M. Aminabhavi. 2021. Recent advances and viability in biofuel production. *Energy Convers. Manag. X* 10: 100070.

Mamtani, K., K. Shahbaz, M.M. Farid. 2021. Glycerolysis of free fatty acids: A review. *Renew. Sustain. Energy Rev.* 137: 110501.

Merklein, K., S.S. Fong, & Y. Deng. 2016. Biomass utilization. In *Biotechnology for Biofuel Production and Optimization*, ed. A.E. Carrie, & T.T. Cong, 291–324. Elsevier.

Mori, T., Y. Tsuboi, N. Ishida, N. Nishikubo, T. Demura, & J. Kikuchi. 2015. Multidimensional high-resolution magic angle spinning and solution-state NMR characterization of ^{13}C-labeled plant metabolites and lignocellulose. *Sci. Rep.* 5: 11848.

Nabaterega, R., R.N. Vesuwe, O.T. Iorhemen, & R.W. Thring. 2026. Biodiesel production unpacked: Data-driven analysis of key operational factors. *Biomass Bioenergy* 205: 108457.

Nogales-Delgado, S. 2025. Biodiesel production and life cycle assessment: Status and prospects. *Energies* 18: 3338.

OECD. 2022. *OECD Economic Outlook, Vol 2022 Issue 2: Confronting the Crisis*. OECD Publishing. doi:10.1787/f6da2159-en

Patel, A.K., R.R. Singhania, S.J. Sim, & A. Pandey. 2019. Thermostable cellulases: Current status and perspectives. *Bioresour. Technol.* 279: 385–392.

Phillips, J.R., H.K. Atiyeh, R.S. Tanner, J.R. Torres, J. Saxena, M.R. Wilkins, & R.L. Huhnke. 2015. Butanol and hexanol production in *Clostridium carboxidivorans* syngas fermentation: Medium development and culture techniques. *Bioresour. Technol.* 190: 114–121.

Procentese, A., F. Raganati, G. Olivieri, M.E. Russo, L. Rehmann, & A. Marzocchella. 2017. Low-energy biomass pretreatment with deep eutectic solvents for bio-butanol production. *Bioresour. Technol.* 243: 464–473.

Rabemanolontsoa, H., & S. Saka. 2016. Various pretreatments of lignocellulosics. *Bioresour. Technol.* 199: 83–91.

Ramos, M., A.P.S. Dias, J.F. Puna, J. Gomes, & J.C. Bordado. 2019. Biodiesel production processes and sustainable raw materials. *Energies* 12: 4408.

Rathnam, V.M., J.M. Modak, & G. Madras. 2020. Non-catalytic transesterification of dry microalgae to fatty acid ethyl esters using supercritical ethanol and ethyl acetate. *Fuel* 275: 117998.

Refaat, A.A. 2012. Biofuels from waste materials. In *Comprehensive Renewable Energy*, ed. A. Sayigh, 217–261. Elsevier.

Research & Markets. 2023. *Global Bioethanol Market Report 2023–2028: Development of Second-Generation Bioethanol Production Presents Opportunities*. Research & Markets.

Rosenbaum, R.K., M.Z. Hauschild, A.-M. Boulay, P. Fantke, A. Laurent, M. Núñez, & M. Vieira. 2017. Life cycle impact assessment. In *Life Cycle Assessment: Theory and Practice*, ed. M.Z. Hauschild, R.K. Rosenbaum, & S.I. Olsen, 167–270. Springer International Publishing.

Sánchez-Ramírez, E., B. Huerta-Rosas, J.G. Segovia-Hernandez, & Juan José Quiroz-Ramírez. 2023. Developments in biobutanol industrial production. In *Advances and Developments in Biobutanol Production*, ed. J.G. Segovia-Hernandez, S. Behera, & E. Sanchez-Ramirez, 47–78. Woodhead Publishing.

Selvam, C.D., Y. Devarajan, & T. Raja. 2025. Sewage sludge as a sustainable feedstock for biodiesel: Advances in conversion technologies and catalytic applications. *Results Eng.* 25: 104000.

Sharma, V.K., G. Monteleone, G. Braccio, C.N. Anyanwu, & N.N. Aneke. 2025. A comprehensive review of green energy technologies: Towards sustainable clean energy transition and global net-zero carbon emissions. *Processes* 13: 69.

Shokravi, H., M. Heidarrezaei, Z. Shokravi, H.C. Ong, W.J. Lau, M.F. Md Din, & A.F. Ismail. 2022. Fourth generation biofuel from genetically modified algal biomass for bioeconomic development. *J. Biotechnol.* 360: 23–36.

Singh, A., R.R. Singhania, S. Soam, C.-W. Chen, D. Haldar, S. Varjani, J.-S. Chang, C.-D. Dong, & A.K. Patel. 2022. Production of bioethanol from food waste: Status and perspectives. *Bioresour. Technol.* 360: 127651.

Solis, C.M., J.L.S. Juan, J.G. Uy, & J.L.S. Juan. 2024. Multi-objective optimization of algal biofuel production integrating resource recirculation and quality considerations. *Clean Technol. Environ. Policy* 26: 1107–1133.

Sundaram, G.A., & G. Raj Muniyandi 2025. Advancements in renewable energy: Innovations and sustainable solutions for a greener future. *Next Res.* 2: 100482.

Tabatabaei, M., M. Aghbashlo, M. Dehhaghi, H.K.S. Panahi, A. Mollahosseini, M. Hosseini, & M.M. Soufiyan. 2019. Reactor technologies for biodiesel production and processing: A review. *Prog. Energy Combust. Sci.* 74: 239–303.

Usha, V.C., & K.N. Nwaigwe. 2016. Economic importance of edible and non-edible almond fruit as bioenergy material: A review. *Am. J. Energy Sci.* 3: 31–39.

Veza, I., M.F.M. Said, & Z.A. Latiff. 2021. Recent advances in butanol production by acetone-butanol-ethanol (ABE) fermentation. *Biomass Bioenergy.* 144: 105919.

Wang, Y., S.-H. Ho, H.-W. Yen, D. Nagarajan, N.-Q. Ren, S. Li, Z. Hu, D.-J. Lee, A. Kondo, & J.-S. Chang. 2017. Current advances on fermentative biobutanol production using third generation feedstock. *Biotechnol. Adv.* 35: 1049–1059.

Wen, P., T. Zhang, Y. Xu, & J. Zhang. 2020. Coproduction of xylo-oligosaccharides and monosaccharides from poplar by a two-step acetic acid and sodium chlorite pretreatment. *Ind. Crops Prod.* 152: 112500.

Wu, W., Y.-C. Lei, & J.-S. Chang. 2019. Life cycle assessment of upgraded microalgae-to-biofuel chains. *Bioresour. Technol.* 288: 121492.

Yaqoob, H., Y.H. Teoh, M.A. Jamil, T. Rasheed, & F. Sher. 2020. An experimental investigation on tribological behavior of tire-derived pyrolysis oil blended with biodiesel fuel. *Sustainability* 12: 9975.

9 Anaerobic Digestion for Biogas and Bioelectricity Generation

Oluwadamilola B. Oyinloye, Adekunle A. Adeleke, Jude A. Okolie, Petrus Nzerem, Seun Jesuloluwa, and Chizoma N. Adewumi

9.1 INTRODUCTION

It is anticipated that the demand for energy would increase by approximately 50% by 2050. However, most of these energy requirements are produced from fossil-based resources. The combustion of fossil fuels accounts for about 88% of global energy, which produces carbon dioxide (CO_2). As a result, the atmosphere is experiencing a sharp increase in the amount of greenhouse gases (GHGs) emitted, with CO_2 being the primary contributor. The current situation has prompted a study into renewable energy generation (Naik et al. 2010). Sustainability is another factor pushing for the use of renewable energy sources. According to some authors, traditional sources have a finite supply and will eventually run out completely (Ighravwe & Babatunde 2018).

The commonly investigated renewable energy sources include biomass, solar, wind, hydro, geothermal energy, and biofuels. Biogas production is one of the processes that has recently attracted a lot of attention. The most efficient alternatives for meeting energy needs are provided by biogas, which is regarded as a low-carbon fuel source. For example, Molino et al. (2018) reported that in 2020, the EU used biofuels as 10% of the energy in the transportation sector. It was recorded that the USA produced roughly 36 billion gallons of biofuels yearly in 2022 (Molino et al. 2018). Nations with an enormous area of fertile cultivable land, favorable climate, as well as water resources can invest in the planting of biomass plants for energy generation. Such agricultural plants include sugarcane, cassava, corn starch, among others. Biomass is classified as non-lignocellulosic or lignocellulosic in nature and exists in various forms such as woody, herbaceous, aquatic debris, farming manure, and byproducts (Osman et al. 2019).

Biofuels can be produced from biomass using various methods, with the biochemical and thermochemical processes being the two most commonly utilized methods. Biochemical processes include fermentation, anaerobic digestion (AD) and combustion, while thermochemical conversion includes pyrolysis, carbonation, combustion, liquefaction, gasification, and supercritical oxidation. The focus of this chapter is on biomass conversion into energy through AD. A brief description of the

DOI: 10.1201/9781003650751-9

process would be provided in subsequent sections, followed by reaction pathways and feedstocks.

9.1.1 Definition of Anaerobic Digestion

In the biological process known as AD, microbes break down organic waste without oxygen to create biogas, which is mainly made up of CO_2 and methane (CH_4). Biogas is composed of 50%–70% CH_4, 30%–50% CO_2, and 1%–5% hydrogen (H_2) (Ramaraj & Dussadee 2015). It has numerous financial and environmental advantages. By providing a rich and sustainable energy source that can be transformed into heat and electricity, it also addresses the problems associated with disposing of historic waste streams.

AD is a complex process, which contains of many rate-limiting factors such as ammonium accumulation, incomplete biomass conversion, and prolonged lag phases, affecting the process efficiency (Adekunle & Okolie 2015). The process consists of four fundamental phases, which are hydrolysis, acidogenesis, acetogenesis, and methanogenesis. AD relies on the interactions of many bacteria capable of performing the four processes outlined above. It can be subdivided into AD of single substrates and anaerobic codigestion. The majority of AD operations are single stage (one stage) processes (Figure 9.1), in which all the biological reactions occur within single reactor. This process presents some challenges such as process instabilities and reactor acidification, which often result in partial breakdown of the reaction (Feng et al. 2020; Mari et al. 2020).

As a result, anaerobic codigestion (Figure 9.2) could be a promising option to overcome the challenges of single digestion and improve the process efficiency. This process separates the acidic and methanogenic phases. It is an interesting alternative for increasing the stability and overall performance of the process (Luo et al. 2011; Schievano et al. 2014). Despite the advantages of anaerobic codigestion such as minimizing the pH fluctuation, ammonium accumulation, year-round availability, and low energy yields, the process possesses some challenges. These include toxic chemicals removal because of toxicity from different feedstock, digester stability, and the ability of microorganisms to acclimatize (Mumme et al. 2014). During the AD process, it takes about 21 d for the microorganisms to familiarize with the new environment, as a result of changes in temperature and substrate. This reaction is said to be a slow process.

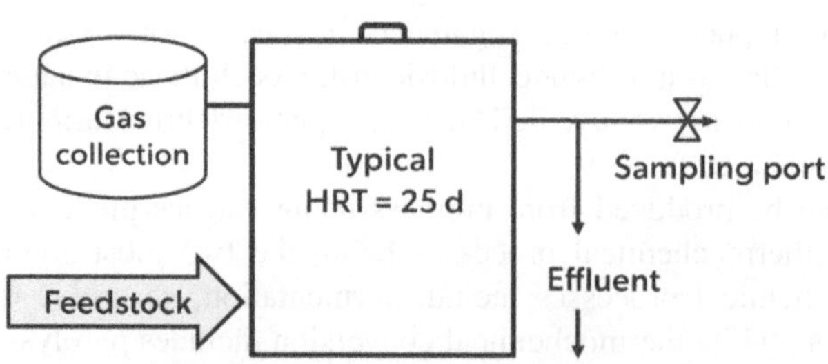

FIGURE 9.1 Single-stage anaerobic digestion.

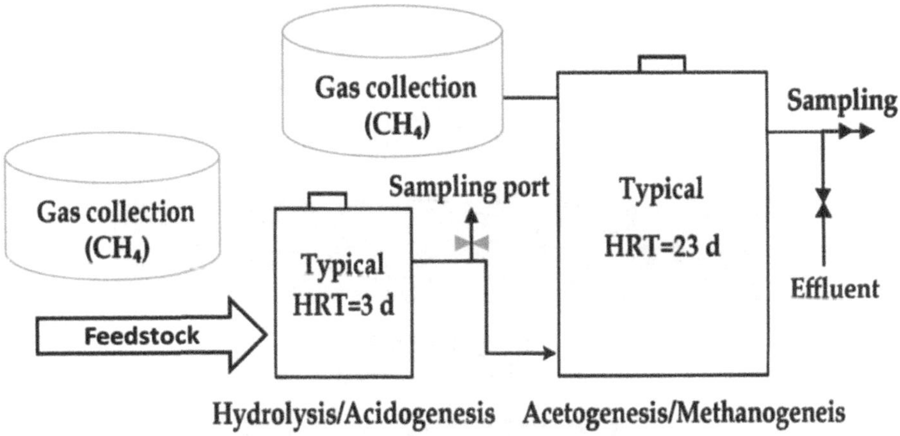

FIGURE 9.2 Anaerobic codigestion.

9.1.2 Importance of Biogas and Bioelectricity

Biogas is a fuel, an intermediate energy carrier (mainly CH_4), while bioelectricity is an energy form, specifically electricity, which can be produced from biogas or other biomass sources. Biogas and bioelectricity are vital components of renewable energy systems gotten through AD, offering significant environmental, economic, and social benefits. Biogas has a positive impact because of its flexibility and relatively cheap energy storage compared to other technologies. Studies have indicated that biogas in municipal sewage waste-based biogas-solar-wind systems cost 22% less over the project lifetime as they are less dependent on fuel price due to locally available substrate. In addition, biogas has a positive impact on CO_2 mitigation, the reduction of waste mismanagement, and the balance of uncertain energy production from intermittent renewable energy sources. Biogas also has the potential to be converted into biomethane via upgrading or used for combined heat and power generation through combustion, making biogas a promising and sustainable approach for bioelectricity. Biogas upgrading to produce biomethane can achieve a high energy conversion efficiency, resulting in biomethane with a purity of more than 95%–97%.

9.2 PRINCIPLES OF ANAEROBIC DIGESTION

AD process is multi-step and involves complex interactions between multiple microbes. The complex biomass polymers are broken down at various stages by diverse microbial populations, resulting in a gaseous mixture. As illustrated in Figure 9.3, the AD processes can be separated into four different stages: hydrolysis, acidogenesis, acetogenesis, and methanogenesis. At particular stages, some bacteria groups are more active than others. Bacteria responsible for hydrolysis and acid fermentation include both facultative and obligate anaerobes (Zamorano-López et al. 2020). Some bacteria found in anaerobic digesters include *Clostridium,*

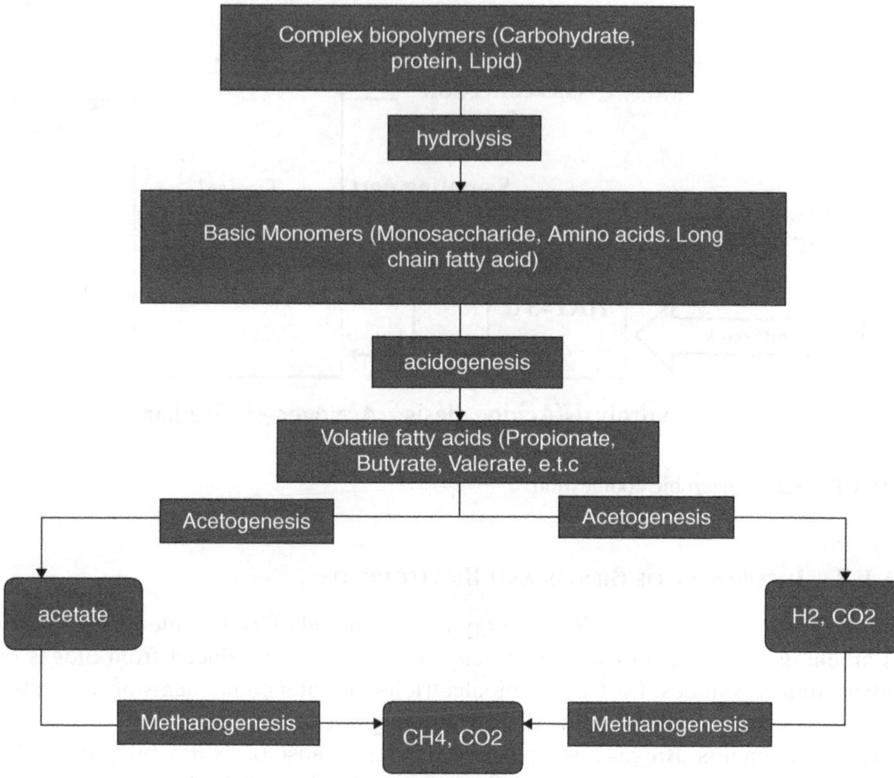

FIGURE 9.3 Stages of the anaerobic digestion process.

Corynebacterium, Actinomyces, Staphylococcus, and Escherichia. The microorganisms responsible for CH_4 conversion include *Methanosarcina, Methanothrix, Methanococcus, Methanobacterium, and Methanobacillus. Methanosarcina and Methanothrix* utilize acetate (CH_3COOH) to produce CH_4 and CO_2, while *Methanococcus, Methanobacterium, and Methanobacillus* oxidize H_2 with CO_2 as an electron acceptor (Rivière et al. 2009; Senés-Guerrero et al. 2019; Walter et al. 2019). The products of one stage are used as the input for the later stages. Each stage is described in the following section.

9.2.1 Hydrolysis

This is essentially the initial stage of the AD process. Water and extracellular enzymes break down the complex polymeric structure of cellulose, starch, proteins and transform them into their respective simple units (monomers or oligomers) such as glucose, fatty acids, and amino acids. The general hydrolysis phase is represented by Equation 9.1.

$$\left(C_{10}H_{10}O_5\right)_n + nH_2O \rightarrow nC_6H_{12}O_6 + nH_2 \tag{9.1}$$

During hydrolysis, an acidic catalyst is required to convert the biopolymers into low-molecular substrates. Fermenting bacteria that disintegrate biomass polymers into soluble components are known as bacteroides, clostridia, and bifidobacteriales (Kour et al. 2019). The primary byproducts of the hydrolysis phase are CH_3COOH and H_2. These are used by methanogens in subsequent phases of AD (Begum et al. 2018). Feedstock hydrolysis is the rate-limiting step in AD of lignocellulosic biomass (Meegoda et al. 2018; Yang et al. 2018). It is typically the cause of AD's lengthy residence time, which is why its effective implementation is necessary (Kiran et al. 2014). The kind of macronutrients (such as the biodegradation of amorphous materials), substrate concentration, particle size, pH (ideally between 5 and 7), and temperature (ideally in the range of 30°C–50°C) determines the hydrolysis rate.

9.2.2 Acidogenesis

The products of hydrolysis are further broken down in the acidogenesis step by acidogenic bacteria. The primary products of acidogenesis are alcohol (ethanol, methanol), ketones (glycerol and acetone), and short-chain volatile fatty acids (VFAs) (acetic acid, propionic acid, formic acid, and lactic acid). As byproducts, CO_2, H_2, ammonia (NH_3), alcohols, and trace amounts of other chemicals are also produced (Kamusoko et al. 2022). At the last stage, methanogens can easily use certain products like CO_2, H_2, CH_3COOH, etc. In order to produce CH_4, other compounds must undergo additional breakdown. Acidogenesis typically proceeds extremely quickly, and if it is not adequately regulated. There is a chance that VFAs will build up in the digester and cause digester toxicity. The chemical reactions in the acidogenesis stage are shown in Equations 9.2–9.4.

$$C_6H_{12}O_6 \leftrightarrow 2CH_3CH_2OH + 2CO_2 \tag{9.2}$$

$$C_6H_{12}O_6 + 2H_2 \leftrightarrow 2CH_3CH_2COOH + 2H_2O \tag{9.3}$$

$$C_6H_{12}O_6 \rightarrow 3CH_3COOH \tag{9.4}$$

9.2.3 Acetogenesis

The acetogenesis stage is thermodynamically feasible due to a syntrophic relation, in which some bacteria are fed from the products of other bacteria. Acetogenic bacteria convert the products of the acidogenesis stage and some of the long-chain fatty acids from the hydrolysis stage into CH_3COOH, CO_2, and H_2 (Kamusoko et al. 2022). Reactions in the acetogenesis stage are not thermodynamically spontaneous, if the partial pressure of H_2 is higher than 10^{-4} atm. However, methanogenic bacteria can lower this partial pressure by consuming the generated H_2. The transfer of H_2 across species is equivalent to the transfer of electrons since H_2 is basically a proton (H^+) plus an extra electron. The rate of digestion as a whole can be strongly impacted by the speed of this electron transfer. The acetogenesis stage's chemical reactions are depicted in Equations 9.5–9.7.

$$CH_2COO^- + 3H_2O \leftrightarrow CH_3COO^- + H + HCO_3^- + 3H_2 \tag{9.5}$$

$$C_6H_{12}O_6 + 2H_2O \leftrightarrow 2CH_3COOH + 2CO_2 + 4H_2 \quad (9.6)$$

$$CH_3CH_2OH + 2H_2O \leftrightarrow CH_3COO^- + 3H_2 + H^+ \quad (9.7)$$

9.2.4 Methanogenesis

Methanogenesis is the last stage, during which all of the intermediate products from the earlier stages are converted to CH_4. Since methanogenic bacteria cannot thrive in the presence of oxygen, this stage is strictly anaerobic. Two distinct bacterial types, such as acetophilic and hydrogenophilic, transform CH_3COOH and H_2 into CO_2 and CH_4. The CH_3COOH is converted into CH_4 and CO_2 by acetophilic bacteria, whereas H_2 and CO_2 are converted to CH_4 by hydrogenophilic bacteria. Equations 9.8–9.10 provide illustrations of the reactions in this stage.

$$CH_3COOH \rightarrow CH_4 + CO_2 \quad (9.8)$$

$$CO_2 + 4H_2 \rightarrow CH_4 + 2H_2O \quad (9.9)$$

$$2CH_3CH_2OH + CO_2 \rightarrow CH_4 + 2CH_3COOH \quad (9.10)$$

9.3 COMPOSITION AND PROPERTIES OF BIOGAS

Biogas is primarily composed of CH_4 gas (40%–75%), CO_2 (15%–60%), and trace amounts of impurities like nitrogen, water vapor, hydrogen sulfide (H_2S), and other gases (Table 9.1). The volume of each gas varies depending on the digestion temperature and feedstock type. CH_4 gas is a vital component of biogas, and increasing its concentration leads to higher quality and cost-effective production. This can be achieved by upgrading biogas using chemical, physical, and biological means. Biogas is a translucent gas with a distinct odor analogous to rotten eggs. It is tasteless and, like CO_2, has a greenhouse effect. Biogas burns with a blue flame at 800°C. This gas is mixed with

TABLE 9.1
Composition and Characteristics of Biogas

Constituents	Concentration (v/v)	Properties
Methane	40%–75%	Combustible
Carbon dioxide	15%–60%	Non- combustible
Water vapor	1%–5%	Non- combustible
Nitrogen	0%–5%	Non- combustible
Hydrogen	Traces	Combustible
Hydrogen sulfide	0–5,000 ppm	Combustible
Oxygen	˂2%	Non- combustible
Trace gases	˂2%	
Ammonia	0–500 ppm	

Sources: Mursec et al. (2009); Baciocchi et al. (2013); Surendra et al. (2014).

air at a 1:20 ratio and has a high ignition speed. The required and optimal pressure for cooking with biogas is in the range of 5–20 cm of water column. Burning I m^3 of biogas can generate 5,500–6,500 kcal of heat, corresponding to running a one-horsepower engine for 2 h. A cubic meter of biogas is equivalent to 0.4 kg of diesel fuel, 0.6 kg of oil and 0.8 kg of coal (Mursec et al. 2009; Baciocchi et al. 2013; Surendra et al. 2014).

9.4 PROCESS CONDITIONS FOR THE ANAEROBIC DIGESTION

Due to the various microorganisms involved in the production of CH_4 gas, a number of parameters such as temperature, pH, carbon/nitrogen (C/N) ratio, hydraulic retention time (HRT), particle size, organic loading rate (OLR), inoculum, and mixing are in charge of the optimal functioning and operation of the AD technology (Figure 9.4). These parameters affect the performance of the biodigesters and the quality of the biogas being produced.

9.4.1 Temperature

One important factor affecting the production of biogas is temperature. Temperature typically ranges from 0°C to roughly 97°C in order to generate CH_4. Microorganisms can withstand a wide range of temperatures. Variations in temperature denature the activity of enzymes, which hinders the system's ability to digest food. Bacteria that are capable of AD operate at various temperatures. Thermophilic bacteria function

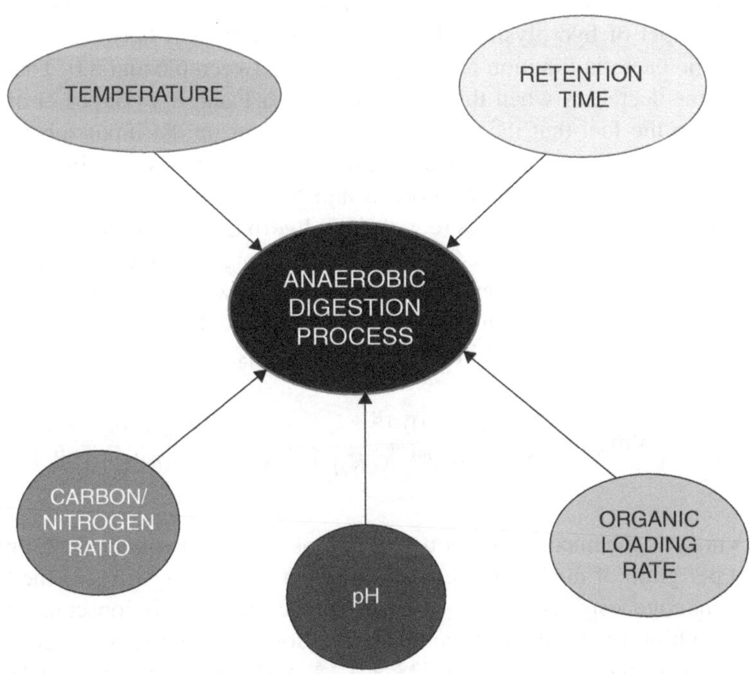

FIGURE 9.4 Process parameters of the anaerobic digestion process.

at temperatures between 55°C and 60°C, mesophilic bacteria tolerate at temperatures ranging from 35°C to 40°C, and psychrophilic bacteria operate at temperatures between 10°C and 20°C. Thermophilic digestion produces more biogas at a faster rate but requires more energy input. But frequently, the process is unsteady. Thermophilic temperature shortens the HRT by increasing the rate of feed breakdown. Additionally, it produces a digestate of superior quality with fewer microorganisms. The majority of commercial digesters operate at mesophilic temperature range. Contrarily, mesophilic digestion produces less biogas than thermophilic digestion. However, it delivers a steadier operation with fewer operational costs. The reaction rate and biogas yield increase as the operating temperature increases. Recent research has demonstrated that temperature affects the microbial communities and metabolic pathways involved in AD, with their metabolic activities increasing significantly as the temperature rises. However, at higher thermophilic temperatures, AD and the dewatering properties of the digested feedstock are reduced (Deago et al., 2023). Studies have shown a correlation between temperature and kinetic rate constant and HRT, with biogas production peaks occurring on different days at different temperatures.

9.4.2 pH

pH is a crucial parameter in the synthesis of biomethane. Different pH values are needed by the microorganisms of various groups participating in AD. In biogas digesters, pH often signifies stability and maintains system homeostasis. Biogas digesters with fluctuating pH values are often unstable and accumulate acid. This may lead to system failure and slow down the rate at which CH_4 is produced. While the favorable pH of hydrolysis and acidogenesis bacteria is between 5.5 and 6.5, methanogenic bacteria function best at pH values between 6.6 and 8.0. The production of biogas decreases when the pH is between 6.1 and 8.3 (Hagos et al. 2017). This is due to the fact that the pH value is dependent on the input substrates. pH drops as a result of the quick acidification process that occurs with substrates high in carbs. It will rise if there are more proteins and NH_3 in the substrate. Prior research has employed chemical substances (e.g. sodium hydroxide, bicarbonate, and NH_3) to increase the pH's alkalinity. Table 9.2 shows the pH ranges for the four stages of AD. The mathematical equation given as Equation 9.11 shows how pH is related to CH_4 production. The model equation can be used to estimate the highest and cumulative CH_4 rate, when pH is at optimal in different conditions.

$$\frac{dCH_4}{dt} = \left(Vm_{max} X_m \frac{Ac^{-1} \times 10^{-pH}}{Ac^{-1} \times 10^{-pH+} K_a K_M} \right) \times \left(\frac{K_{im} K_a}{K_{im} K_a + Ac^{-1} \times 10^{-pH}} \right), \quad (9.11)$$

where, Vm_{max} is the maximum production rate of biogas (in volume at °C and 1 atm pressure) per gram of methanogenesis bacteria per day (L/day), X_m is the biomass of the methanogenesis (g/L), K_M is the methane yield saturation constant (g/L), K_{im} is the inhibition constant of methane yield for CH_3COOH, K_a is constant for CH_3COOH dissociation given as 1.728×10^{-5}, and Ac is the ionized CH_3COOH concentration (g/L).

TABLE 9.2
pH Values for Different Stages of the Anaerobic Digestion Process

Anaerobic Digestion Stage	pH
Hydrolysis	5.5–6.5
Acetogenesis	5.5–6.5
Acidogenesis	6.6–.0
Methanogenesis	6.8–8.0

Source: Hagos et al. (2017).

9.4.3 Carbon/Nitrogen Ratio

Another important component of a balanced AD process is the substrate C/N ratio. Microorganisms use carbon as their energy source, but they also need a particular amount of nitrogen for development or metabolism. Depending on the properties of the substrate, the ideal C/N ratio for the majority of AD falls between 20 and 30. A high nitrogen concentration, which is represented by a low C/N ratio, will ultimately result in excess NH_3. Excess NH_3 lowers the biogas output by inhibiting the digestion process and increasing digester alkalinity. Conversely, a high C/N ratio indicates deficit of nitrogen in a substrate. This nitrogen will be quickly consumed by microorganisms, leaving insufficient nitrogen for their metabolism. In order to attain the ideal C/N ratio, the AD process frequently involves adding carbon or nitrogen-rich organic material to the substrate in a codigestion system (Kamusoko & Mukumba 2024).

Ghaleb et al. (2021) performed a study to investigate the codigestion of oily-biological sludge (low C/N) with sugarcane bagasse (high C/N) to improve biogas production. Results showed that increasing the C/N ratio of oily-sludge significantly improved biogas and CH_4 yield. At C/N of 30 (cosubstrate VS/inoculum VS = 0.18), maximum yields were achieved: 200.6 mL biogas/g VS removed and 65.1 mL CH_4/g VS removed, which was 70% and 67% higher, respectively, compared to C/N of 20. The improvement was linked to high volatile solids content, balanced nutrient ratios, and stable pH (6.9–7.3), which favored microbial activity.

9.4.4 Hydraulic Retention Time and Residence Time

HRT is defined as the amount of time the liquid phase (wastewater, slurry, and digestate) stays inside the biogas digester before it is released. The average amount of time the substrate spends in the digester is determined by HRT. To guarantee that most organic materials are converted into biogas, substrates should be left in the digester for an adequate amount of time. While residence time (sometimes called solid retention time) refers to the average length of time organic solids (substrate/biomass) remain inside the anaerobic digester, residence time affects the extent of organic matter breakdown, microbial population stability, and CH_4 yield.

Longer residence times generally allow for more complete digestion of organic material but require larger reactor volumes. However, as Equation 9.12 demonstrates, a longer HRT necessitates a larger digester size, raising the digester's capital cost. The amount of OLR and HRT determines the volume of the biogas digester. In mesophilic conditions, biomass deteriorates over a period of at least 10–30 d, whereas in thermophilic environments, retention duration is fewer than 10 d. It takes between 20 and 30 d with temperatures around 30°C to produce biogas (Garfí et al. 2011).

$$\text{Digester volume}\left(\text{m}^3\right) = \text{HRT}\left(\text{day}\right) \times \text{Substrate input flow rate}\left(\text{m}^3/\text{day}\right) \quad (9.12)$$

9.4.5 Organic Loading Rate

The OLR describes the consumption rate of organic material per unit volume of the digester. The OLR can be defined as:

$$OLR = Q \times COD / V \quad (9.13)$$

where COD, a measurement of the amount of organic material in any substrate, is expressed in kilograms COD/m^3, Q is the feed flow rate in meters per day, and V is the reactor volume in meters.

OLR is influenced by the amount of organic matter in the substrate, and the ideal rate is found through experimentation. Above-optimal OLR can be hazardous to the digester and reduce methanogenic activity. Conversely, biogas production is decreased by an OLR that is below ideal.

9.4.6 Particles Size

The impact of particle size is recognized as a very essential aspect in the AD technology. Its effect in terms of speed and stability depends on particle size distribution. For instance, the success of the AD technology is claimed to be dependent on the equipment used for processing the substrate, such as a rotary cutter, wet macerator, and the combination of the shear shredder (Zhang & Banks 2013). Such equipment has been utilized in different investigations to produce different particle sizes. The outcome demonstrates that particle size distribution affects the system's performance behavior rather than biogas production (Zhang & Banks 2013). The increase in surface area brought about by tiny particles affects methanogenic bacteria, which speeds up the production of biogas and the pace at which the feedstock breaks down. Accordingly, tiny particle sizes are recommended for the AD to function well. Overly large particle sizes have the drawback of clogging or obstructing biogas digesters.

9.4.7 Inoculum

Since biogas-producing bacteria are few in number, particularly in fresh organic waste, biogas generation occurs when there are enough microorganisms in the biogas digester. Gas output is increased by around 30%, when fresh organic waste is injected. It is beneficial to seed biogas digesters with active inoculums, ideally of low temperature.

9.4.8 Mixing

Mixing is another major factor to be considered in AD systems. Its main purpose is to ensure uniformity of the fluid and stability within the reactor. The activities of bacteria within the digester are dependent on the rate of mixing. For instance, slow mixing causes improper mixing and short circulation. Also, the contents of the digestion mixture affect the mixing taking place in the AD reactor. The kinetics of AD increase with respect to mixing to ensure an even distribution of the anaerobes and close contact with their substrate. The substrate needs to be stirred continuously. Another process related to mixing is agitation. It is also considered an important factor for the successful operation of the AD process. Stirring increases the rate of CH_4 production and minimizes or eliminates decomposition of solids inside the biodigesters. Inadequate agitation may lead to the formation of layers in the digesters, causing the digested sludge to be separated. Some examples of mixing devices used for the stirring mechanism are propeller and scraper. The best frequency for agitation depends on the type of reactor to be used. The impact of mixing and scale on performance and hydrodynamics of AD, stating their type of mixing and respective biogas production and CH_4 yield are shown in Table 9.3.

9.5 FEEDSTOCK FOR ANAEROBIC DIGESTION

There are various feedstocks that can be used for AD. These feedstocks include agricultural residues, food waste, sewage sludge, energy crops, and algae (Jiang et al. 2018). The properties of various feedstocks and their potential for biogas production through AD are discussed in this section.

TABLE 9.3
Type of Mixing with Their Values for Biogas Production, Methane Yield, Total Solids, and Volatile Solids

Type of Mixing	Biogas Production Rate (L/d)	Methane Yield (L/g VS Loaded)	Total Solids (%)	Volatile Solid Reduction (%)
Unmixed	0.92	0.19	41	35
Gas mixed	1.07	0.21	49	39
Impeller mixed	1.14	0.23	47	41
Slurry recirculation	1.20	0.24	45	35

9.5.1 Agricultural Waste

The AD of agricultural byproducts offers a reliable means of lowering GHG emissions and using waste for the creation of clean energy. Agricultural residues are materials that are frequently left over on farmed land after harvesting of crops or fiber byproducts. These include rice husks, wheat straw, corn stover, and sugarcane bagasse (Jingura & Matengaifa 2018; Kamusoko et al. 2021). Traditionally, as they burn, CO_2 is released into the atmosphere. Agricultural residues can be employed in the production of pulp, paper, textiles, and composites (Okolie et al. 2020). They might also be ploughed back into the ground, where their breakdown reduces the soil acidity and promotes leaching, or they can be used as bedding. Agricultural waste can be in the form of primary or secondary residues. Secondary residues are produced during processing, including coconut shell, saw dust, and palm oil shell (Rowan et al. 2022). Primary residues are field-based residues, often generated from crop harvesting. Examples include fruit bunches, palm oil frond, maize stalks, soybean straw, etc.

Agricultural residues are highly lignocellulosic, composed of cellulose (35–55 wt%), hemicellulose (20–40 wt%), and lignin (10–25 wt%), as main building blocks (Okolie et al. 2021). Table 9.4 shows the lignocellulosic composition of different types of agricultural residues. The most plentiful, widely distributed, and conveniently accessible organic substance on the planet is lignocellulosic biomass (Sahoo et al. 2022). However, biogas from lignocellulosic biomass is not being produced on a wide scale due to the high expenses of turning it into gaseous fuel. The integration and fundamental understanding of pretreatment is necessary for cost-competitive conversion of lignocellulose into biogas.

9.5.2 Energy Crops

Energy cover crops (ryegrass, miscanthus, switchgrass, etc.) are dedicated plants that are cultivated to produce high biomass yields for biofuel production. These crops

TABLE 9.4
Characteristics of Major Crop Agricultural Residues

Crop Residue	Lignin (%)	Cellulose (%)	Hemicellulose (%)	References
Barley straw	16–21	34–38	30–37	Sun et al. (2011); Meyer & van Rensburg (2022)
Corn stover	12–19	37–41	22–31	Mensah et al. (2021)
Rice straw	10–30	30–50	10–35	Chen et al. (2021)
Wheat straw	11–26	32–45	20–45	Zhang et al. (2022)
Sugar cane straw	19–43	29–44	27–31	Bilatto et al. (2020); Pagliuso et al. (2021)

are generally seeded and harvested between two cash crops, that is, in systems using double cropping or growing three crops in two years. They are neither grown as cash crops nor do they compete with food crops because they develop over a period that is usually too short to grow food crops. However, this period is already being increasingly used to produce supplementary fodder (Andersen et al. 2020; Binder et al. 2020) or even food crops due to climate change (Meza et al. 2008; Sandler et al. 2015). In Western Europe, energy cover crops are planted either during the summer fallow period (June–October) or during the winter fallow period (September–May). They function as multiservice cover crops. The biomass can be harvested and digested anaerobically to produce biogas. The fact that energy cover crops are harvested for biogas rather than being incorporated into the soil is what differentiates them from conventional multiservice cover crops (Graß et al. 2013).

9.5.3 Food Waste

Food processing industries annually produce a large amount of food waste, which mostly consists of waste from dairy, fruits, and vegetables, consumer or domestic garbage, perished or damaged items, general waste, and compost (Mirabella et al. 2014). In some nations, the retail industry is the primary source of nearly half of the food waste, as premium, high-energy food is discarded because of marketing mandates and overindulgent customers. For instance, about 2.6% (1.7 Mt/yr) of consumer industry products and 4% (17 Mt/yr) of agri-food products are thrown away in Italy. Moreover, around 5 Mt of food are lost from stores annually, with 40% of that quantity consisting of fruits and vegetables with high sugar content (Mirabella et al. 2014). Conventional waste management techniques are labor-intensive, exhibit ineffective closed-loop sustainability, and are harmful to biodegradability and CH_4 leaching, use of land area, and yield no financial gains (Bong et al. 2018). It is predicted that two tons of CO_2 emissions are generated for every ton of food waste produced, making the need for sustainable valorization routes for food waste (Mirabella et al. 2014). Food waste is becoming more important as an AD feedstock, not only for its harvestable energy value, but for its potential to help meet EU circular economy targets. High moisture content, low pH, biodegradability, and high dry matter content are characteristics of food waste that are associated with high energy potential and biogas output. Even though there is a chance to produce a lot of economic energy yields, the heterogeneity of food waste and the threshold limit of VFA inhibition may all work against it (Bong et al. 2018).

9.5.4 Sewage Sludge

Sewage sludge is largely produced in wastewater treatment plants during biological, chemical, and physical treatments. It is mainly composed of dewatered microbial biomass. Additionally, sewage sludge contains pathogens, heavy metals, and other hazardous materials (Li et al. 2018). Sewage sludge represents approximately 2% of the treated sewage volume (Zhen et al. 2017). Its management and treatment is expensive, accounting for nearly one-half of the total sewage treatment cost (Neumann et al. 2016). Basic disposal practices include agricultural use, landfills, composting, AD, recycling as a construction material, and incineration (Zhen et al. 2017; Hanum et al. 2019; Zhu et al. 2019). Most of these management practices have specific practical constraints related to human and environmental health (Zheng et al. 2020). AD is considered an effective, economical, and eco-friendly way to stabilize sludge, remove odor and pathogens, and generate CH_4 gas. However, the complex organic matter composition of sludge creates obstacles for efficient AD. The presence of a complex floc structure extracellular polymeric substances, recalcitrant cell walls, and other high molecular weight compounds in sludge have been identified to hinder the hydrolysis step (Kinnunen et al. 2015; Kumaran et al. 2016; Zhen et al. 2017; Xu & Dai 2020).

9.6 FEEDSTOCK PRETREATMENT

There are different pretreatment technologies necessary for AD to generate sufficient amounts of biogas. The volume of biogas generated and the AD performance can be impacted by the structure and composition of the waste feedstock (Angelidaki et al. 2018). Additionally, particle size of the feed material and its organic content affect the AD process. Materials with low organic content might not provide enough nutrients for efficient digestion, whereas larger particles are more difficult for bacteria to break down (Izumi et al. 2010). The structural complexity of lignocellulosic biomass poses some challenges during AD. This is due to the presents of cellulose, hemicellulose, and lignin, which are combined to create a matrix that is resistant to biodegradation (Carrere et al. 2016). Feed with complex composition and high lignocellulosic material needs an additional pretreatment step to promote digestion (Khalid et al. 2011). Pretreatment techniques such as mechanical, chemical, thermal, biological, plasma, and combinatorial techniques have been developed in recent decades to combat these biodegradation limits affecting AD (Lymperatou et al. 2023). By making biomass more biodegradable, these techniques can boost the efficiency of AD conversion. The primary goal of pretreatment is to extract lignin from materials that are lignocellulosic. The most challenging aspect of pretreatment is choosing an appropriate method without causing inhibition.

9.6.1 Mechanical Pretreatment

This process involves cutting, squeezing, and sorting waste to submit it to severe mechanical stress. The fundamental procedure entails particle size reduction of the organic waste. Mechanical methods aim to increase the surface area for microbial attack and transform the substrates promptly. The success of mechanical pretreatment influenced by the feed material, technique, and AD system parameters (Mata-Alvarez et al. 2000). Strong and dependable equipment is recommended to avoid harming the object and property, since wear and tear of the machinery is an issue to consider in mechanical pretreatment. Table 9.5 illustrates the different mechanical pretreatment methods and their impact on biogas yield for different feedstocks. It shows an increase in biogas yield after exposure of feedstocks to mechanical stress.

TABLE 9.5
Different Mechanical Pretreatment Technologies and Their Effect on Biogas Yield

Pretreatment	Feedstock	Biogas Yield Increase (%)	Reference
Microwave	OFMSW	78	Shahriari et al. (2012)
	Rice straw	71	Kainthola et al. (2019)
	Food waste	59	Yue et al. (2021)
	Herbal	72	Cheng & Liu (2010)
Electromagnetic mill	OFMSW	25 and 50	Litti et al. (2018)
	Wheat straw	4–29	Wahid et al. (2015)
Extrusion	Barley straw	70	Hjorth et al. (2011)
	Wheat straw	23	Hjorth et al. (2011)

OFMSW, Organic fraction of municipal solid waste.

9.6.2 Thermal Pretreatment

In this procedure, the substrate is heated, usually between 125°C and 190°C, under pressure for up to 1 h. Thermal pretreatment alters the physicochemical characteristics of the waste material to improve the digestion process. It is an energy-intensive process that can hydrolyze and solubilize complex organic components, including proteins, carbohydrates, and lipids into simpler ones that bacteria can readily digest. Furthermore, it enhances the waste biodegradability and biogas generation by releasing nutrients, destroying the cell walls of undesirable microorganisms, and reducing inhibiting substances (Yan et al. 2022). Operating temperature and treatment time have the most profound effects on the efficacy of thermal pretreatment. For instance, long treatment periods and high temperatures encourage the solubilization of organic matter, which may have an impact on the digestate composition (Xue et al. 2015). There are different types of thermal pretreatment such as vapothermal, microwave, steam explosion, hydrothermal, etc., and their effects on the substrate varies as shown in Table 9.6. Each thermal technology has its own advantages and disadvantages that determines its utility. Steam explosion at 180°C–240°C and 1–3.5 MPa seems to be the most promising thermal pretreatment that can enhance CH_4 production in AD.

9.6.3 Biological Pretreatment

The most promising, cost-effective, and ecologically friendly approach for producing biogas is biological pretreatment. It is a method that utilizes microorganisms (mostly bacteria and fungi) and their cellular components (i.e. enzymes) to disintegrate cross-linked structures in complex substrates, as shown in Table 9.7. Recent studies are putting emphasis on microbial consortium pretreatment, which is considered to be more effective than single microorganism pretreatment. Biological

TABLE 9.6
Different Types of Thermal Pretreatment Technologies

Pretreatment	Effects	Advantages	Disadvantages	State-of-the Art
Hydrothermal	- Increases pore volume and accessible surface area - Reduces crystallinity - Elutes degradable substances	- Capable of continuous operation - Recycling of waste heat - Easy use of dissolved degradable substances	- High energy consumption - Very low pH may cause inhibitions	- Large-scale applications for sanitation purposes - Few large-scale plants are already in operation
Vapothermal	- Dissolves organic substances creating larger pore volume - Improves grindability	- Waste heat can be reused - Relatively low energy use - Applicable for chunky or fibrous material	- Low water content is needed to minimize energy use - Precipitation could block the accessible surface	- Demonstration plants
Microwave	- Breaks down the educt matrix, making the material more easily digestible - Elutes degradable substances	- Rapid heat-up time - Small processing equipment	- Need for electrical energy - Scalability is costly	- Laboratory-scale
Steam explosion	- Tears the educt's structure - Hydrolyzes organic components	- Very short residence time - Low energy use when using dry substrates	- Necessity of steam condensation to use dissolved degradable substances - Higher temperature requirements	- Large-scale operations for lignocellulosic biomass, municipal solid wastes and sludge

pretreatment changes the content of organic matter by breaking down large organic chemicals into simple and more easily biodegradable molecules. The synthesis of sugars, pH stabilization, hydrolysis rate, adaptability, and substrate conversion rate can be increased by biological pretreatment in a mild environment without generating inhibitory compounds (Mishra et al. 2018). Biological pretreatment boosts biogas generation by lowering pollutants (e.g. sulfide and NH_3) that kill anaerobes

TABLE 9.7
Biological Pretreatment of Biomass for Anaerobic Digestion

Pretreatment	Feedstock	Biogas Yield Increase (%)	Reference
Mixed microbial community	Corn straw	131.6	Li et al. (2020)
Microbial consortium	Municipal solid waste	6.1–105.1	Yuan et al. (2014)
Microbial consortium	Cotton stalk	39.6–120.4	Yuan et al. (2016)
Cellulolytic microbial consortium	Crop residues	60.9–84.7	Kamusoko & Mukumba (2025b)
Rumen flora	Coconut fibers	46–76	Kivaisi & Eliapenda (1994)
Rumen flora	Bagasse	22–44	Kivaisi & Eliapenda (1994)
Enzymatic	Chicken manure	35	Kucuker et al. (2020)
Microaerobic pretreatment with *Bacillus subtilis*	Corn straw	17.4	Xu et al. (2018)
Phanerochaete chrysosporium and *Aspergillus niger*	Paper waste	226 and 73, respectively	Ofon et al. (2025)

(Harirchi et al. 2022). The material to be treated should, however, determine the choice of the biological pre-treatment method. Pretreatment conditions should be tailored for any relevant enzyme or microbial functions. The ability to operate at lower temperatures without the need for energy or chemicals is the primary benefit of the biological technology over other pretreatment methods. However, its drawbacks include the potential for slower processing times compared to the non-organic approach and the potential need for more extensive pretreatment and storage facilities (Kamusoko & Mukumba 2024).

9.6.4 Chemical Pretreatment

Chemical pretreatment is used to break down complex organic substances and alter their structure and pH. This makes a substrate more accessible and favorable for microbial degradation. For example, acidifying feed can dissolve organic matter, improving the biological breakdown of waste for increased biogas (Woiciechowski et al. 2020). The type of waste, chemical dosage and cost, and duration may affect the effectiveness of a particular pretreatment. Using strong chemicals in pretreatment could have a detrimental effect on process stability, microbial activity, and the environment (Hvid Hansen et al. 1998). As a result, several considerations should be taken into account, when choosing the chemical for biomass pretreatment. Alkaline pretreatment solubilizes lignin using alkaline solutions, including H_2O_2, $Ca(OH)_2$, NaOH, Na_2CO_3, and CaO (Baruah et al. 2018). With a variety of

TABLE 9.8
Effect of Chemical Pretreatment of Biogas Production

Pretreatment	Chemical	Feedstock	Biogas Yield Increase (%)	Reference
Alkaline	NaOH	Corn stover	37	Zhu et al. (2010)
	$Ca(OH)_2$	Sugarcane bagasse	23.3	Mustafa et al. (2018)
	NaOH	Pulp and paper sludge	183.5	Lin et al. (2009)
	NaOH	Rice straw	50	Sabeeh et al. (2020)
Acid	Peracetic acid	Sewage sludge	40.4	Ren et al. (2024)
	Acetic acid	Mixed agro-industrial waste	205	Nava-Valente et al. (2023)
	H_2SO_4	*Salvinia molesta*	71.3–81.8	Syaichurrozi et al. (2019)
	H_2SO_4	*Water hyacinth*	131.5	Sarto et al. (2019)
Ozonation	O_3	*Food waste*	5.0–.5	Ariunbaatar et al. (2014)
Fenton process	$Fe2^{+}/H_2O_2$	*Slaughterhouse sludge*	31	Anwer et al. (2025)

substrates such as corn cob, corn straw, wheat straw, pearl millet straw, and floral waste, alkaline pretreatment has been found to enhance biogas by up to 360%. Another technique for raising the yield of biomass's enzymatic hydrolysis is ammonia fiber expansion (AFEX). This procedure involves treating organic materials with liquid NH_3 at 325 psi, then rapidly release the pressure to increase the fiber structure and biodegradability. For example, Rojas-Sossa et al. (2019) showed that AFEX-treated maize stover codigestion systems produced 22% higher biogas and improved the archaea community. Ozonation and Fenton processes have demonstrated encouraging outcomes in the pretreatment of rice stover and straw, respectively. Table 9.8 shows the potential of various chemicals to treat biomass for enhanced biogas production.

9.7 BIOGAS FOR ELECTRICITY GENERATION

Using biogas in industrial settings offers environmental sustainability, waste management, and economic benefits. Cogeneration systems, also known as combined heat and power systems (CHP), are commonly used in commercial and industrial setups to maximize biogas efficiency (Kabeyi & Olanrewaju 2022). These systems can generate power and capture waste heat simultaneously, improving energy efficiency in several industrial operations. Industrial AD relies on CHP engine. It is the most vital aspect of generating renewable electricity; hence, it creates revenue for

AD. About 1 m^3 of CH_4 can generate 9.9 kWh of energy. During the process of producing electricity, two types of heat become available. These are low-grade hot water (70°C–90°C) and high-grade engine exhaust gas (450°C).

The CHP plant is advantageous because it improves energy efficiency and services, creates local jobs and global competitiveness, lowers GHGs and environmental degradation, and cuts funding for regional power plants. CHP units can attain up to 90% efficiency, surpassing stand-alone plants (~20%–45%) (Kamusoko & Mukumba 2025a). It can provide both electricity and heat, with electricity contributing about 60% of the total output. It is critical to clean H_2O and H_2S from biogas before feeding it into engines to avert condensation and corrosion. Biogas power plants in industrial sites can connect to the electrical grid (Liu et al. 2023). Surplus electricity can be transmitted to the grid, increasing overall power supply and possibly creating income through feed-in tariffs or power purchase agreements (Hou et al. 2022). Biogas is extensively utilized in industries, including food processing, agriculture, and wastewater treatment, to generate vast amounts of power (Afridi et al. 2023).

It is possible to use biogas in heating without the need for electricity generation. The process involves burning raw biogas in industrial boilers or heaters (Figure 9.5). Such heat has an efficiency of 75%–85% or higher (Ullah Khan et al. 2017). This heat can be used as a source of thermal energy in businesses like food processing, manufacturing, and chemical production (Sun et al. 2015).

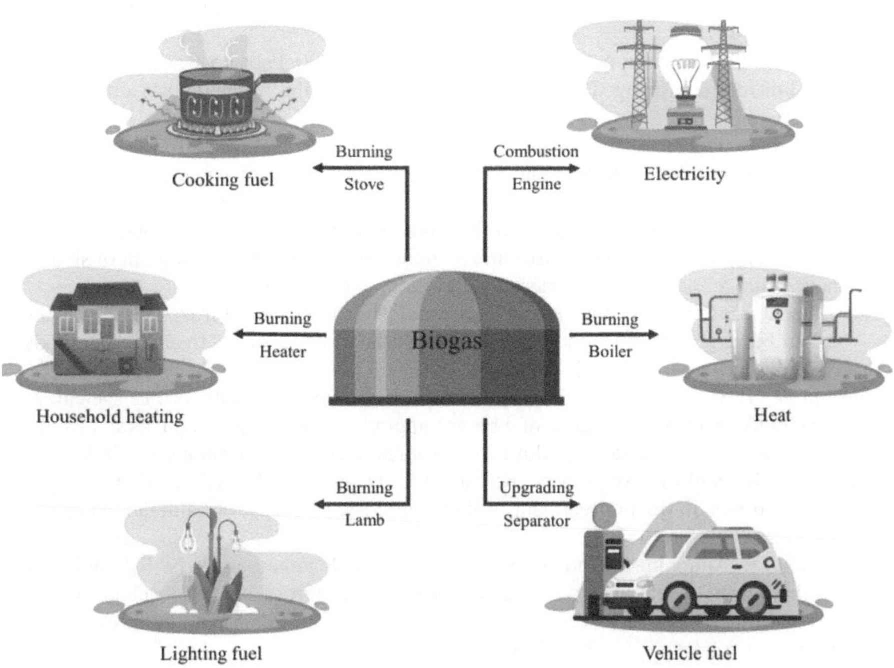

FIGURE 9.5 Biogas utilization and application.

9.8 CONCLUSION

AD for the generation of biogas and bioelectricity has been thoroughly examined in this chapter. It is a well-established, eco-friendly, and sustainable technology essential to the world's paradigm shift to a circular bioeconomy and renewable energy. AD involves the microbial degradation of organic matter (food waste, agricultural waste, sewage sludge, etc.) in the absence of oxygen, releasing a digestate rich in plant nutrients and biogas. Biogas is mostly made up of CO_2 and CH_4, although a few impurities may be present. The technology not only offers the twin advantages of energy recovery and waste management, but it also makes a significant contribution to the reduction of GHGs. AD promotes two primary energy-generating outputs: biogas, which can be converted into renewable natural gas or burned directly to generate heat and bioelectricity using CHP systems. Energy poverty can be improved, particularly in rural and decentralized areas, by using the electricity produced locally or feeding it into national grids. In conclusion, anaerobic digestion stands out as a critical green technology in the energy-to-waste interface. In line with international objectives for climate resilience and sustainable development, it provides a viable route for resource circularity, environmental preservation, and energy recovery when correctly executed.

REFERENCES

Adekunle, K.F., & J.A. Okolie. 2015. A review of biochemical process of anaerobic digestion. *Adv. Biosci. Biotechnol.* 6: 205–212.

Afridi, Z.U.R., K. Ullah, M.F. Mustafa, H. Saleem, B. Shaker, N. Ashraf, & S. Aslam. 2023. Biogas as sustainable approach for social uplift in South East Asian region. *Energy Rep.* 10: 4808–4818.

Andersen, B.J., D.P. Samarappuli, A. Wick, & M.T. Berti. 2020. Faba bean and pea can provide late-fall forage grazing without affecting maize yield the following season. *Agronomy* 10: 80.

Angelidaki, I., L. Treu, P. Tsapekos, G. Luo, S. Campanaro, H. Wenzel, & P.G. Kougias. 2018. Biogas upgrading and utilization: Current status and perspectives. *Biotechnol. Adv.* 36: 452–466.

Anwer, M., M.A.N. Shaikh, S.U. Khan, S. Mirkhalafi, M.S. Mahtab, I.H. Farooqi, S. Ayub, & M.H. Dehghani. 2025. Response surface modelling of Fenton pretreatment of slaughterhouse sludge for enhanced anaerobic digestion. *Sci Rep.* 15: 31584.

Ariunbaatar, J., A. Panico, L. Frunzo, G. Esposito, P.N.L. Lens, & F. Pirozzi. 2014. Enhanced anaerobic digestion of food waste by thermal and ozonation pretreatment methods. *J. Environ. Manage.* 146: 142–149.

Baciocchi, R., E. Carnevale, G. Costa, R. Gavasci, L. Lombardi, T. Olivieri, L. Zanchi, & D. Zingaretti. 2013. Performance of a biogas upgrading process based on alkali absorption with regeneration using air pollution control residues. *Waste Manage.* 33: 2694–2705.

Baruah, J., B.K. Nath, R. Sharma, S. Kumar, R.C. Deka, D.C. Baruah, & E. Kalita. 2018. Recent trends in the pretreatment of lignocellulosic biomass for value-added products. *Front. Energy Res.* 6: 141.

Begum, S., G.R. Anupoju, S. Sridhar, S.K. Bhargava, V. Jegatheesan, & N. Eshtiaghi. 2018. Evaluation of single and two stage anaerobic digestion of landfill leachate: Effect of pH and initial organic loading rate on volatile fatty acid (VFA) and biogas production. *Bioresour. Technol.* 251: 364–373.

Bilatto, S., J.M. Marconcini, L.H.C. Mattoso, & C.S. Farinas. 2020. Lignocellulose nanocrystals from sugarcane straw. *Ind. Crops Prod.* 157: 112938.

Binder, J.M., H.D. Karsten, D.B. Beegle, & C.J. Dell. 2020. Manure injection and rye double cropping increased nutrient recovery and forage production. *Agron. J.* 112: 2968–2977.

Bong, C.P.C., L.Y. Lim, C.T. Lee, J.J. Klemeš, C.S. Ho, & W.S. Ho. 2018. The characterization and treatment of food waste for improvement of biogas production during anaerobic digestion – A review. *J. Clean. Prod.* 172:1545–1558.

Carrere, H., G. Antonopoulou, R. Affes, F. Passos, A. Battimelli, G. Lyberatos, & I. Ferrer. 2016. Review of feedstock pretreatment strategies for improved anaerobic digestion: From lab-scale research to full-scale application. *Bioresour Technol.* 199: 386–397.

Chen, C., X. Deng, W. Kong, M.F. Qaseem, S. Zhao, Y. Li, & A.M. Wu. 2021. Rice straws with different cell wall components differ on abilities of saccharification. *Front. Bioeng. Biotechnol.* 8: 624314.

Cheng, X.Y., & C.Z. Liu. 2010. Enhanced biogas production from herbal-extraction process residues by microwave-assisted alkaline pretreatment. *J. Chem. Technol. Biotechnol.* 85: 127–131.

Deago, E., Ramírez, M., Espino, K., Nieto, D., Barragán, M., García, M., & Guevara-Cedeño, J. (2023). Optimizing anaerobic digestion at ambient temperatures: energy efficiency and cost reduction potential in panama. *Water (Switzerland), 15*(14). https://doi.org/10.3390/w15142653

Feng, K., Q. Wang, H. Li, Y. Zhang, Z. Deng, J. Liu, & X. Du. 2020. Effect of fermentation type regulation using alkaline addition on two-phase anaerobic digestion of food waste at different organic load rates. *Renew. Energy* 154: 385–393.

Garfí, M., L. Ferrer-Martí, I. Perez, X. Flotats, & I. Ferrer. 2011. Codigestion of cow and guinea pig manure in low-cost tubular digesters at high altitude. *Ecol. Eng.* 37: 2066–2070.

Ghaleb, A.A.S., S.R.M. Kutty, G.H.A. Salih, A.H. Jagaba, A. Noor, V. Kumar, N.M.Y. Almahbashi, A.A.H. Saeed, & B.N. Saleh Al-dhawi. 2021. Sugarcane bagasse as a cosubstrate with oil-refinery biological sludge for biogas production using batch mesophilic anaerobic codigestion technology: Effect of carbon/nitrogen ratio. *Water* 13: 590.

Graß, R., F. Heuser, R. Stülpnagel, H.P. Piepho, & M. Wachendorf. 2013. Energy crop production in double-cropping systems: Results from an experiment at seven sites. *Eur. J. Agron.* 51: 120–129.

Hagos, K., J. Zong, D. Li, C. Liu, & X. Lu. 2017. Anaerobic codigestion process for biogas production: Progress, challenges and perspectives. *Renew. Sustain. Energy Rev.* 76: 1485–1496.

Hanum, F., L.C. Yuan, H. Kamahara, H.A. Aziz, Y. Atsuta, T. Yamada, & H. Daimon. 2019. Treatment of sewage sludge using anaerobic digestion in Malaysia: Current state and challenges. *Front. Energy Res.* 7: 19.

Harirchi, S., S. Wainaina, T. Sar, S.A. Nojoumi, M. Parchami, M. Parchami, S. Varjani, S.K. Khanal, J. Wong, M.K. Awasthi, & M.J. Taherzadeh. 2022. Microbiological insights into anaerobic digestion for biogas, hydrogen or volatile fatty acids (VFAs): A review. *Bioengineered* 13: 6521–6557.

Hjorth, M., K. Gränitz, A.P.S. Adamsen, & H.B. Møller. 2011. Extrusion as a pretreatment to increase biogas production. *Bioresour. Technol.* 102: 4989–4994.

Hou, X., S. Zhong, & J. Zhao. 2022. A critical review on decarbonizing heating in China: Pathway exploration for technology with multi-sector applications. *Energies* 15: 1183.

Hvid Hansen, K., I. Angelidaki, & B.K. Ahring. 1998. Anaerobic digestion of swine manure: Inhibition by ammonia. *Water Res.* 32: 5–12.

Ighravwe, D.E., & M.O. Babatunde. 2018. Determination of a suitable renewable energy source for mini-grid business: A risk-based multicriteria approach. *J. Renew. Energy* 2018: 2163262.

Izumi, K., Y. Okishio, N. Nagao, C. Niwa, S. Yamamoto, & T. Toda. 2010. Effects of particle size on anaerobic digestion of food waste. *Int. Biodeterior. Biodegrad.* 64: 601–608.

Jiang, J., L. Li, M. Cui, F. Zhang, Y. Liu, Y. Liu, J. Long, & Y. Guo. 2018. Anaerobic digestion of kitchen waste: The effects of source, concentration, and temperature. *Biochem. Eng. J.* 135: 91–97.

Jingura R.M., & R. Matengaifa. 2018. The potential for energy production from crop residues in Zimbabwe. *Biomass Bioenergy* 32: 1287–1292.

Kabeyi, M.J.B., & O.A. Olanrewaju. 2022. Biogas production and applications in the sustainable energy transition. *J. Energy* 2022: 8750221.

Kainthola, J., M. Shariq, A.S. Kalamdhad, & V.V. Goud. 2019. Enhanced methane potential of rice straw with microwave assisted pretreatment and its kinetic analysis. *J. Environ. Manage.* 232:188–196.

Kamusoko, R., R.M. Jingura, Z. Chikwambi, & W. Parawira. 2021. Strategies for valorization of crop residues into biofuels and other value-added products. *Biofuels Bioprod. Bioref.* 15: 1950–1964.

Kamusoko, R., R.M. Jingura, W. Parawira, & Z. Chikwambi. 2022. Biogas: Microbiological research to enhance efficiency and regulation. In *Handbook of Biofuels*, ed. S. Sahay, 485–497. Elsevier.

Kamusoko, R., & P. Mukumba. 2024. Potential of wheat straw for biogas production by anaerobic digestion in South Africa: A review. *Energies* 17: 4662.

Kamusoko, R., & P. Mukumba. 2025a. Valorization of maize stover into biogas for heat and power generation: A South African perspective. *Fermentation* 11: 338.

Kamusoko, R., & P. Mukumba. 2025b. Effect of biological pretreatment with a cellulolytic bacteria consortium on biogas production from crop residues. Paper presented at the *42nd Johannesburg International Conference on Chemical, Biological and Environmental Engineering (JCBEE-25)*, 20–21 November 2025, Johannesburg, South Africa.

Khalid, A., M. Arshad, M. Anjum, T. Mahmood, & L. Dawson. 2011. The anaerobic digestion of solid organic waste. *Waste Manage.* 31: 1737–1744.

Kinnunen, V., A. Ylä-Outinen, & J. Rintala. 2015. Mesophilic anaerobic digestion of pulp and paper industry biosludge-long-term reactor performance and effects of thermal pretreatment. *Water Res.* 87:105–111.

Kiran, E.U., A.P. Trzcinski, W.J. Ng, & Y. Liu. 2014. Bioconversion of food waste to energy: A review. *Fuel* 134: 389–399.

Kivaisi, A.K., & S. Eliapenda. 1994. Pretreatment of bagasse and coconut fibers for enhanced anaerobic degradation by rumen microorganisms. *Renew. Energy* 5: 791–795.

Kour, D., K.L. Rana, N. Yadav, A.N. Yadav, A.A. Rastegari, C. Singh, P. Negi, K. Singh, & A.K. Saxena. 2019. Technologies for biofuel production: Current development, challenges, and future prospects. In *Prospects of Renewable Bioprocessing in Future Energy Systems. Biofuel and Biorefinery Technologies*, vol. 10, ed. A. Rastegari, A. Yadav, & A. Gupta. Springer.

Kucuker, M.A., B. Demirel, & T.T. Onay. 2020. Enhanced biogas production from chicken manure via enzymatic pretreatment. *J. Mater. Cycles Waste Manag.* 22: 1521–1528.

Kumaran, P., D. Hephzibah, R. Sivasankari, N. Saifuddin, & A.H. Shamsuddin. 2016. A review on industrial scale anaerobic digestion systems deployment in Malaysia: Opportunities and challenges. *Renew. Sustain. Energy Rev.* 56: 929–940.

Li, H., D. Si, C. Liu, K. Feng, & C. Liu. 2018. Performance of direct anaerobic digestion of dewatered sludge in long-term operation. *Bioresour. Technol.* 250: 355–364.

Li, Y., D. Hua, H. Xu, F. Jin, H. Mu, Y. Zhao, & X. Fang. 2020. Acidogenic and methanogenic properties of corn straw silage: Regulation and microbial analysis of two-phase anaerobic digestion. *Bioresour. Technol.* 307: 123180.

Lin, Y., D. Wang, S. Wu, C. Wang. 2009. Alkali pretreatment enhances biogas production in the anaerobic digestion of pulp and paper sludge. *J. Hazard. Mater.*170: 366–373.

Litti, Y., D. Kovalev, A. Kovalev, I. Katraeva, Y. Russkova, & A. Nozhevnikova. 2018. Increasing the efficiency of organic waste conversion into biogas by mechanical pretreatment in an electromagnetic mill. *J. Phys. Conf. Ser.* 1111: 012013.

Liu, Z., G. Ming, & J. Su. 2023. Performance testing and case studies of rural household biogas micro-CHP system. *E3S Web Conf.* 466: 01005.

Luo, G., L. Xie, Q. Zhou, & I. Angelidaki. 2011. Enhancement of bioenergy production from organic wastes by two-stage anaerobic hydrogen and methane production process. *Bioresour. Technol.* 102: 8700–8706.

Lymperatou, A., T.K. Engelsen, I.V. Skiadas, & H.N. Gavala. 2023. Prediction of methane yield and pretreatment efficiency of lignocellulosic biomass based on composition. *Waste Manage.* 155: 302–310.

Mari, A.G., C.L. Andreani, T.U. Tonello, L.C.C. Leite, J.R. Fernandes, D.D. Lopes, J.A.D. Rodrigues, & S.D. Gomes. 2020. Biohydrogen and biomethane production from cassava wastewater in a two-stage anaerobic sequencing batch biofilm reactor. *Int. J. Hydrog. Energy* 45: 5165–5174.

Mata-Alvarez, J., S. Macé, & P. Llabrés. 2000. Anaerobic digestion of organic solid wastes. An overview of research achievements and perspectives. *Bioresour. Technol.* 74: 3–16.

Meegoda, J.N., B. Li, K. Patel, & L.B. Wang. 2018. A review of the processes, parameters, and optimization of anaerobic digestion. *Int. J. Environ. Res. Public Health* 15: 2224.

Mensah, M.B., H. Jumpah, N.O. Boadi, & J.A.M. Awudza. 2021. Assessment of quantities and composition of corn stover in Ghana and their conversion into bioethanol. *Sci. Afr.* 12: e00731.

Meyer, G.V.O., & E. van Rensburg. 2022. A rumen based anaerobic digestion approach for lignocellulosic biomass using barley straw as feedstock. *S. Afr. J. Chem. Eng.* 41: 98–104.

Meza, F.J., D. Silva, & H. Vigil. 2008. Climate change impacts on irrigated maize in Mediterranean climates: Evaluation of double cropping as an emerging adaptation alternative. *Agric. Syst.* 98: 21–30.

Mirabella, N., V. Castellani, & S. Sala. 2014. Current options for the valorization of food manufacturing waste: A review. *J. Clean. Prod.* 65: 28–41.

Mishra, S., P.K. Singh, S. Dash, & R. Pattnaik. 2018. Microbial pretreatment of lignocellulosic biomass for enhanced biomethanation and waste management. *3 Biotech* 8: 458.

Molino, A., V. Larocca, S. Chianese, & D. Musmarra. 2018. Biofuels production by biomass gasification: A review. *Energies* 11: 811.

Mumme, J., F. Srocke, K. Heeg, & M. Werner. 2014. Use of biochars in anaerobic digestion. *Bioresour. Technol.* 164: 189–197.

Mursec, B., P. Vindis, M. Janzekovic, M. Brus, & F. Cus. 2009. Analysis of different substrates for processing into biogas. *J. Achiev. Mater. Manuf. Eng.* 37: 652–659.

Mustafa, A.M., H. Li, A.A. Radwan, K. Sheng, & X. Chen. 2018. Effect of hydrothermal and $Ca(OH)_2$ pretreatments on anaerobic digestion of sugarcane bagasse for biogas production. *Bioresour. Technol.* 259: 54–60.

Naik, S. N., V.V. Goud, P.K. Rout, & A.K. Dalai. 2010. Production of first and second generation biofuels: A comprehensive review. *Renew. Sustain. Energy Rev.* 14: 578–597.

Nava-Valente, N., E. Hernández-Aguilar, A. Alvarado-Lassman, & J.M. Méndez-Contreras. 2023. Effect of acid pretreatment on the anaerobic codigestion of sewage sludge, chicken litter, and sugarcane waste (SCW) for biogas production. *Bioenerg. Res.* 16: 1889–1901.

Neumann, P., S. Pesante, M. Venegas, & G. Vidal. 2016. Developments in pretreatment methods to improve anaerobic digestion of sewage sludge. *Rev. Environ. Sci. Biotechnol.* 15: 173–211.

Ofon, U.A., U.U. Ndubuisi-Nnaji, A.A. Adegoke, N.-A.O. Offiong, O.P. Ewenike, & S.E. Shaibu. 2025. Bioenergy potential of paper waste: Fungal pretreatment and kinetics modelling. *Next Sustain.* 5: 100086.

Okolie, J.A., S. Nanda, A.K. Dalai, & J.A. Kozinski. 2020. Hydrothermal gasification of soybean straw and flax straw for hydrogen-rich syngas production: Experimental and thermodynamic modeling. *Energy Convers. Manage.* 208: 112545.

Okolie, J.A., S. Nanda, A.K. Dalai, & J.A. Kozinski. 2021. Chemistry and specialty industrial applications of lignocellulosic biomass. *Waste Biomass Valor.* 12: 2145–2169.

Osman, A.I., A. Abdelkader, C. Farrell, D. Rooney, & K. Morgan. 2019. Reusing, recycling and up-cycling of biomass: A review of practical and kinetic modelling approaches. *Fuel Process. Technol.* 192: 179–202.

Pagliuso, D., A. Grandis, C.R. de Sousa, A.P. de Souza, C. Driemeier, & M.S. Buckeridge. 2021. The effect of sugarcane straw aging in the field on cell wall composition. *Front. Plant Sci.* 12: 652168.

Ramaraj, R., & N. Dussadee. 2015. Biological purification processes for biogas using algae cultures: A review. *Int. J. Precis. Eng. Manuf. Green Technol.* 4: 20–32.

Ren, W., Y. Zhang, X. Liu, S. Li, H. Li, & Y. Zhai. 2024. Peracetic acid pretreatment improves biogas production from anaerobic digestion of sewage sludge by promoting organic matter release, conversion and affecting microbial community. *J. Environ. Manage.* 349: 119427.

Rivière, D., V. Desvignes, E. Pelletier, S. Chaussonnerie, S. Guermazi, J. Weissenbach, T. Li, P. Camacho, & A. Sghir. 2009. Towards the definition of a core of microorganisms involved in anaerobic digestion of sludge. *ISME J.* 3: 700–714.

Rojas-Sossa, J.P., Y. Zhong, F. Valenti, J. Blackhurst, T. Marsh, D. Kirk, D. Fang, B. Dale, & W. Liao. 2019. Effects of ammonia fiber expansion (AFEX) treated corn stover on anaerobic microbes and corresponding digestion performance. *Biomass Bioenergy* 127: 105263.

Rowan, M., G.C. Umenweke, E.I. Epelle, I.C. Afolabi, P.U. Okoye, B. Gunes, & J.A. Okolie. 2022. Anaerobic codigestion of food waste and agricultural residues: An overview of feedstock properties and the impact of biochar addition. *Digit. Chem. Eng.* 4: 100046.

Sabeeh, M., S. Zeshan, R. Liaquat, & A. Maryam. 2020. Effect of alkaline and alkaline-photocatalytic pretreatment on characteristics and biogas production of rice straw. *Bioresour. Technol.* 309: 123449.

Sahoo, A., S. Kumar, & K. Mohanty. 2022. A comprehensive characterization of non-edible lignocellulosic biomass to elucidate their biofuel production potential. *Biomass Convers. Biorefin.* 12: 5087–5103.

Sandler, L., K.A. Nelson, & C. Dudenhoeffer. 2015. Winter wheat row spacing and alternative crop effects on relay-intercrop, double-crop, and wheat yields. *Int. J. Agron.* 2015: 369243.

Sarto, S., H. Raudati, & I. Syaichurrozi. 2019. Effect of chemical pretreatment using sulfuric acid on biogas production from water hyacinth and kinetics. *Renew. Energy* 132: 335–350.

Schievano, A., A. Tenca, S. Lonati, E. Manzini, & F. Adani. 2014. Can two-stage instead of one-stage anaerobic digestion really increase energy recovery from biomass?. *Appl. Energy* 124: 335–342.

Senés-Guerrero, C., F.A. Colón-Contreras, J.F. Reynoso-Lobo, B. Tinoco-Pérez, J.H. Siller-Cepeda, & A. Pacheco. 2019. Biogas-producing microbial composition of an anaerobic digester and associated bovine residues. *Microbiology Open* 8: e00854.

Shahriari, H., M. Warith, M. Hamoda, & K.J. Kennedy. 2012. Anaerobic digestion of organic fraction of municipal solid waste combining two pretreatment modalities, high temperature microwave and hydrogen peroxide. *Waste Manage.* 32: 41–52.

Sun, Q., H. Li, J. Yan, L. Liu, Z. Yu, & X. Yu. 2015. Selection of appropriate biogas upgrading technology – A review of biogas cleaning, upgrading and utilization. *Renew. Sustain. Energy Rev.* 51: 521–532.

Sun, X.F., Z. Jing, P. Fowler, Y. Wu, & M. Rajaratnam. 2011. Structural characterization and isolation of lignin and hemicelluloses from barley straw. *Ind. Crops Prod.* 33: 588–598.

Surendra, K.C., D. Takara, A.G. Hashimoto, & S.K. Khanal. 2014. Biogas as a sustainable energy source for developing countries: Opportunities and challenges. *Renew. Sustain. Energy Rev.* 31: 846–859.

Syaichurrozi, I., P.K. Villta, N. Nabilah, & R. Rusdi. 2019. Effect of sulfuric acid pretreatment on biogas production from *Salvinia molesta*. *J. Environ. Chem. Eng.* 7: 102857.

Ullah Khan, I., M.H.D. Othman, H. Hashim, T. Matsuura, A.F. Ismail, M. Rezaei-DashtArzhandi, & I. Wan Azelee. 2017. Biogas as a renewable energy fuel – A review of biogas upgrading, utilization and storage. *Energy Convers. Manage.* 150: 277–294.

Wahid, R., M. Hjorth, S. Kristensen, & H.B. Møller. 2015. Extrusion as pretreatment for boosting methane production: Effect of screw configurations. *Energy Fuels* 29: 4030–4037.

Walter, A., M. Probst, I.H. Franke-Whittle, C. Ebner, S.M. Podmirseg, M. Etemadi-Shalamzari, S. Hupfauf, & H. Insam. 2019. Microbiota in anaerobic digestion of sewage sludge with and without cosubstrates. *Water Environ. J.* 33: 214–222.

Woiciechowski, A.L., C.J.D. Neto, L.P. de Souza Vandenberghe, D.P. de Carvalho Neto, A.C.N. Sydney, L.A.J. Letti, S.G. Karp, L.A.Z. Torres, & C.R. Soccol. 2020. Lignocellulosic biomass: Acid and alkaline pretreatments and their effects on biomass recalcitrance – Conventional processing and recent advances. *Bioresour. Technol.* 304: 122848.

Xu, W., S. Fu, Z. Yang, J. Lu, & R. Guo. 2018. Improved methane production from corn straw by microaerobic pretreatment with a pure bacteria system. *Bioresour. Technol.* 259: 18–23.

Xu, Y., & X. Dai. 2020. Integrating multi-state and multi-phase treatment for anaerobic sludge digestion to enhance recovery of bioenergy. *Sci. Total Environ.* 698: 134196.

Xue, Y., H. Liu, S. Chen, N. Dichtl, X. Dai, & N. Li. 2015. Effects of thermal hydrolysis on organic matter solubilization and anaerobic digestion of high solid sludge. *Chem. Eng. J.* 264:174–180.

Yan, W., H. Xu, D. Lu, & Y. Zhou. 2022. Effects of sludge thermal hydrolysis pretreatment on anaerobic digestion and downstream processes: Mechanism, challenges and solutions. *Bioresour. Technol.* 344: 126248.

Yang, G., Y. Sun, L. Li, P. Lv, X. Kong, & D. Huang. 2018. Hydrolysis dynamics for batch anaerobic digestion of elephant grass. *RSC Adv.* 8: 22670–22675.

Yuan, X., L. Ma, B. Wen, D. Zhou, M. Kuang, W. Yang, & Z. Cui. 2016. Enhancing anaerobic digestion of cotton stalk by pretreatment with a microbial consortium (MC1). *Bioresour. Technol.* 207: 293–301.

Yuan, X., B. Wen, X. Ma, W. Zhu, X. Wang, S. Chen, & Z. Cui. 2014. Enhancing the anaerobic digestion of lignocellulose of municipal solid waste using a microbial pretreatment method. *Bioresour. Technol.* 154: 1–9.

Yue, L., J. Cheng, S. Tang, X. An, J. Hua, H. Dong, & J. Zhou. 2021. Ultrasound and microwave pretreatments promote methane production potential and energy conversion during anaerobic digestion of lipid and food wastes. *Energy* 228: 120525.

Zamorano-López, N., L. Borrás, A. Seco, & D. Aguado. 2020. Unveiling microbial structures during raw microalgae digestion and codigestion with primary sludge to produce biogas using semi-continuous AnMBR systems. *Sci. Total Environ.* 699: 134365.

Zhang, L., A. Larsson, A. Moldin, & U. Edlund. 2022. Comparison of lignin distribution, structure, and morphology in wheat straw and wood. *Ind. Crops Prod.* 187: 115432.

Zhang, Y., & C.J. Banks. 2013. Impact of different particle size distributions on anaerobic digestion of the organic fraction of municipal solid waste. *Waste Manage.* 33: 297–307.

Zhen, G., X. Lu, H. Kato, Y. Zhao, & Y.Y. Li. 2017. Overview of pretreatment strategies for enhancing sewage sludge disintegration and subsequent anaerobic digestion: Current advances, full-scale application and future perspectives. *Renew. Sustain. Energy Rev.* 69: 559–577.

Zheng, X., Z. Jiang, Z. Ying, J. Song, W. Chen, & B. Wang. 2020. Role of feedstock properties and hydrothermal carbonization conditions on fuel properties of sewage sludge-derived hydrochar using multiple linear regression technique. *Fuel* 271: 117609.

Zhu, J., C. Wan, & Y. Li. 2010. Enhanced solid-state anaerobic digestion of corn stover by alkaline pretreatment. *Bioresour. Technol.* 101: 7523–7528.

Zhu, X., W. Yuan, M. Lang, G. Zhen, X. Zhang, & X. Lu. 2019. Novel methods of sewage sludge utilization for photocatalytic degradation of tetracycline-containing wastewater. *Fuel* 252: 148–156.

10 Biotechnology of Single Cell Protein Production from Biomass Feedstock

Thembekile Ncube, Nerve Zhou, Tawanda P. Makopa, and Ngonidzashe Mangoma

10.1 INTRODUCTION

The increasing global demand for food protein has led to the expansion of livestock production and aquaculture, creating a shortage of feed protein, hence the need for sustainable innovative strategies to meet dietary protein requirements (Kim et al. 2019). The global demand for alternative protein sources is projected to reach US$ 290 billion by 2035 (Severo et al. 2024), and recent studies have considered several potentially sustainable substitutes, including plant-based protein, insect-based protein, and cell-based/laboratory-cultured meat (de Oliveira Padilha et al. 2022; Li et al. 2024; Liceaga et al. 2022; Nirmal et al. 2025). While plant-based protein may be demanding on resources such as land, water, and energy (Wu et al. 2014), the other two alternatives are promising but are faced by significant consumer acceptance and safety concerns hurdles (Megido et al. 2016; Pakseresht et al. 2022). One emerging alternative that appears to bear significant potential is single cell protein (SCP), a type of cheap protein derived from microbial cells. SCP has gained traction due to its relatively high nutritional value and low environmental impact (D'Almeida & de Albuquerque 2025; Fasihi et al. 2025).

This chapter will discuss the current diversity of microbial strains used in SCP production, the various biomass feedstocks, and the advances in the pretreatment technologies. Furthermore, this chapter will delve into the biotechnological improvements seen through metabolic engineering and process optimization, highlighting the current challenges and future perspectives in SCP production.

10.1.1 What Is Single Cell Protein?

SCP refers to dried microbial cells belonging to bacteria, fungi, or microalgae that are formulated and used as a protein supplement in human food and livestock feed (Reihani & Khosravi-Darani 2019). Derived from microorganisms

 DOI: 10.1201/9781003650751-10

such as bacteria, filamentous fungi, microalgae, and yeasts, SCP offers high protein content, essential amino acids, and can be produced rapidly with minimal land and water use (Ledbetter 2025; Raziq et al. 2020). The main advantage of microbial sources of protein is their rapid growth on cheap and easily accessible growth media, making proteins available for humans and/or livestock on short notice (Ritala et al. 2017). Moreover, SCP is characterized by high protein content, which generally exceeds that of most plant- and meat-based sources. It is comparable to insect protein in terms of dry weight (D'Almeida & de Albuquerque 2025). Among SCP sources, bacterial and microalgae typically exhibit the highest protein content (50%–80% and 60%–70% dry weight, respectively), while fungi (30%–50% dry weight) tend to have slightly lower protein content (Ritala et al. 2017). Although all SCP sources contain complete essential amino acid profiles, quantitative differences may exist as observed in fungi, which often exhibit low sulfur-containing amino acids such as methionine (Puyol et al. 2017). The final formulation and presentation of SCP comes in various forms such as bars, tablets or capsules, supplements or may be incorporated into other foods such as cereal bars or other protein-rich foods.

Indeed, SCP production could play a significant role toward the achievement of several sustainable development goals (SDGs), including food security, good health, clean energy, and promoting inclusive and sustainable industrialization (Akenji & Bengtsson 2014; Sekoai et al. 2024). Although the diversity of microbial species with potential in SCP production is a key driving force for advancement in this field, they still remain largely underexplored (Dong et al. 2025). In addition to their desirable capacity to use diverse raw materials, they can utilize readily available carbon sources such as biomass feedstock, significantly reducing the cost of production (Muniz et al. 2024). Biomass feedstocks are central to making SCP, a sustainable, cost-effective, and globally relevant solution to protein scarcity (Sharifian et al. 2025). While SCP production offers significant profit potential compared to traditional protein production methods, there is a need for more techno-economic studies on the upcycling of specific biomass feedstocks into high-value SCP (Vlaeminck et al. 2023). The ongoing advances in the biotechnology of SCP from biomass feedstocks are aimed at improved microbial strains, efficient pretreatment technologies, and optimized conditions for microbial cell accumulation (Sharifian et al. 2025).

10.1.2 History of Single Cell Protein

Research into SCP began in the early 20th century as scientists explored microbial sources of protein to combat food shortages, particularly during times of war (Ritala et al. 2017). In the 1910s, yeast was studied in Germany as a potential food supplement, and by World War II, various countries were using microbial biomass such as *Candida utilis* to address nutritional deficiencies. The post-war era saw increased interest due to population growth and protein scarcity, leading to large-scale industrial research by the 1970s onwards (Onyeaka et al. 2022). The British Petroleum and Imperial Chemical Industries developed SCP from bacteria using hydrocarbons, while the National Aeronautics and Space Administration (NASA) investigated its use in closed-loop life support systems for astronauts (Spalvins et al. 2023). In the 1980s, the United Kingdom introduced Quorn, the first commercially successful SCP product derived from the fungus *Fusarium venenatum* (Whittaker et al. 2020). Although early efforts faced challenges related to cost and public acceptance, advances in biotechnology and sustainability concerns have revitalized SCP research in recent decades, positioning it as a promising solution for global food security.

10.2 SELECTION AND IMPROVEMENT OF MICROBIAL STRAINS

A wide range of microbial species is employed in SCP production due to their high growth rates, protein content, and adaptability to diverse feedstocks (Bajpai 2017). Bioconversion of fermentable sugars in biomass feedstocks into SCP can be done by facultative anaerobes or obligate aerobes (Kasperski 2023). These microorganisms can simultaneously be employed as cocultures for improved efficiency and productivity. While microorganisms are mostly used for cell biomass production, recent studies have employed them in the pretreatment of biomass feedstocks for their ability to break down complex lignocellulosic compounds (Munir & Levin 2016). In addition to unique physiological attributes such as wide carbon assimilation profiles and robust industrial stress tolerance (Makopa et al. 2024), the selection of these microbial species is based on safety and pathogenicity risk (Hadi & Brightwell 2021).

Several studies have reported the potential in bioprospecting microbial species from their natural environments for novel species with unique attributes acquired through adaptation (Antranikian et al. 2005; Niehaus et al. 1999; Nwaefuna et al. 2021; Rammala & Zhou 2022). Thus, targeted isolation of extremophilic microbial species from biomass feedstocks could lead to the discovery of more efficient microbial strains. Table 10.1 shows the diversity of microbial strains used in SCP production, although some are multicellular organisms.

Yeasts such as *Saccharomyces cerevisiae* are widely used in food and feed industries for their high nutritional value and ability to grow on agro-industrial waste and molasses (Ravindra 2000). Although generally regarded as safe, there have been several reports of clinical isolates of *S. cerevisiae*, and thus, handling of any microbial strains for food and feed should be done cautiously (Munoz et al. 2005; Murphy & Kavanagh 1999; Strope et al. 2015). Several methylotrophic, ethanol-utilizing, and fat- and hydrocarbon-utilizing yeasts, including species of *Candida, Ogataea, Pichia*, and *Trichosporon* have also been used for biomass production (Kurtzman et al. 2011). While methylotrophs are usually preferred because of their ability to use inexpensive substrate, methanol, other non-methylotrophic yeasts have been utilized, including *C. utilis* (Qi et al. 2024). Yeasts possess unique flocculation abilities and relatively large cell sizes, making it easier to harvest the biomass from fermentation liquor. Moreover, unlike bacteria, many yeasts contain low quantities of nucleic acids, which can have detrimental nutritional effects (Ravindra 2000). However, some yeasts used as food supplements have been suspected of causing various health problems such as allergies, vaginitis, irritable

TABLE 10.1
Current Microbial Species Used in Single Cell Protein Production

Microorganism	Type	Substrate	Applications	Reference
Candida utilis	Yeast	Molasses, wood hydrolysate, industrial waste	Animal feed, flavor enhancer in food	Kurcz et al. (2018)
Saccharomyces cerevisiae	Yeast	Starch hydrolysate, agricultural waste	Livestock feed, human nutritional supplements	Gervasi et al. (2018)
Fusarium venenatum	Fungus	Glucose, starch-rich substrates	Mycoprotein (Quorn) for human consumption	Whittaker et al. (2020)
Aspergillus niger	Fungus	Food waste, agro-industrial residues	Enzyme production, feed supplements	Oshoma & Eguakun-Owie (2018)
Rhizopus oligosporus	Fungus	Soybeans, starch waste	Fermented single cell protein, traditional tempeh	Farooq et al. (2025)
Spirulina platensis	Cyano-bacterium	Carbon dioxide, sunlight, mineral salts	Human dietary supplements, cosmetics	Becker (2013)
Chlorella vulgaris	Microalga	Carbon dioxide, light, nutrients	Food supplements, aquaculture feed	Becker (2013)
Scenedesmus obliquus	Microalga	Carbon dioxide, sunlight, waste nutrients	Aquafeed, protein-rich biomass	Barka & Blecker (2016)
Methylococcus capsulatus	Bacterium	Methane	Fish and livestock feed	Øverland et al. (2010)
Methylophilus methylotrophus	Bacterium	Methanol	Animal feed	Mahmoudi et al. (2024)
Cupriavidus necator	Bacterium	Carbon dioxide, hydrogen	Future sustainable food systems, space food	Matassa (2016)
Hydrogenobacter thermophilus	Bacterium	Hydrogen, carbon dioxide	Autotrophic single cell protein production	Watanabe et al. (2023)

bowel syndrome, immunodeficiency, and other maladies (Airola et al. 2006; Mazzanti et al. 2017; Smith et al. 2024). This is thought to be due to the presence of yeast (fungal) antigens or to high levels of chromium and salicylates, and possibly contaminants within the processes (Airola et al. 2006; Raziq et al. 2020; Smith et al. 2024).

Fungi like *F. venenatum* are notable for producing mycoprotein, commercialized under the brand Quorn and used as a meat substitute in vegetarian diets (Wiebe 2002). Algae, including *Spirulina platensis* and *Chlorella vulgaris*, are rich in proteins, vitamins, and antioxidants. They are used in dietary supplements and functional foods (Becker 2013). Bacteria such as *Methylococcus capsulatus* and *Methylophilus methylotrophus* are employed in large-scale SCP production from methane and methanol, respectively. They are primarily used in aquaculture and livestock feeds (Gundupalli et al. 2024; Øverland et al. 2010). Emerging research explores the use of hydrogen-oxidizing bacteria like *Cupriavidus necator*, which can convert carbon dioxide and hydrogen into high-protein biomass offering sustainable food options, even in resource-limited settings (Matassa 2016).

While many organisms have been evaluated for their potential use in SCP production in their native state, scope exists for the improvement of these organisms using techniques such as random or targeted mutagenesis or genetic engineering (Balagurunathan et al. 2022). Through metabolic engineering, microorganisms are genetically modified to enhance protein yield, improve amino acid profiles, and broaden substrate utilization. This enables them to grow on unconventional feedstocks such as methanol, carbon dioxide, or lignocellulosic biomass. A number of engineering studies have been done using yeasts as models. For example, Xie et al. (2025) systematically engineered the cell wall of *S. cerevisiae* to improve SCP production. In non-conventional yeasts, which are rapidly expanding, a strain of *Yarrowia lipolytica* was engineered to simultaneously produce lipase and SCP from agro-industrial waste (Yan et al. 2018). Other microbial species have been improved by metabolic engineering in enhanced protein biomass production such as *Corynebacterium glutamicum* (Wolf et al. 2021), *Komagataella phaffii* (Carneiro et al. 2022), and *Escherichia coli* (Blandón et al. 2023). Metabolic engineering has been shown to improve various aspects related to SCP production, including tolerance to inhibitors generated during biomass pretreatment and low levels of unwanted byproducts, channeling more metabolic energy toward protein synthesis (Ling et al. 2014).

10.3 DIVERSITY AND AVAILABILITY OF BIOMASS FEEDSTOCKS FOR SINGLE CELL PROTEIN

Biomass feedstocks are essential in SCP production, serving as the primary source of carbon and energy, and sometimes nitrogen, required for microbial growth (Nasseri et al. 2011). The cost of production and conflict of interest associated with conventional feedstocks demand that we opt for readily available and renewable biomass feedstocks (Bardhan et al. 2015; Melero et al. 2012), hence the highly sustainable second generation biorefineries (Bajpai 2017). The versatility of microbial strains allows SCP production to be tailored to locally available resources, making it scalable and adaptable across regions (Sharifian et al. 2025). Moreover, by integrating biomass feedstocks into a circular economy model, SCP production significantly lowers its environmental footprint compared to traditional protein sources (Matassa 2016). In Figure 10.1, the various known biomass feedstocks are classified in three broad groups (i.e. primary, secondary, and tertiary) based on their degree of processing and position within the production value chain applications. Although cost and availability are important practical considerations, the classification adopted here is based primarily on processing hierarchy rather than economic value, which may vary across regions.

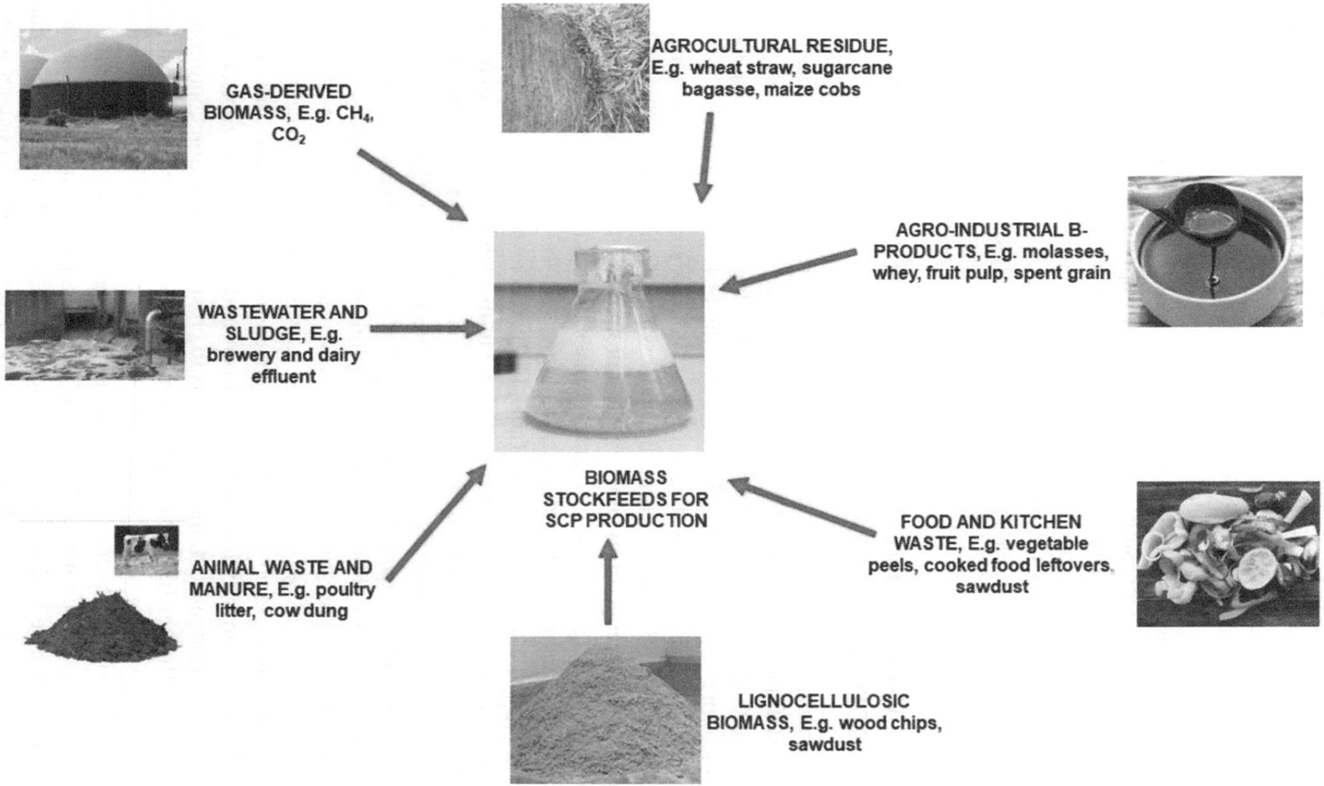

FIGURE 10.1 Classification of biomass feedstock with potential in single cell protein production.

Primary feedstocks comprise raw agricultural and natural biomass such as crop residues and lignocellulosic materials, which are abundant but often require pretreatment. Secondary feedstocks include agro-industrial byproducts and organic waste streams such as molasses, whey, spent grain, food waste, and animal manure, representing partially processed materials with enhanced nutrient accessibility. Tertiary feedstocks encompass industrial effluents and gaseous substrates, including wastewater, sludge, carbon dioxide, and methane. These require technological recovery but offer strong potential for waste mitigation and carbon valorization.

The diversity of available substrates offers strong opportunities for bioprospecting robust microbial species from understudied and resource-limited environments shaped by extreme selective pressures. Therefore, expanding the search to unconventional feedstocks, including toxic wastes, microbe- and insect-derived biomass, and gaseous emissions, can further diversify SCP inputs, while enhancing sustainability in SCP production.

10.4 ADVANCES IN PRETREATMENT AND FERMENTATION PROCESS OPTIMIZATION

Various biomass feedstock processing technologies are critical in preparing substrates for SCP production, enabling the efficient conversion of raw materials into nutrients suitable for microbial growth. These technologies include mechanical milling, chemical pretreatment (e.g. acid or alkaline hydrolysis), enzymatic hydrolysis, and thermal processing to break down complex biomass such as lignocellulosic residues, food waste, and agro-industrial byproducts into fermentable sugars or gases. Table 10.2 shows a range of methods used to pretreat biomass for SCP production. According to the stage of development, biomass pretreatment is divided into two broad categories: ancient and modern technologies.

Considering the examples given in Table 10.2, crop residues like corn stover and wheat straw must undergo pretreatment to release fermentable sugars like maltose, glucose, and xylose for microbes like *S. cerevisiae* and *C. utilis*. Methane-rich biogas or methanol produced from organic waste can support the growth of methanotrophs and methylotrophs such as *M. capsulatus* and *M. methylotrophus*. Algae such as *Spirulina* and *Chlorella* can utilize carbon dioxide and sunlight in photobioreactors, while fungi like *F. venenatum*, used in Quorn production, thrive on starch- and sugar-based hydrolysates. Some bacteria, like *Hydrogenobacter* spp., can grow autotrophically on hydrogen and carbon dioxide, enabling protein production from non-organic carbon sources.

Once the right starter culture microorganism and the most appropriate growth substrate have been identified, the next task is to create fermentation conditions that allow the most efficient biomass conversion into microbial biomass, most importantly protein. Depending on the specific starter microbe being used and the nature of the substrate, fermentation for the purpose of SCP production may be either submerged or solid-state. Submerged fermentation entails processing the substrate in a fluidized reactor, where both the substrate and microbe are submerged in an aqueous fermentation medium. On the other hand, solid-state fermentation occurs when the microbe acts on moistened solid media. Table 10.3 shows the different fermentation techniques used in SCP production.

TABLE 10.2
Pretreatment Methods Used to Prepare Substrate for Single Cell Protein Production

Type	Technique	Description	Reference
Ancient	Mechanical grinding	- Reduces particle size to increase surface area	Nigam & Pandey (2009)
	Steam cooking	- Thermal softening of plant material	Wang et al. (2015)
	Alkaline soaking (lime)	- Used traditionally to treat agricultural residues	Sun & Cheng (2005)
Modern	Ionic liquid pretreatment	- Disrupts lignin-cellulose structure using ionic liquids	Sun et al. (2022)
	Deep eutectic solvents	- Ecofriendly lignin solubilization	Francisco et al. (2012)
	Ammonia fiber expansion	- Non-oxidative method using liquid ammonia	Balan et al. (2012)
	Supercritical carbon dioxide explosion	- Ruptures biomass using high-pressure carbon dioxide	Ramesh et al. (2018)
	Enzymatic hydrolysis	- Breaks down cellulose using enzymes	Zhang et al. (2015)

Despite the significant progress in microbial strain development and the availability of inexpensive feedstocks, the production of SCP is still more costly than traditional protein production. Hence, biotechnological advances in process optimization could play a crucial role in advancing process efficiency and reducing cost. Process optimization refines operational parameters such as pH, temperature, and aeration to maximize microbial growth and protein output. Innovations in bioreactor design, including fed-batch and continuous systems as well as the use of immobilized cells, have improved scalability and consistency. Real-time monitoring using artificial intelligence and sensor technologies further enhances process control, while improvements in downstream processing ensure high recovery and purity of SCP. Combining metabolic engineering and process optimization will synergistically enhance SCP viability. This makes SCP a more competitive, scalable, and sustainable alternative protein source for food, feed, and specialized applications.

TABLE 10.3
Fermentation Systems Used in Single Cell Protein Production

Type	Technique	Description	Reference
Ancient	Natural fermentation	- Uses native microbes on substrates	Steinkraus (1997)
	Batch fermentation	- Closed-system microbial cultivation	Azwar et al. (2021)
	Solid-state fermentation	- Fermentation on moist solids without free water	Artola et al. (2024)
Modern	Fed-batch fermentation	- Nutrient feeding during cultivation	Gupta & Shukla (2017)
	Continuous fermentation	- Steady-state operation for high productivity	Vlaeminck et al. (2023)
	Immobilized cell systems	- Cells bound to supports for reuse	Singh & Mishra (2019)
	Co-culture systems	- Multiple microbes in synergistic production	Singh et al. (2019)
	Artificial intelligence-controlled fermentation	- Real-time adjustments using sensors and artificial intelligence	Al Noor & Anshika (2025)

10.5 CURRENT APPLICATIONS, CHALLENGES, AND FUTURE PERSPECTIVES

SCP has a wide range of applications across various sectors due to its high protein content, fast production rates, and sustainability. In animal feed, SCP derived from yeast, bacteria, algae, and fungi is used as a high-quality protein supplement for livestock, poultry, aquaculture, and pets, offering an alternative to fishmeal and soybean meal (Raziq et al. 2020). In the human food industry, SCP is incorporated into food products as a meat substitute, protein enhancer, or dietary supplement, particularly for vegetarians and in regions with protein scarcity (Bratosin et al. 2021). The nutritional profile of SCP, being rich in essential amino acids, vitamins, and minerals, makes it suitable for nutraceuticals and functional foods. SCP technology also finds use in bioremediation and waste valorization, where microbes convert organic waste or industrial byproducts into valuable biomass (Bratosin et al. 2021). Moreover, SCP is gaining attention in space exploration and extreme environments, with organisms like *Spirulina* being studied by NASA and the European Space Agency for their potential to produce food, oxygen, and recycle waste in closed-loop life support systems (Alvarado 2020). These diverse applications highlight SCP's potential to contribute to food security, environmental sustainability, and innovative biotechnological solutions.

The use of biomass feedstocks in SCP production presents several challenges, including variability in composition, the presence of microbial inhibitors, low nutrient availability, and the need for complex pretreatment processes. In addition, concerns related to high nucleic acid content in microbial biomass may lead to uric acid accumulation and associated health risks if not adequately reduced during downstream processing (Schulz & Oslage 1976; Yakubu et al. 2024). Additionally, logistical issues such as seasonal availability, transportation, and storage further complicate consistent supply. However, these challenges can be addressed through strategies like optimizing pretreatment methods to reduce inhibitors, blending and standardizing feedstocks, supplementing with essential nutrients, and engineering robust microbial strains that can tolerate toxic compounds. Implementing integrated biorefinery approaches and improving supply chain logistics also enhance economic viability. Together, these solutions pave the way for sustainable and scalable SCP production from diverse biomass sources.

10.6 CONCLUSION

In a world plagued by several challenges such as climate change, growing human and livestock population, a gradual shift in human behavior and urbanization, and conflict, SCP emerges as a viable alternative to traditional protein sources such as animal products and legume seeds. The ubiquity of starter microbes and fermentable substrates creates fertile ground for the production and adoption of SCP as a human and livestock protein source. More research in areas such as starter culture selection and development, optimization of substrate, and fermentation technologies is needed in order to achieve meaningful growth in this area.

REFERENCES

Airola, K., L. Petman, & S. Mäkinen-Kiljunen. 2006. Clustered sensitivity to fungi: Anaphylactic reactions caused by ingestive allergy to yeasts. *Ann. Allergy Asthma Immunol.* 97: 294–297.

Akenji, L., & M. Bengtsson. 2014. Making sustainable consumption and production the core of sustainable development goals. *Sustainability* 6: 513–529.

Al Noor, A., & Anshika. 2025. Role of single cell proteins in bioremediation. In *Recent Advances in Bioremediation and Phytoremediation*, ed. A. Kuanar, D. Kar, A.P. Das, D. Bhanja, 63–91. Cham: Springer.

Alvarado, K.A. 2020. *Efficient Alternative Food Systems for Earth and Space*. MSc diss, Alaska: University of Alaska Fairbanks.

Antranikian, G., C.E. Vorgias, & C. Bertoldo. 2005. Extreme environments as a resource for microorganisms and novel biocatalysts. In *Marine Biotechnology I. Advances in Biochemical Engineering/Biotechnology*, vol. 96, ed. R. Ulber, & Y. Le Gal, 219–262. Berlin, Heidelberg: Springer.

Artola, A., X. Font, J. Moral-Vico, & A. Sánchez. 2024. The role of solid-state fermentation to transform existing waste treatment plants based on composting and anaerobic digestion into modern organic waste-based biorefineries, in the framework of circular bioeconomy. *Front. Chem. Eng.* 6:1463785.

Azwar, A., M. Mukhlishien, A. Muslim, P. Hadissa, U.H. Ningsih, M. Zanil, & J.M. Ali. 2021. Production of single cell protein from banana peel waste in batch fermentation using *Saccharomyces cerevisiae*. *J. Bahan Alam Terbarukan* 10: 104–112.

Bajpai, P. 2017. *Single Cell Protein Production from Lignocellulosic Biomass*. Singapore: Springer.

Balagurunathan, B., H. Ling, W.J. Choi, & M.W. Chang. 2022. Potential use of microbial engineering in single cell protein production. *Curr. Opin. Biotechnol.* 76: 102740.

Balan, V., L. Sousa, S.P.S. Chundawat, J.F. Humpula, & B.E. Dale. 2012. Overview to ammonia pretreatments for lignocellulosic biorefineries. *Dyn. Biochem. Process Biotechnol. Mol. Biol.* 6: 1–11.

Bardhan, S.K., S. Gupta, M.E. Gorman, & M.A. Haider. 2015. Biorenewable chemicals: Feedstocks, technologies and the conflict with food production. *Renew. Sustain. Energy Rev.* 51: 506–520.

Barka, A. & C. Blecker. 2016. Microalgae as a potential source of single cell proteins. A review. *Biotechnol. Agron. Soc. Environ.* 20: 427–436.

Becker, E.W. 2013. Microalgae for human and animal nutrition. In *Handbook of Microalgal Culture: Applied Phycology and Biotechnology*, ed. A. Richmond, & Q. Hu, 461–503. Hoboken, NJ: Wiley-Blackwell.

Blandón, J.F.V., C.P.S. Henao, J.E.Z. Montoya, & S. Ochoa. 2023. l-Threonine production from whey and fish hydrolysate by *E. coli* ATCC® 21277TM. *Heliyon* 9: e18744.

Bratosin, B. C., S. Darjan, & D.C. Vodnar. 2021. Single cell protein: A potential substitute in human and animal nutrition. *Sustainability* 13: 9284.

Carneiro, C.V.G.C., L.A. Serra, T.F. Pacheco, L.M.M. Ferreira, L.T.D. Brandão, M.N.D.M. Freitas, D. Trichez, & J.R.M. de Almeida. 2022. Advances in *Komagataella phaffii* engineering for the production of renewable chemicals and proteins. *Fermentation* 8: 575.

D'Almeida, A.P., & T.L. de Albuquerque. 2025. Is it possible to produce meat without animals? The potential of microorganisms as protein sources. *Fermentation* 11: 24.

de Oliveira Padilha, L.G., L. Malek, & W.J. Umberger. 2022. Consumers' attitudes towards lab-grown meat, conventionally raised meat and plant-based protein alternatives. *Food Qual. Prefer.* 99: 104573.

Dong, L., Y. Wu, M. Li, C. Zhang, J. Cao, R. Ledesma-Amaro, W. Zhao, & D. Kang. 2025. Exploring the fermentation potential of *Kluyveromyces marxianus* NS127 for single cell protein production. *Fermentation* 11: 70.

Farooq, M.H., A. Naureen, A. Sarwar, R. Nelofer, M. Nadeem, T. Aziz, M.M. Alomran, A. Shami, F. Al-Asmari, Al-Joufi, A.M. Nahari, & R.S. Jalal. 2025. Optimization of fermentation parameters for enhanced and low-cost production of single cell protein from rice polishing using *Rhizopus oligosporus*. *Ital. J. Food Sci.* 37: 219–232.

Fasihi, M., F. Jouzi, P. Tervasmäki, P. Vainikka, & C. Breyer. 2025. Global potential of sustainable single cell protein based on variable renewable electricity. *Nat. Commun.* 16: 1496.

Francisco, M., A. van den Bruinhorst, & M.C. Kroon. 2012. New natural and renewable low transition temperature mixtures (LTTMs): Screening as solvents for lignocellulosic biomass processing. *Green Chem.* 14: 2153–2157.

Gervasi, T., V. Pellizzeri, G. Calabrese, G. Di Bella, N. Cicero, & G. Dugo. 2018. Production of single cell protein (SCP) from food and agricultural waste by using *Saccharomyces cerevisiae*. *Nat. Prod. Res.* 32: 648–653.

Gundupalli, M.P., S. Ansari, J.P.V. da Costa, F. Qiu, J. Anderson, M. Luckert, & D.C. Bressler. 2024. Bacterial single cell protein (BSCP): A sustainable protein source from *Methylobacterium* species. *Trends Food Sci. Technol.* 147: 104426.

Gupta, S.K. & P. Shukla. 2017. Sophisticated cloning, fermentation, and purification technologies for an enhanced therapeutic protein production: A review. *Front. Pharmacol.* 8: 419.

Hadi, J., & G. Brightwell. 2021. Safety of alternative proteins: Technological, environmental and regulatory aspects of cultured meat, plant-based meat, insect protein and single cell protein. *Foods* 10: 1226.

Kasperski, S.D. 2023. *Bioconversion of Steam-Pretreated Sugarcane Bagasse to Single Cell Protein*. PhD diss, Stellenbosch: Stellenbosch University.

Kim, S.W., J.F. Less, L. Wang, T. Yan, V. Kiron, S.J. Kaushik, & X.G. Lei. 2019. Meeting global feed protein demand: Challenge, opportunity, and strategy. *Annu. Rev. Anim. Biosci.* 7: 221–243.

Kurcz, A., S. Błażejak, A.M. Kot, A. Bzducha-Wrobel, & M. Kieliszek. 2018. Application of industrial wastes for the production of microbial single-cell protein by fodder yeast *Candida utilis*. *Waste Biomass Valor.* 9: 57–64.

Kurtzman, C.P., J.W. Fell, & T. Boekhout. 2011. Definition, classification and nomenclature of the yeasts. In *The Yeasts: A Taxonomic Study*, ed. C.P. Kurtzman, J.W. Fell, & T. Boekhout, 3–5. Burlington: Elsevier.

Ledbetter, F. 2025. Single cell protein for feed and food. *EuroBiotech J.* 9: 117–127.

Li, Y., N. Xiang, Y. Zhu, M. Yang, C. Shi, Y. Tang, W. Sun, K. Sheng, D. Liu, & X. Zhang. 2024. Blue source-based food alternative proteins: Exploring aquatic plant-based and cell-based sources for sustainable nutrition. *Trends Food Sci. Technol.* 147: 104439.

Liceaga, A.M., J.E. Aguilar-Toalá, B. Vallejo-Cordoba, A.F. González-Córdova, & A. Hernández-Mendoza. 2022. Insects as an alternative protein source. *Annu. Rev. Food Sci. Technol.* 13: 19–34.

Ling, H., W. Teo, B. Chen, S.S.J. Leong, & M.W. Chang. 2014. Microbial tolerance engineering toward biochemical production: from lignocellulose to products. *Curr. Opin. Biotechnol.* 29: 99–106.

Mahmoudi, S., J.M. Mosaabadi, R. Khalilzadeh, S.M. Robatjazi, & F. Rezaie. 2024. Evaluation of nutritional value of microbial protein produced with methanol using *Methylophilus methylotrophus* bacteria. *Res. J. Livest. Sci.* 37: 57–70.

Makopa, T.P., T. Ncube, S. Alwasel, T. Boekhout, & N. Zhou. 2024. Yeast-insect interactions in Southern Africa: Tapping the diversity of yeasts for modern bioprocessing. *Yeast* 41: 330–348.

Matassa, S. 2016. *Exploring Protein Production by Hydrogen-Oxidizing Microbiomes*. PhD diss, Ghent: Ghent University.

Mazzanti, G., P.A. Moro, E. Raschi, R. Da Cas, & F. Menniti-Ippolito. 2017. Adverse reactions to dietary supplements containing red yeast rice: Assessment of cases from the Italian surveillance system. *Br. J. Clin. Pharmacol.* 83: 894–908.

Megido, R.C., C. Gierts, C. Blecker, Y. Brostaux, É. Haubruge, T. Alabi, & F. Francis. 2016. Consumer acceptance of insect-based alternative meat products in Western countries. *Food Qual. Prefer.* 52: 237–243.

Melero, J.A., J. Iglesias, & A. Garcia. 2012. Biomass as renewable feedstock in standard refinery units. Feasibility, opportunities and challenges. *Energy Environ. Sci.* 5: 7393–7420.

Munir, R., & D.B. Levin. 2016. Enzyme systems of anaerobes for biomass conversion. In *Anaerobes in Biotechnology. Advances in Biochemical Engineering/Biotechnology*, vol. 156, ed. R. Hatti-Kaul, G. Mamo, & B. Mattiasson, 113–138. Cham: Springer.

Muniz, E.d.N., R.T. de Queiroz Montenegro, D.N. da Silva, A.P. D'Almeida, L.R.B. Gonçalves, & T.L. de Albuquerque. 2024. Advances in biotechnological strategies for sustainable production of non-animal proteins: Challenges, innovations, and applications. *Fermentation* 10: 638.

Munoz, P., E. Bouza, M. Cuenca-Estrella, J.M. Eiros, M.J. Pérez, M. Sánchez-Somolinos, C. Rincón, J. Hortal, & T. Peláez. 2005. *Saccharomyces cerevisiae* fungemia: An emerging infectious disease. *Clin. Infect. Dis.* 40: 1625–1634.

Murphy, A., & K. Kavanagh. 1999. Emergence of *Saccharomyces cerevisiae* as a human pathogen: Implications for biotechnology. *Enzyme Microb. Technol.* 25: 551–557.

Nasseri, A. T., S. Rasoul-Amini, M.H. Morowvat, & Y. Ghasemi. 2011. Single cell protein: Production and process. *Am. J. Food Technol.* 6: 103–116.

Niehaus, F., C. Bertoldo, M. Kähler, & G. Antranikian. 1999. Extremophiles as a source of novel enzymes for industrial application. *Appl. Microbiol. Biotechnol.* 51:711–729.

Nigam, P.S.-N., & A. Pandey. 2009. Solid-state fermentation technology for bioconversion of biomass and agricultural residues. In *Biotechnology for Agro-Industrial Residues Utilisation: Utilization of Agro-Residues*, ed. P. Nigam, & A. Pandey, 197–221. Dordrecht: Springer.

Nirmal, N., C.F. Anyimadu, A.C. Khanashyam, A.E. Bekhit, & B.K. Dhar. 2025. Alternative protein sources: Addressing global food security and environmental sustainability. *Sustain. Dev.* 33: 3958–3969.

Nwaefuna, A.E., K. Rumbold, T. Boekhout, & N. Zhou. 2021. Bioethanolic yeasts from dung beetles: Tapping the potential of extremophilic yeasts for improvement of lignocellulolytic feedstock fermentation. *Biotechnol. Biofuels* 14: 86.

Onyeaka, H., C.K. Anumudu, C. Okpe, A. Okafor, F. Ihenetu, T. Miri, O.A. Odeyemi, & A. Anyogu. 2022. Single cell protein for foods and feeds: A review of trends. *Open Microbiol. J.* 16: e187428582206160.

Oshoma, C., & S.O. Eguakun-Owie. 2018. Conversion of food waste to single cell protein using *Aspergillus niger*. *J. Appl. Sci. Environ. Manage.* 22: 350–355.

Øverland, M., A.-H. Tauson, K. Shearer, & A. Skrede. 2010. Evaluation of methane-utilizing bacteria products as feed ingredients for monogastric animals. *Arch. Anim. Nutr.* 64: 171–189.

Pakseresht, A., S.A. Kaliji, & M. Canavari. 2022. Review of factors affecting consumer acceptance of cultured meat. *Appetite* 170: 105829.

Puyol, D., D.J. Batstone, T. Hülsen, S. Astals, M. Peces, & J.O. Krömer. 2017. Resource recovery from wastewater by biological technologies: Opportunities, challenges, and prospects. *Front. Microbiol.* 7: 2106.

Qi, H., R. Wang, C. Wang, R. Wang, J. Shen, H. Fang, & J. Zhang. 2024. Integrated microbiome and metabolomics analysis of the effects of dietary supplementation with corn-steep-liquor-derived *Candida utilis* feed on black pigs. *Animals* 14: 306.

Ramesh, D., I.K. Muniraj, K. Thangavelu, & S. Karthikeyan. 2018. Pretreatment of lignocellulosic biomass feedstocks for biofuel production. In *Advanced Solid Catalysts for Renewable Energy Production*, ed. S. González-Cortés, & F.E. Imbert, 33–60. Hershey, PA: IGI Global Scientific Publishing,

Rammala, B., & N. Zhou. 2022. Looking into the world's largest elephant population in search of ligninolytic microorganisms for biorefineries: A mini-review. *Biotechnol. Biofuels Bioprod.* 15: 64.

Ravindra, P. 2000. Value-added food: Single cell protein. *Biotechnol. Adv.* 18: 459–479.

Raziq, A., M. Lateef, A. Ullah, H. Ullah, & M.W. Khan. 2020. 02. Single cell protein (SCP) production and potential substrates: A comprehensive review. *Pure Appl. Biol.* 9: 1743–1754.

Reihani, S.F.S., & K. Khosravi-Darani. 2019. Influencing factors on single cell protein production by submerged fermentation: A review. *Electron. J. Biotechnol.* 37: 34–40.

Ritala, A., S.T. Häkkinen, M. Toivari, & M.G. Wiebe. 2017. Single cell protein - State-of-the-art, industrial landscape and patents 2001–2016. *Front. Microbiol.* 8: 2009.

Schulz, E., & H.J. Oslage. 1976. Composition and nutritive value of single cell protein (SCP). *Anim. Feed Sci. Technol.* 1: 9–24.

Sekoai, P.T., Y. Roets-Dlamini, F. O'Brien, S. Ramchuran, & V. Chunilall. 2024. Valorization of food waste into single cell protein: An innovative technological strategy for sustainable protein production. *Microorganisms* 12: 166.

Severo, I.A., G.S. de Lira, R.R. Ambati, R.A. Gokare, J.V.C. Vargas, J. Ordonez, & A.B. Mariano. 2024. Disruptive potential of microalgae proteins: Shaping the future of the food industry. *Future Foods* 9: 100318.

Sharifian, P., A. Amiri, D.J. McClements, M. Kubiak, L. Favaro, S. Maggo, A. Singh, & A. Pratap-Singh. 2025. Single cell proteins: Fermentation pathways, nutritional quality and digestibility, and a techno-economic and environmental outlook for sustainable scale-up. *Crit. Rev. Food Sci. Nutr.* 29: 1–29.

Singh, R., J. Ryu, & S.W. Kim. 2019. Microbial consortia including methanotrophs: Some benefits of living together. *J. Microbiol.* 57: 939–952.

Singh, V., & V. Mishra. 2019. Bioremediation of nutrients and heavy metals from wastewater by microalgal cells: Mechanism and kinetics. In *Microbial Genomics in Sustainable Agroecosystems*, ed. V. Tripathi, P. Kumar, P. Tripathi, A. Kishore, & M. Kamle, 319–357. Singapore: Springer.

Smith, K., A.W. Watson, M. Lonnie, W.M. Peeters, D. Oonincx, N. Tsoutsoura, G. Simon-Miquel, K. Szepe, N. Cochetel, A.G. Pearson, O.C. Witard, A.M. Salter, M. Bennett, & B.M. Corfe. 2024. Meeting the global protein supply requirements of a growing and ageing population. *Eur. J. Nutr.* 63: 1425–1433.

Spalvins, K., Z. Kusnere, & S. Raita. 2023. Sustaining a mars colony through integration of single cell oil in biological life support systems. *Environ. Clim. Technol.* 27: 339–367.

Steinkraus, K. H. 1997. Classification of fermented foods: Worldwide review of household fermentation techniques. *Food Control* 8: 311–317.

Strope, P.K., D.A. Skelly, S.G. Kozmin, G. Mahadevan, E.A. Stone, P.M. Magwene, F.S. Dietrich, & J.H. McCusker. 2015. The 100-genomes strains, an *S. cerevisiae* resource that illuminates its natural phenotypic and genotypic variation and emergence as an opportunistic pathogen. *Genome Res.* 25: 762–774.

Sun, J., R. Ding, & J. Yin. 2022. Pretreatment of corn cobs and corn stalks with tetrabutyl phosphate hydroxide ionic liquid to enhance enzymatic hydrolysis process. *Biochem. Eng. J.* 177: 108270.

Sun, Y., & J.J. Cheng. 2005. Dilute acid pretreatment of rye straw and bermudagrass for ethanol production. *Bioresour. Technol.* 96: 1599–1606.

Vlaeminck, E., E. Uitterhaegen, K. Quataert, T. Delmulle, S.-S. Kontovas, N. Misailidis, R.G. Ferreira, D. Petrides, K.D. Winter, & W.K. Soetaert. 2023. Single cell protein production from industrial off-gas through acetate: Techno-economic analysis for a coupled fermentation approach. *Fermentation* 9: 771.

Wang, K., J. Chen, S.-N. Sun, & R.-C. Sun. 2015. Steam explosion. *Pretreatment of Biomass: Processes and Technologies*, ed. A. Pandey, S. Negi, P. Binod, & C. Larroche, 75–104. Amsterdam: Academic Press.

Watanabe, Y., W. Aoki, & M. Ueda. 2023. Ammonia production using bacteria and yeast toward a sustainable society. *Bioengineering* 10: 82.

Whittaker, J.A., R.I. Johnson, T.J.A. Finnigan, S.V. Avery, & P.S. Dyer. 2020. The biotechnology of Quorn mycoprotein: Past, present and future challenges. In *Grand Challenges in Fungal Biotechnology*, ed. H. Nevalainen, 59–79. Cham: Springer.

Wiebe, M. 2002. Myco-protein from *Fusarium venenatum*: A well-established product for human consumption. *Appl. Microbiol. Biotechnol.* 58: 421–427.

Wolf, S., J. Becker, Y. Tsuge, H. Kawaguchi, A. Kondo, J. Marienhagen, M. Bott, V.F. Wendisch, & C. Wittmann. 2021. Advances in metabolic engineering of *Corynebacterium glutamicum* to produce high-value active ingredients for food, feed, human health, and well-being. *Essays Biochem.* 65: 197–212.

Wu, G., F.W. Bazer, & H.R. Cross. 2014. Land-based production of animal protein: Impacts, efficiency, and sustainability. *Ann. N. Y. Acad. Sci.* 1328: 18–28.

Xie, L., S. Tian, Z. Jin, T. Zou, M. Chang, H. Tang, T. Yu, & Z. Zhuang. 2025. Systematic engineering of cell wall for improving single cell protein (SCP) production. *J. Biotechnol.* 410: 173–183.

Yakubu, M., H.H. Laminu, & M.A. Milala. 2024. Production of low nucleic acid containing single cell protein from banana (*Musa* spp.) peels using the yeast (*Saccharomyces cerevisiae*). *Sys. Rev. Pharm.* 15: 244–250.

Yan, J., B. Han, X. Gui, G. Wang, L. Xu, Y. Yan, C. Madzak, D. Pan, Y. Wang, Zha, G., & L. Jiao. 2018. Engineering *Yarrowia lipolytica* to simultaneously produce lipase and single cell protein from agro-industrial wastes for feed. *Sci. Rep.* 8: 758.

Zhang, J., W. Liu, Q. Hou, J. Chen, N. Xu, & F. Ji. 2015. Effects of different preextractions combining with chemi-thermomechanical treatments on the enzymatic hydrolysis of wheat straw. *Bioresour. Technol.* 175: 75–81.

11 Biomass Feedstock in the Manufacture of Biochemicals, Bioplastics, and Biobased Materials

Lanrewaju I. Fajimi, Matthew A. Onu, Farooq Sher, Rishen Roopchund, and Naadhira Seedat

11.1 INTRODUCTION

The global shift toward sustainability has intensified the search for alternatives to fossil-based resources, particularly in the production of chemicals, plastics, and materials (Sharma et al. 2025). As climate change, plastic pollution, and resource depletion continue to challenge conventional industrial systems, biomass feedstock has emerged as a promising foundation for a biobased economy (Ingle et al. 2025). Derived from organic sources, such as agricultural residues, forestry byproducts, and organic waste, biomass offers a renewable and versatile platform for manufacturing biochemicals, bioplastics, and biobased materials, with reduced environmental footprints (Guo et al. 2024). Recent years have seen a growing emphasis on valorizing biomass into high-value products, driven by advances in biotechnology, materials science, and process engineering (Ning et al. 2021). Biomass feedstocks are typically classified into first-generation (food crops), second-generation (lignocellulosic residues), and third-generation (algae and aquatic biomass) (Singh et al. 2023). Among these, lignocellulosic biomass, composed of cellulose, hemicellulose, and lignin, has garnered particular interest due to its abundance and non-competition with food supply (Fajimi et al. 2022). However, its structural complexity necessitates robust pretreatment and conversion technologies to unlock its full potential (Juneja et al. 2023).

The transformation of biomass into platform chemicals, such as lactic acid, succinic acid, and furfural, has paved the way for the development of bioplastics and specialty chemicals (Galán 2024; Kumar et al. 2024). These intermediates serve as precursors for materials like polylactic acid (PLA) and polyhydroxyalkanoates (PHAs), which are biodegradable and increasingly adopted in packaging, agriculture, and biomedical applications (Naser et al. 2021). PLA, derived from microbial fermentation of sugars, offers a viable alternative to petroleum-based plastics, while PHAs,

DOI: 10.1201/9781003650751-11

produced by bacterial synthesis, are gaining traction for their versatility and environmental compatibility (Shekhar & Mondal 2024). In parallel, biobased materials, such as nanocellulose, biochar, and biocomposites, are redefining material performance and sustainability (Naser et al. 2021). Nanocellulose, extracted from plant fibers, exhibits exceptional mechanical strength, barrier properties, and biocompatibility, making it suitable for applications in electronics, coatings, and drug delivery systems (Norizan et al. 2022; Kunnath et al. 2024). Biochar, produced via pyrolysis of biomass, serves as a soil amendment, carbon sink, and precursor for activated carbon and construction additives (Shyam et al. 2025). Biocomposites, which combine natural fibers with biobased polymers, are being explored for automotive, aerospace, and building materials due to their lightweight and renewable nature (Ullah et al. 2025).

The economic and environmental viability of biomass-based production is influenced by feedstock availability, conversion efficiency, market dynamics, and policy frameworks (Dunn et al. 2018). In regions such as Southern Africa, agricultural residues like sugarcane bagasse, fruit peels, and invasive alien plants offer untapped potential for decentralized biomanufacturing and rural development (Khan et al. 2015). However, challenges such as feedstock variability, high capital costs, and limited infrastructure persist, necessitating the use of techno-economic assessment (TEA) and life cycle analysis (LCA) to guide strategic decisions (Nair et al. 2020). Technological innovations are rapidly improving the feasibility of biomass conversion. Engineered microbial strains now enable high-yield production of biochemicals from mixed sugar streams, while novel catalysts facilitate selective transformation of lignin into aromatic compounds (Carruthers & Lee 2020). These advances are supported by digital tools for process simulation, optimization, and sustainability modeling, which are critical for scaling up laboratory breakthroughs to industrial applications (Batista et al. 2023). Importantly, the transition to biobased manufacturing is not solely a technical challenge, but it is also a socio-economic transformation. It requires inclusive innovation, capacity building, and equitable access to technologies that empower communities and foster resilience. Public engagement, consumer education, and policy alignment are essential to ensure that biobased solutions are not only environmentally sound but also socially just.

This chapter explores the evolving role of biomass feedstock in the manufacture of biochemicals, bioplastics, and biobased materials, with a focus on recent technological advances, regional opportunities, and sustainability frameworks. A novel contribution of this work is the integration of open-source TEA tools with localized biomass data to support context-specific decision-making for biobased manufacturing in resource-constrained environments.

11.2 BIOCHEMICAL PRODUCTION FROM BIOMASS

11.2.1 Fermentation-Based Biochemicals

Fermentation-based biochemicals represent a cornerstone of biobased manufacturing, offering a sustainable alternative to petrochemical-derived products. Through microbial fermentation, a wide array of valuable compounds, including organic acids, alcohols, amino acids, and biopolymers, can be produced from renewable biomass. This approach leverages the metabolic capabilities of microorganisms, such as *Saccharomyces cerevisiae*, *Escherichia coli*, and *Corynebacterium glutamicum*, which have been extensively engineered to enhance yield, substrate tolerance, and product specificity (Williams 2025). Recent advances in metabolic engineering and synthetic biology have enabled the development of robust microbial strains capable of converting mixed sugar streams derived from lignocellulosic biomass. These innovations are particularly relevant for Southern Africa, where agricultural residues like sugarcane bagasse and fruit peels offer abundant feedstock. The use of extremophilic microbes, adapted to high temperatures or acidic conditions, improves process stability and reduces contamination risks (Williams 2025). Feedstock diversification is a critical driver of economic viability. Traditional fermentation relies heavily on food-grade substrates, such as glucose and corn syrup, raising concerns about food versus fuel trade-offs. In response, researchers are exploring lignocellulosic hydrolysates, glycerol from biodiesel production, and other industrial byproducts as low-cost carbon sources. Pretreatment technologies, including dilute acid hydrolysis and enzymatic saccharification, have improved the conversion of complex biomass into fermentable sugars, supporting circular economy goals (Ewing et al. 2022). Bioprocess optimization plays a vital role in scaling fermentation technologies. Modern systems increasingly employ continuous or semi-continuous operations, supported by Process Analytical Technologies (PAT) that monitor pH, oxygen levels, and substrate concentrations in real time. Innovations in bioreactor design, such as improved aeration, agitation, and temperature control, enable high-cell-density cultures and consistent product yields. Immobilized cell systems and biofilm reactors are gaining traction for their ability to enhance productivity and reduce operational costs (Williams 2025).

Electro-fermentation is an emerging hybrid technology that integrates microbial metabolism with electrochemical control. By using electrodes as electron donors or acceptors, electro-fermentation can modulate redox conditions, improve product titers, and reduce byproduct formation. This approach has shown promise in enhancing the synthesis of butyric acid and other value-added compounds from lignocellulosic feedstocks (Pilania et al. 2025). The spectrum of fermentation-based biochemicals continues to expand, encompassing applications in biofuels, food additives, pharmaceuticals, and biodegradable plastics. For example, lactic acid and succinic acid serve as precursors for bioplastics, while ethanol and butanol are key biofuels. Amino acids like lysine and glutamate are widely used in animal feed and food industries. To ensure long-term impact, fermentation-based production must be embedded within inclusive innovation frameworks. This includes capacity building, equitable access to technologies, and community engagement, especially in rural and underserved regions.

11.2.2 Enzymatic and Chemical Conversion Pathways

Enzymatic and chemical conversion pathways are pivotal in the valorization of biomass into fuels, chemicals, and materials within modern biorefineries. These approaches offer complementary advantages, for example, enzymatic routes provide high specificity and operate under mild conditions, while chemical methods enable rapid and scalable transformations of complex substrates. Enzymatic conversion utilizes biocatalysts, primarily proteins, which accelerate reactions such as hydrolysis, oxidation, and isomerization. Enzymes like cellulases, hemicellulases, and ligninases are employed to depolymerize lignocellulosic biomass into fermentable sugars and aromatic compounds. Their selectivity reduces unwanted byproducts and energy demands, making them ideal for sustainable processing. Recent advances in enzyme engineering, directed evolution, and immobilization have enhanced catalytic efficiency, thermostability, and substrate range (Jatoi & Iqbal 2025). Enzymatic cascades are being explored for CO_2 conversion into methanol and formic acid, offering eco-friendly routes for carbon capture and utilization (Markandan et al. 2023).

Chemical conversion pathways involve acid/base catalysis, oxidation, hydrogenation, and dehydration to transform biomass-derived intermediates into platform chemicals. For instance, glucose can be dehydrated into 5-hydroxymethylfurfural, which serves as a precursor for bioplastics and fuels. Similarly, lignin can be depolymerized into phenolic monomers using metal oxide catalysts or ionic liquids (Chatterjee et al. 2015). These processes often require elevated temperatures and pressures but offer high throughput and compatibility with continuous flow systems. Chemo-enzymatic hybrid systems are gaining traction for their ability to merge the precision of enzymes with the robustness of chemical catalysis. Such integration minimizes purification steps and enhances the overall yield. Computational tools like minChemBio have been developed to design optimal hybrid pathways by minimizing transitions between enzymatic and chemical steps (Anand et al. 2025). Globally, these technologies are being adopted across sectors, from pharmaceuticals to biofuels, driven by the need for low-carbon, circular production systems. However, challenges remain, including enzyme cost, feedstock heterogeneity, and process integration (Kang et al. 2021). As synthetic biology, AI-driven pathway design, and green chemistry continue to evolve, enzymatic and chemical conversion pathways will play a central role in shaping the future of sustainable biomanufacturing (Kang et al. 2021).

11.2.3 Case Studies in Biochemical Manufacturing

11.2.3.1 Case Study 1: NatureWorks LLC – PLA Bioplastics from Lactic Acid

NatureWorks LLC, headquartered in Minnesota, USA, is widely recognized as a pioneer in the commercial production of PLA, a biodegradable polymer derived from renewable resources. Their flagship product, Ingeo®, is manufactured through a multistep process that begins with the fermentation of dextrose, typically sourced from corn starch, into lactic acid. The lactic acid is then converted into lactide, a cyclic dimer, which undergoes ring-opening polymerization to form PLA. The company has invested heavily in optimizing microbial strains and fermentation conditions to achieve high yields and purity, which are critical for downstream polymerization and product performance. The environmental benefits of PLA are significant. According to Vink et al. (2003), the production of PLA results in up to 60% lower greenhouse gas (GHG) emissions compared to conventional petroleum-based plastics, such as polyethylene terephthalate (PET). Additionally, PLA is compostable under industrial conditions, making it suitable for applications in packaging, textiles, and disposable items.

NatureWorks has conducted extensive LCA to quantify the environmental impact of its process, including energy use, water consumption, and land use. Their findings support the claim that PLA offers a more sustainable alternative to fossil-derived polymers, particularly when produced from second-generation feedstocks. From a techno-economic perspective, NatureWorks has demonstrated that PLA can be cost-competitive at scale. Its TEA models incorporate variables, such as feedstock cost, capital investment, and market dynamics. The company has also explored diversification into non-food biomass sources, including cellulosic sugars, to mitigate concerns around land use and food security. This strategic shift aligns with the broader goals of circular economy and feedstock flexibility. NatureWorks' success illustrates how biochemical manufacturing can achieve both environmental and economic viability through process innovation, strategic partnerships, and rigorous sustainability assessment (Vink et al. 2003).

11.2.3.2 Case Study 2: BASF and Corbion – Biobased Succinic Acid

Baden Aniline and Soda Factory (BASF), one of the world's largest chemical companies, partnered with Corbion to develop a biobased succinic acid production process that leverages renewable sugars and proprietary bacterial strains. Succinic acid is a versatile platform chemical used in the manufacture of polyurethanes, solvents, biodegradable plastics, and food additives. The joint venture, known as Succinity GmbH, aimed to demonstrate the commercial viability of producing succinic acid through microbial fermentation rather than petrochemical synthesis. The process begins with the fermentation of glucose derived from sugar beet or sugarcane using anaerobic strains of *Actinobacillus succinogenes*. The fermentation is conducted under controlled pH and temperature conditions to maximize yield and productivity. Downstream processing involves crystallization and purification steps to isolate high-purity succinic acid suitable for industrial applications. The pilot plant in Pomacle, France, and the subsequent commercial facility in Montmeló, Spain, served as proof-of-concept for scaling biobased C4 chemical production. TEA study revealed that the biobased process could achieve production costs in the range of US$ 2.50–3.00 per kilogram, making it competitive with petrochemical alternatives under favorable market conditions. LCA indicated a 50%–70% reduction in GHG emissions compared to conventional routes, along with lower energy and water usage. However, the venture faced challenges related to market volatility, feedstock price sensitivity, and limited policy incentives. In 2018, BASF and Corbion dissolved the joint venture, but the technological insights gained continue to inform BASF's broader sustainability strategy (Song & Lee 2006). This case underscores the importance of integrating TEA and LCA into early-stage development to identify risks and opportunities. It also highlights the need for adaptive business models and supportive policy frameworks to ensure the long-term success of biobased chemical ventures.

11.2.3.3 Case Study 3: BioAmber Industrial-Scale Biosuccinic Acid

BioAmber Inc., a Canadian-American startup, was among the first companies to commercialize biosuccinic acid at an industrial scale. Their facility in Sarnia, Ontario, Canada, represented a significant milestone in biochemical manufacturing, demonstrating the feasibility of producing high-value platform chemicals from renewable feedstocks. The process utilized genetically engineered strains of *E. coli* to ferment glucose into succinic acid, followed by ion exchange and crystallization for purification. The Sarnia plant had a production capacity of 30,000 metric tons per year and was designed to integrate renewable energy sources to further reduce its environmental footprint. The engineered *E. coli* strains exhibited enhanced flux through the C4 pathway, achieving yields of up to 0.9 grams of succinic acid per gram of glucose. The hybrid fermentation process combined aerobic and anaerobic phases to optimize productivity and minimize byproduct formation. TEA study estimated the capital cost of the facility at approximately US$ 150 million, with operating costs ranged from US$ 1.80 to 2.20 per kilogram of succinic acid. These figures positioned BioAmber's product as a competitive alternative to petrochemical succinic acid, particularly in markets demanding sustainable solutions. The company targeted applications in polybutylene succinate (PBS), coatings, resins, and food additives, leveraging the growing demand for biobased materials. Despite its technological success, BioAmber filed for bankruptcy in 2018 due to a combination of market saturation, feedstock price fluctuations, and insufficient policy support (Li & Mupondwa 2021; Mancini et al. 2022). The case serves as a cautionary tale about the complexities of scaling biochemical innovations, emphasizing the need for robust TEA, strategic partnerships, and flexible business strategies. BioAmber's legacy continues through technology licensing and knowledge transfer, contributing to the advancement of biobased chemical manufacturing.

11.2.3.4 Novamont – Integrated Biorefinery for Mater-Bi® Bioplastics

Novamont, an Italian company based in Novara, exemplifies a holistic approach to biochemical manufacturing through its integrated biorefinery model. The company's flagship product, Mater-Bi®, is a compostable bioplastic derived from starch, vegetable oils, and agricultural residues. Novamont's strategy emphasizes regional development, feedstock sustainability, and circular economy principles, making it a model for place-based innovation in biochemical production. The production process involves enzymatic hydrolysis of starch to produce fermentable sugars, which are then converted into monomers through microbial fermentation. These monomers are polymerized into biodegradable plastics suitable for packaging, agriculture, and consumer goods. Novamont collaborates with local farmers to source non-food biomass, thereby supporting rural economies and reducing environmental impact. Its biorefinery in Terni integrates multiple unit operations, including fermentation, chemical synthesis, and polymerization, within a closed-loop system. LCA conducted by Novamont indicated significant reductions in GHG emissions, water use, and land footprint compared to conventional plastics. The company also invests in social sustainability, including education, job creation, and community engagement. TEA study reveals that Mater-Bi® can achieve cost parity with fossil-based plastics under supportive regulatory environments, particularly in regions with composting infrastructure and plastic bans. Novamont's success illustrates the potential of biochemical manufacturing to align with environmental, economic, and social goals. Its integrated approach demonstrates how innovation can be rooted in local contexts while addressing global sustainability challenges (Briassoulis et al. 2013). The company's ongoing research into new feedstocks and process optimization further strengthens its position as a leader in sustainable materials.

11.2.3.5 DSM and Veramaris – Omega-3 Fatty Acids from Algal Biomass

Royal DSM, a Dutch multinational corporation, has developed a fermentation-based process for producing omega-3 fatty acids using marine microalgae. Traditionally sourced from fish oil, omega-3s are essential for human health but raise sustainability concerns due to overfishing and marine ecosystem degradation. DSM's approach eliminates dependency on marine resources by cultivating *Schizochytrium* species in controlled bioreactors, followed by lipid extraction and purification. The process begins with the cultivation of microalgae under heterotrophic conditions using glucose as a carbon source. The algae accumulate high levels of docosahexaenoic acid (DHA) and eicosapentaenoic acid (EPA), which are extracted using solvent-based or supercritical CO_2 methods. The resulting oil is refined and formulated for use in aquaculture, human nutrition, and pet food. DSM's partnership with Evonik led to the formation of Veramaris, a joint venture that supplies algal omega-3s to major seafood producers and nutrition companies. Veramaris has demonstrated the scalability of algal omega-3 production through its large-scale facility in Nebraska, USA, capable of supplying global aquaculture markets. The product has been adopted by major seafood brands seeking sustainable feed alternatives, reducing reliance on wild-caught fish oil. DSM's innovation addresses both environmental and ethical concerns, offering a vegan-friendly, traceable source of essential fatty acids. LCA confirms significant reductions in carbon emissions, water use, and biodiversity impact (Colombo et al. 2018; Santigosa et al. 2023). This case exemplifies how biochemical manufacturing can deliver high-value nutritional products, while advancing marine conservation and aligning with global sustainability goals.

11.3 BIOPLASTICS: SYNTHESIS AND APPLICATIONS

In order to be classified as bioplastic, a polymeric material should be biobased or composed of renewable resources that can decompose. Bioplastics are specific types of polymeric materials derived from biological resources, considered economically and environmentally feasible alternatives to traditional petrochemical plastics (Nanda et al. 2022). Because bioplastics are naturally biodegradable, replacing synthetic plastics with them is an effective way to address the current global issue of plastic waste accumulation in the environment. The development of materials with improved qualities, made possible by advances in bioplastic research, has enabled their use in various applications across key commercial industries. Numerous natural sources, including microbes, plants, and animals, are used to make bioplastics. A biopolymer produced by bacteria, microbial fermentation, PHA has mechanical and physicochemical properties similar to those of synthetic polymers (Thomas et al. 2023). Bioplastics have grown significantly during the last ten years as a result of industry investments in research and development to improve qualities like processability, durability, and thermal insulation. Nowadays, bioplastics are widely used

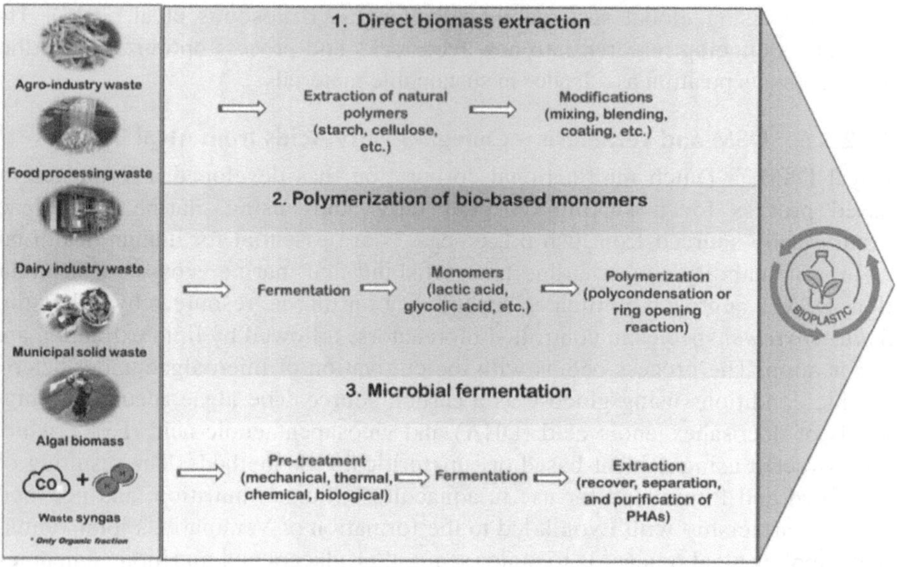

FIGURE 11.1 Synthesis techniques for bioplastics from organic waste.

(Adapted from Thomas et al. 2023.)

in a variety of products, such as textiles, consumer goods, automotive parts, and packaging (Thomas et al. 2023). Figure 11.1 shows the methods used in bioplastics synthesis.

11.3.1 Types of Bioplastics

Biobased chemicals and polymers provide an alternative to conventional, comparable products derived from fossil fuels and their byproducts. Biobased chemicals and plastics frequently provide sustainability advantages over their fossil fuel-based equivalents, such as a lower carbon footprint, less reliance on fossil fuels, and more alternatives for disposal at the end of their useful lives. In many instances, such as with 3D printing filaments, bioplastics can also provide better product performance and functionality (Lovett et al. 2016). The various types of bioplastics are comprehensively discussed in this section.

11.3.1.1 Starch-Based

Many foods, including corn, potatoes, wheat, rice, sorghum, and others, are good sources of starch. Starch-based bioplastics are intricate mixtures of starch and biodegradable plastic ingredients, such as PHAs, PLA, PBS, polybutylene adipate terephthalate, and polycaprolactone (PCL) (Rao et al. 2022; Gurunathan et al. 2025). These mixtures improve the starch-derived bioplastics' mechanical, thermal, and water-resistance qualities. Anitha et al. (2024) in their study used various starch sources, glycerol, and water to produce two distinct bioplastic films (Anitha et al.

2024). According to a similar study by Rao et al. (2022) used naturally biodegradable materials, rice and corn starch, along with varying amounts of other elements, such as ethanol, sorbital, rice oil, and olive oil, to create the degradable bioplastic blends (Rao et al. 2022). The structural properties of glucans and granules, as well as the chemical makeup of starches derived from different sources, vary because of genetic, environmental, and nutritional influences that plants are exposed to throughout their development. To prevent water absorption, lipids, phosphorylated residues, and lengthy lateral chains of amylopectin interact with amylose granules on the interior of starch. On the other hand, hydration through freer hydrogen bonding is made possible by the large percentage of amylopectin, particularly with short lateral chains. This leads to a gel that is prone to retrogradation after extended storage. Although the surface pores and channels on the surface of starch granules facilitate water uptake, smaller starch grains have a bigger superficial area that permits quick hydration. Thus, the starch granules' ability to expand, their viscosity, and their gelatinization are all increased by the higher hydration rate (Anugrahwidya et al. 2021).

11.3.1.2 Cellulose-Based

Plant-based cellulose is a polysaccharide that is used to make cellulose-based bioplastics. Commercial cellulose is derived from weeds, bamboo, wood, and other forest and agricultural waste. It is quite challenging to produce pure cellulose bioplastic due to its robust and well-organized network of intermolecular hydrogen connections. For this reason, cellulose derivatives are frequently produced industrially to create cellulose-based bioplastics. The two primary categories of cellulose polymers are regenerated cellulose and cellulose derivatives, such as cellulose ethers and esters. "Regenerated cellulose" is defined as cellulose that has undergone physical and chemical dissolution before being reassembled into fibers or films. Cellulose-based bioplastics are made from cellulose esters, including cellulose acetates, nitrates, butyrate, and propionate. Starches are mixed with cellulosic bioplastics to create bioplastics with better mechanical qualities (Sharma et al. 2025). One of the earliest polymers utilized in industry was cellulose-based plastics (celluloid and cellophane), which are now utilized as a framework for compostable polymer composites to enhance their mechanical and hydrophobic qualities (Álvarez-Castillo et al. 2021).

11.3.1.3 Protein-Based

Bioplastics made of proteins derived from soybeans (38%–45% protein), wheat gluten (containing 75%–80% protein), and corn protein zein are recognized as sustainable, biodegradable, and renewable substitutes for petroleum-derived plastics (Sharma et al. 2025). However, the majority of studies have concentrated on using casting, compression, or injection molding to prepare soy protein isolates or concentrates. Although soy protein bioplastics have generally shown satisfactory mechanical qualities, their low water resistance, which is linked to their high aspartic and glutamic acid residue content, has been the primary source of criticism. Despite this, the field of superabsorbent materials has taken advantage of this characteristic (Álvarez-Castillo et al. 2021). Because proteins are composed of different amino acids that enable significant intermolecular interaction, they offer a greater range of functional possibilities than polysaccharides and lipids. Although protein-based packaging has numerous advantages, certain problems, such as poor mechanical qualities, limited moisture resistance, and microbiological contamination, still restrict its widespread use (Qazanfarzadeh & Kumaravel 2023).

11.3.2 Feedstock Selection for Bioplastic Production

The sources of the materials used to produce bioplastics can be categorized using a biofuel categorization system that divides feedstocks into first-, second-, and third-generation categories. Sugarcane, whey, and maize are examples of edible biomass that are utilized as first-generation feedstocks. Non-edible biomasses from lignocellulosic sources, including urban waste, agricultural, forest, and animal processing waste, make up the second-generation. The most uncommon source is algal biomass, which is listed as a third-generation (Coppola et al. 2021). The availability, renewable nature, and inedibility of lignocellulosic materials make them extremely promising for the development of environmentally benign bioplastics. It is possible to extract cellulose and lignin from these feedstocks. By altering their surfaces and undergoing additional chemical derivatizations, both materials can be readily adjusted to yield various bioplastics. Biopolyethylene, PLA, PHAs, polyurethanes, and starch-based nanocellulosic bioplastics are examples of common biobased plastics that can be made from cellulose or lignin (Reshmy et al. 2021). Table 11.1 shows the three-generation feedstocks for bioplastic development. Potato starch and banana peels are utilized to create bioplastics due to their high starch content, which is composed of two distinct polymer chains, termed amylose and amylopectin. These chains are composed of adjacent glucose molecules that are connected to make bioplastic (Beevi et al. 2020). Soy, corn, and sugarcane feedstocks have been investigated due to their widespread application in the production of bioplastics and biofuels (Grabowski et al. 2015).

TABLE 11.1
A Summary of Feedstock Generations for Bioplastics Production

First-Generation	Second-Generation	Third-Generation
Corn	Wood	Biomass from algae
Wheat	Wheat straw	Municipal waste
Rice	Sugar cane bagasse	Industrial waste
Sugar cane	Corn stover	Byproducts from the food industry
Sugar beet	Palm fruit bunches	
Potato	Switch grass	

Source: Wellenreuther & Wolf (2020).

11.3.3 Processing Techniques and Market Trends

11.3.3.1 Direct Biomass Extraction

This process modifies natural polymers like proteins, cellulose, pullulan, and starch. Because of their easy availability, affordability, and renewable nature, waste substrates that are high in these biopolymers could be employed as an efficient feedstock (Thomas et al. 2023). The polysaccharide corn starch, which is taken from corn kernels, offers a flexible and sustainable foundation for the manufacturing of bioplastics. By using the inherent qualities of maize starch, this method produces biodegradable plastics that can drastically lessen the environmental impact of plastic waste. There are usually several important phases involved in the manufacture of bioplastics based on corn starch. Corn kernels are first processed to extract corn starch. The kernels are broken down into component parts, protein, fiber, and starch, by soaking them in hot water and sulfur dioxide. After being extracted and refined, the starch is used as the main raw ingredient to make bioplastics (Dino et al. 2025).

11.3.3.2 Polymerization of Biobased Monomers

Bioplastics can be produced using bottom-up processes, such as monomer-based polymerization and biotechnological conversion of renewable resources. The manufacture of PLA in industrial settings is one such example. Lignocellulosic substrates are the primary source of lactic acid, which is required for PLA polymerization. Polysaccharide-based natural products are typically fermented in batches for 3–5 d at 43°C and a pH range of 6.0–7.0 using lactic acid bacteria like *Lactobacillus bulgaricus*. To make PLA, lactic acid must be polymerized through condensation or lactide production (Haq et al. 2025). According to Thomas et al. (2023), the bioplastic production process involves the fermentation of raw materials to form monomers, followed by polymerization (PLA and polyglycolic acid-PGA). The process of making PLA involves first turning the source of carbohydrates into dextrose, which is subsequently fermented to lactic acid and then polymerized. It is possible to use agricultural waste, cotton, and coffee as substrates to provide the synthesis with carbohydrates (Thomas et al. 2023).

11.3.3.3 Microbial Fermentation

Microbial synthesis of bioplastics is a novel and environmentally friendly method of employing microorganisms to create biodegradable plastics. Microbial colonies taken from natural habitats can be cultured and adapted or created by genetic engineering to make PHAs, a type of bioplastic (Thomas et al. 2023). Through microbial fermentation, carbohydrates from these renewable resources are transformed into lactic acid, which is subsequently polymerized to create PLA. The manufacture of PLA is frequently accomplished by the microorganisms *Lactobacillus* and *Lactococcus* (Bhatt 2023). Selection of the substrate, selection of microorganisms, fermentation, accumulation of bioplastics, harvesting, extraction, and processing are the steps involved in the microbial manufacture of bioplastics (Bhatt 2023).

11.3.3.4 Market Trends

The market for bioplastics is expanding quickly as a result of increased environmental concerns, laws against single-use plastics, and consumer desire for sustainable materials. By 2025, the bioplastics market was worth US$ 17 billion, up from US$ 6.8 billion in 2020. In 2025, it is projected that the market will expand 19.9% year-over-year. The market is anticipated to grow at a Compound Annual Growth Rate (CAGR) of 20.1% during the assessment period (2025–2035), reaching US$ 106 billion by the end of 2035 (Future Market Insights 2025). Over the past few years, the market has grown steadily, indicating a promising growth trajectory. Market sales increased significantly during the historical era due to consistent demand from a variety of end-use sectors. It is anticipated that this increasing trend will continue, with additional growth anticipated in the upcoming years. The market is expected to increase steadily each year, building on past momentum, thanks to better industry prospects and changing market dynamics. The market is expected to achieve a significant valuation in the coming years, with a promising forecast for further growth. The market is expected to continue growing in the future due to rising demand in several important industries. The market is expected to grow steadily, which boosts confidence in long-term growth. The market's potential for continuous expansion throughout the course of the evaluation period is highlighted by the projected growth trajectory (Future Market Insights 2025).

11.4 BIOBASED MATERIALS: INNOVATIONS AND CHALLENGES

11.4.1 Composite Materials from Biomass

Natural fibers or natural resins are used to create biocomposite materials rather than synthetic fibers (carbon, glass, etc.) or resins (polyvinyl alcohol, epoxy, etc.). Jute, sisal, flax, hemp, bamboo, hair, wool, silk, and other bio-based fibers are derived from either plants or animals. Additionally, plants are used to make natural matrix materials like polyester, natural rubber, etc. Many researchers have been drawn to biocomposites because of their benefits, which include simplicity of disposal, the ability to compost after the expiration date, which is typically not achievable with standard synthetic materials, and their sustainability and renewable nature (Yıldızhan et al. 2018). A novel class of materials called biopolymer-based composites from agricultural waste biomass is produced, when biopolymers are combined with biomass from agricultural waste. Biopolymers are a class of naturally occurring polymers, such as proteins, polysaccharides, and nucleic acids, which are produced by living things. "Agricultural waste" refers to biomass that comes from agricultural byproducts. This includes byproducts, such as sawdust, animal manure, and crop remnants (Gowda et al. 2023).

11.4.2 Nanocellulose and Lignin-Based Materials

Cellulose/lignin composite membranes are effectively made by combining nanocellulose and nanolignin and imbibing them with a metal ion solution, which is driven by the reinforcing mechanisms seen in biomaterials. A strong cross-linking network is produced between cellulose and lignin by hydrogen bonding and metal ion cross-linking.

The composite films outperform widely used commercial petroleum-based plastics and have a tensile strength of 223.8 MPa, more than twice that of pure nanocellulose films (104 MPa) (Xu et al. 2025). With a high potential for manufacturing useful biomaterials, lignocellulosic biomass, which is mostly made up of cellulose, hemicellulose, and lignin, is a plentiful and renewable biomass resource. The inefficiencies of conventional processing techniques and their intricate structure, however, make their sustainable conversion difficult. There are now more ways to convert lignocellulosic biomass into valuable products, such as nanomaterials, biofuels, nanocellulose, charcoal, and nanofertilizers, thanks to recent developments in nanotechnology (Ingle et al. 2025). Lignin-based materials are frequently utilized in energy storage, the environment, electronic devices, and other disciplines since they are typically inexpensive and environmentally beneficial. Due to its minimal toxicity, environmental friendliness, and outstanding biocompatibility, lignin is being investigated extensively for high-value products rather than burning. Additionally, because of its aromatic qualities, phenol can be substituted to create phenolic resin adhesives (Ma et al. 2021).

11.4.3 Performance and Sustainability Considerations

A more sustainable future depends on the use of sustainable materials and bioplastics. The goal of advances in material science is to prevent harmful environmental impacts like air pollution and the disposal of vast amounts of waste into the ocean, which were prevalent in the previous century. In order to solve this, industries must reduce the amount of waste and emissions they produce, which is impossible with the common plastics. With a strong emphasis on material research and future innovation, companies have been moving in the direction of a circular economy, which is good for the environment. This adjustment is also aimed at improving the social setting. Marketing tools include eco-friendly packaging, enhancing and building a brand's reputation and image, and boosting networking and competition (Ekeinde & Eruni 2024). The need for sustainability propels the use of bioplastics, which could increase the competitiveness of bioplastics relative to traditional resins based on petrochemicals. There are chances for innovation in the fields of specialty plastics creation, biopolymer mixing, and modification chemistry (Ekeinde & Eruni 2024). Bioplastics are used in many different industries and contribute to efforts to reduce plastic pollution and promote environmental sustainability. Bioplastics are paving the way for sustainable and ecologically friendly products (Sharma et al. 2025). In the fields of biomedical engineering, aircraft, automotive, marine, sports, packaging, electronics, and other fields, biocomposites are rapidly emerging as a viable substitute for metal, traditional reinforcement-based composites, and ceramic-based materials. Biocomposites are also used in electrical, medicinal, magnetic, optical, coating, packaging, and electromagnetic shielding (Andrew & Dhakal 2022).

11.5 CHALLENGES IN BIOMASS CONVERSION PROCESSES AND THE WAY FORWARD

Concerns about the environment and human health have made the conversion of renewable biomass feedstock into value-added goods using bioprocessing platforms appealing. When designing bioprocesses to create desired products from biomass feedstock, process performance and cost competitiveness are important considerations. With the right pretreatment, cellulose can be easily hydrolyzed to monomeric sugars by removing hemicellulose and delignifying the liquid fraction (Khunnonkwao et al. 2024). Because bioplastic production methods are frequently less well-established than those for traditional plastics, production prices are higher. For broader use, these procedures must be optimized for cost and efficiency. Standardization of materials is required to guarantee uniform product quality across various bioplastic varieties. This standardization will boost consumer trust and enable dependable performance for particular applications (Olagunju & Kiambi 2024). The performance attributes (strength, durability, and barrier qualities) of conventional plastics might not always be matched by bioplastics. This may present restrictions for some uses, such as food packaging that needs particular gas or moisture barriers. The goal of ongoing research is to enhance bioplastics' performance using a variety of methods. Material development, the creation of new bioplastic materials with improved qualities via material science breakthroughs, is an important field of study. Certain performance attributes, like strength, barrier qualities, or thermal stability, can be enhanced by adding chemicals or changing the structure of the bioplastic. These performance constraints can be addressed to increase the adaptability and applicability of bioplastics in a greater variety of applications (Olagunju & Kiambi 2024). One of the key elements influencing final properties and different uses is biomass processing. Since then, processing has been greatly hampered by the extremely crystalline, variable reactivity, and restricted solubility of lignocellulosic materials. Often, expensive techniques are employed to break down lignocellulosic materials into energy and usable goods (Usmani et al. 2017).

11.6 CONCLUSION

Valorization of biomass into biochemicals, bioplastics, and biobased materials is no longer a distant vision; it is a necessity for building a sustainable and resilient future. As the world grapples with climate change, resource depletion, and mounting plastic waste, the biobased economy offers a pathway that aligns environmental protection with economic opportunity. Biomass, derived from agricultural residues, forestry byproducts, algae, and organic waste, has the potential to replace fossil-based feedstocks, while generating new value streams for industries and communities. Advances in metabolic engineering, enzymatic catalysis, and hybrid bioprocesses have already transformed the feasibility of these solutions, turning what was once experimental into commercial reality. Success stories, from NatureWorks' large-scale production of PLA to Novamont's biodegradable Mater-Bi® and DSM's algal-derived omega-3 oils, illustrate how innovation, investment, and policy alignment can drive market penetration. Bioplastics, in particular, are emerging as credible alternatives to conventional plastics, with expanding applications in packaging, medicine, and consumer goods. Similarly, biobased materials, such as nanocellulose, lignin-derived composites, and biochar, are demonstrating superior performance in sectors, ranging from construction to energy storage. These examples highlight not only technological progress but also the socio-economic potential of biomass valorization, including job creation, rural empowerment, and new industrial ecosystems. Yet the urgency of this transition cannot be overstated. Without decisive action, the world risks locking into unsustainable systems that deepen environmental crises. Key barriers, such as high capital costs, feedstock variability, and limited infrastructure, can be overcome through targeted investments in research, scalable processing technologies, and supportive regulatory frameworks. Equally vital is fostering consumer trust through education and transparent sustainability labeling. The biobased economy is more than an alternative, but a strategic imperative. It bridges climate goals with industrial innovation, rural livelihoods with global markets, and circularity with economic resilience. By prioritizing biomass valorization today, societies can accelerate the shift from a fossil-dependent past to a bio-driven future that safeguards both people and the planet.

REFERENCES

Álvarez-Castillo, E., C. Bengoechea, M. Felix, & A. Guerrero. 2021. Protein-based bioplastics from biowastes: Sources, processing, properties and applications. In *Bioplastics for Sustainable Development*, ed. M. Kuddus, & Roohi, 137–176. Singapore: Springer.

Anand, M., V. Upadhyay, & C.D. Maranas. 2025. minChemBio: Expanding chemical synthesis with chemo-enzymatic pathways using minimal transitions. *ACS Synth. Biol.* 14: 756–770.

Andrew, J.J., & H. N. Dhakal. 2022. Sustainable biobased composites for advanced applications: Recent trends and future opportunities – A critical review. *Compos. C Open Access* 7: 100220.

Anitha, R., K. Jayakumar, G.V. Samuel, M.E. Joice, M. Sneha, & D.S. Seeli. 2024. Synthesis and characterization of starch-based bioplastics: A promising alternative for a sustainable future. *Eng. Proc.* 61: 30.

Anugrahwidya, R., B. Armynah, & D. Tahir. 2021. Bioplastics starch-based with additional fiber and nanoparticle: Characteristics and biodegradation performance: A review. *J. Polym. Environ.* 29: 3459–3476.

Batista, R.M., A. Converti, J. Pappalardo, M. Benachour, & L.A. Sarubbo. 2023. Tools for optimization of biomass-to-energy conversion processes. *Processes* 11: 854.

Beevi, K.R., A.R.S. Fathima, A.I.T. Fathima, N. Thameemunisa, C.M. Noorjahan, & T. Deepika. 2020. Bioplastic synthesis using banana peels and potato starch and characterization. *Int. J. Sci. Technol. Res.* 9: 1809–1814.

Bhatt, S.M. 2023. Microbial production of bio plastics. *J. Agric. Biotechnol. Appl. Sci.* 1: 1–4.

Briassoulis, D., E. Babou, M. Hiskakis, G. Scarascia, P. Picuno, D. Guarde, & C. Dejean. 2013. Review, mapping and analysis of the agricultural plastic waste generation and consolidation in Europe. *Waste Manag. Res.* 31: 1262–1278.

Carruthers, D.N., & T.S. Lee. 2020. Translating advances in microbial bioproduction to sustainable biotechnology. *Front. Bioeng. Biotechnol.* 10: 968437.

Chatterjee, C., F. Pong, & A. Sen. 2015. Chemical conversion pathways for carbohydrates. *Green Chem.* 17: 40–71.

Colombo, S.M., L.G. Campbell, E.J. Murphy, S.L. Martin, & M.T. Arts. 2018. Potential for novel production of omega-3 long-chain fatty acids by genetically engineered oilseed plants to alter terrestrial ecosystem dynamics. *Agric. Syst.* 164: 31–37.

Coppola, G., M.T. Gaudio, C.G. Lopresto, V. Calabro, S. Curcio, & S. Chakraborty. 2021. Bioplastic from renewable biomass: A facile solution for a greener environment. *Earth Syst. Environ.* 5: 231–251.

Dino, A.A., G. Kishore, S.L. Rokhum, & G. Baskar. 2025. Bioplastic production using waste macroalgal biomass: A holistic review on challenges, prospects, economic viability and sustainability analysis. *J. Environ. Chem. Eng.* 13: 116108.

Dunn, J.B., M. Biddy, S. Jones, H. Cai, P. T. Benavides, J. Markham, L. Tao, E. Tan, C. Kinchin, R. Davis, A. Dutta, M. Bearden, C. Clayton, S. Phillips, K. Rappe, & P. Lamers. 2018. Environmental, economic, and scalability considerations and trends of selected fuel economy-enhancing biomass-derived blendstocks. *ACS Sustain. Chem. Eng.* 6: 561–569.

Ekeinde, E.B., & U. Eruni. 2024. The role of bioplastics and sustainable materials in achieving environmental sustainability. *Int. J. Res. Publ. Rev.* 5: 3969–3978.

Ewing, T.A., N. Nouse, M. van Lint, J. van Haveren, J. Hugenholtz, & D.S. van Es. 2022. Fermentation for the production of biobased chemicals in a circular economy: A perspective for the period 2022–2050. *Green Chem.* 24: 6373–6405.

Fajimi, L.I., J. Chrisostomou, & B.O. Oboirien. 2022. A techno-economic analysis (TEA) of a combined process of torrefaction and gasification of lignocellulose biomass (bagasse) for electricity production and methanol production. *Biomass Conv. Bioref.* 14: 12501–12516.

Future Market Insights. 2025. Bioplastics market: Global industry analysis 2020–2024 and opportunity assessment, 2025–2035. https://www.futuremarketinsights.com/ (accessed on 4 August 2025).

Galán, M.L. 2024. Upstream bioprocesses to biomass-based platform chemicals and derivatives. *Fermentation* 10: 59.

Gowda, Y.T.G., S.B. Nagaraju, M. Puttegowda, A. Verma, S.M. Rangappa, & S. Siengchin. 2023. Biopolymer-based composites: An eco-friendly alternative from agricultural waste biomass. *J. Compos. Sci.* 7: 242.

Grabowski, A., S.E.M. Selke, R. Auras, M.K. Patel, & R. Narayan. 2015. Life cycle inventory data quality issues for bioplastics feedstocks. *Int. J. Life Cycle Assess.* 20: 584–596.

Guo, J., Y. Zhang, J. Fang, Z. Ma, C. Li, M. Yan, N. Qiao, Y. Liu, & M. Bian. 2024. Reduction and reuse of forestry and agricultural biowaste through innovative green utilization approaches: A review. *Forests* 15: 1372.

Gurunathan, M.K., R.J.H. Navasingh, J.D.R. Selvam, & R. Čep. 2025. Development and characterization of starch bioplastics as a sustainable alternative for packaging. *Sci. Rep.* 15: 15264.

Haq, F., M. Kiran, I.A. Khan, S. Mehmood, T. Aziz, & M. Haroon. 2025. Exploring the pathways to sustainability: A comprehensive review of biodegradable plastics in the circular economy. *Mater. Today Sustain.* 29: 101067.

Ingle, A.P., S. Saxena, M.P. Moharil, J.D. Rivaldi, L. Ramos, & A.K. Chandel. 2025. Production of biomaterials and biochemicals from lignocellulosic biomass through sustainable approaches: Current scenario and future perspectives. *Biotechnol. Sustain. Mater.* 2: 3.

Jatoi, A.S., & A. Iqbal. 2025. Enzymatic conversion. In *Integrated Food and Bioprocessing*, ed. A.S. Jatoi, 69–85. Cham: Springer.

Juneja, A., D. Kumar, K. Rajendran, & A. Mittal. 2023. Pretreatment technologies for lignocellulosic biomass refineries. In *Advances in Lignocellulosic Biofuel Production Systems*, ed. P. Moodley, R.C. Ray, & E.B.G. Kana, 81–106. Oxford: Woodhead Publishing.

Kang, K., N.B. Klinghoffer, I. ElGhamrawy, & F. Berruti. 2021. Thermochemical conversion of agroforestry biomass and solid waste using decentralized and mobile systems for renewable energy and products. *Renew. Sustain. Energy Rev.* 149: 111372.

Khan, N., M. le Roes-Hill, P.J. Welz, K.A. Grandin, T. Kudanga, J.S. van Dyk, C. Ohlhoff, W.H. van Zyl, & I. Pletschke. 2015. Fruit waste streams in South Africa and their potential role in developing a bioeconomy. *S. Afr. J. Sci.* 111: 1–11.

Khunnonkwao, P., S. Thitiprasert, P. Jaiaue, K. Khumrangsee, B. Cheirsilp, & N. Thongchul. 2024. The outlooks and key challenges in renewable biomass feedstock utilization for value-added platform chemical via bioprocesses. *Heliyon* 10: e30830.

Kumar, V., P. Kumar, S.K. Maity, D. Agrawal, V. Narisetty, S. Jacob, G. Kumar, S.K. Bhatia, D. Kumar, & V. Vivekanand. 2024. Recent advances in biobased production of top platform chemical, succinic acid: An alternative to conventional chemistry. *Biotechnol. Biofuels Bioprod.* 17: 72.

Kunnath, S.M., A. Padinjareveettil, P.V. Sivankuttynair, & S.K. Gundappa. 2024. Nanocellulose-based composites as advanced materials. *Mat. Int.* 6: 18.

Li, X., & E. Mupondwa. 2021. Empirical analysis of large-scale biosuccinic acid commercialization from a techno-economic and innovation value chain perspective: BioAmber biorefinery case study in Canada. *Renew. Sustain. Energy Rev.* 137: 110587.

Lovett, J., F. De Bie, & D. Visser. 2016. *Sustainable Sourcing of Feedstocks for Bioplastics: Clarifying Sustainability Aspects Around Feedstock Use for the Production of Bioplastics*. Gorinchem: Total Corbion PLA bv.

Ma, C., T. Kim, K. Liu, M. Ma, S. Choi, & C. Si. 2021. Multifunctional lignin-based composite materials for emerging applications. *Front. Bioeng. Biotechnol.* 9: 708976.

Mancini, E., R. Dickson, S. Fabbri, I.A. Udugama, H.I. Ullah, S. Vishwanath, K.V. Gernaey, J. Luo, M. Pinelo, & S.S. Mansouri. 2022. Economic and environmental analysis of biosuccinic acid production: From established processes to a new continuous fermentation approach with in-situ electrolytic extraction. *Chem. Eng. Res. Des.* 179: 401–414.

Markandan, K., R. Sankaran, Y.W. Tiong, H. Siddiqui, M. Khalid, S. Kalik, & S. Rustagi. 2023. A review on the progress in chemo-enzymatic processes for CO_2 conversion and upcycling. *Catalysts* 13: 611.

Nair, S.K., R.M. Emerson, & J.R. Solomon. 2020. Biomass supply chain risk: Towards a better understanding of feedstock availability, cost, variability, and uncertainty to catalyze and de-risk biobased investment. In *Handbook of Biorefinery Research and Technology*, ed. V. Bisaria, 1–28. Dordrecht: Springer.

Nanda, S., B.R. Patra, R. Patel, J. Bakos, & A.K. Dalai. 2022. Innovations in applications and prospects of bioplastics and biopolymers: A review. *Environ. Chem. Lett.* 20: 379–395.

Naser, A.Z., I. Deiab, F. Defersha, & S. Yang. 2021. Expanding poly(lactic acid) (PLA) and polyhydroxyalkanoates (PHAs) applications: A review on modifications and effects. *Polymers* 13: 4271.

Ning, P., G. Yang, L. Hu, J. Sun, L. Shi, Y. Zhou, Z. Wang, & J. Yang. 2021. Recent advances in the valorization of plant biomass. *Biotechnol. Biofuels* 14: 102.

Norizan, M.N., S.S. Shazleen, A.H. Alias, F.A. Sabaruddin, M.R. Asyraf, E.S. Zainudin, N. Abdullah, M.S. Samsudin, S.H. Kamarudin, & M.N.F. Norrrahim. 2022. Nanocellulose-based nanocomposites for sustainable applications: A review. *Nanomaterials* 12: 3483.

Olagunju, O.A., & S.L. Kiambi. 2024. The environmental benefits, technical challenges, and scalability of the bioplastic production methods. In *Biomass-Based Bioplastic and Films: Preparation, Characterization, and Application*, ed. O.J. Gbadeyan, & N. Deenadayalu, 207–216. Cham: Springer.

Pilania, P., K. Bhushan, & U.G. Phutela. 2025. Electro-fermentation for biofuel and biochemical production. *Fermentation* 11: 219.

Qazanfarzadeh, Z., & V. Kumaravel. 2023. Hydrophobization approaches of protein-based bioplastics. *Trends Food Sci. Technol.* 138: 27–43.

Rao, L.S., C.D. Naidu, & S. Tiwari. 2022. Investigation on synthesis, structure and degradability of starch based bioplastics. *Mater. Today: Proc.* 49: 257–261.

Reshmy, R., D. Thomas, E. Philip, S.A. Paul, A. Madhavan, R. Sindhu, R. Sirohi, S. Varjani, A. Pugazhendhi, A. Pandey, & P. Binod. 2021. Bioplastic production from renewable lignocellulosic feedstocks: A review. *Rev. Environ. Sci. Bio/Technol.* 20: 167–187.

Santigosa, E., R.E. Olsen, A. Madaro, V.V. Trichet, & I. Carr. 2023. Algal oil gives control of long-chain omega-3 levels in full-cycle production of Atlantic salmon, without detriment to zootechnical performance and sensory characteristics. *J. World Aquac. Soc.* 54: 861–881.

Sharma, S., G. Kumar, S.K. Bhardwaj, A. Chauhan, A. Rathi, J. Kumar, K. Sahil, & S. Vishvamitera. 2025. Bioplastics as next-generation eco-friendly materials: A comprehensive review. *Clean Technol. Environ. Policy* 27: 5009–5027.

Shekhar, N., & A. Mondal. 2024. Synthesis, properties, environmental degradation, processing, and applications of polylactic acid (PLA): An overview. *Polym. Bull.* 81: 11421–11457.

Shyam, S., S. Ahmed, S.J. Joshi, & H. Sarma. 2025. Biochar as a soil amendment: Implications for soil health, carbon sequestration, and climate resilience. *Discov. Soil.* 2: 18.

Singh, A., P. Prajapati, S. Vyas, V.K. Gaur, R. Sindhu, P. Binod, V. Kumar, R.R. Singhania, M.K. Awasthi, Z. Zhang, & S. Varjani. 2023. A comprehensive review of feedstocks as sustainable substrates for next-generation biofuels. *BioEnergy Res.* 16: 105–122.

Song, H., & S.Y. Lee. 2006. Production of succinic acid by bacterial fermentation. *Enzyme Microb. Technol.* 39: 352–361.

Thomas, A.P., V.P. Kasa, B.K. Dubey, R. Sen, & A.K. Sarmah. 2023. Synthesis and commercialization of bioplastics: Organic waste as a sustainable feedstock. *Sci. Total Environ.* 904: 167243.

Ullah, S., Z. Akhter, A. Palevicius, & G. Janusas. 2025. Natural fiber-based biocomposites for potential advanced automotive applications. *J. Eng. Fiber. Fabr.* 20. 15589250241311468.

Usmani, M.A., I. Khan, A. Haque, A.H. Bhat, D. Mondal, & U. Gazal. 2017. Biomass-based composites from different sources: Properties, characterization, and transforming biomass with ionic liquids. In *Lignocellulosic Fibre and Biomass-Based Composite Materials: Lignocellulosic Fibre and Biomass-Based Composite Materials*, ed. M. Jawaid, P. Md Tahir, & N. Saba, 45–76. Kidlington: Woodhead Publishing.

Vink, E.T.H., K.R. Rabago, D.A. Glassner, & P.R. Gruber. 2003. Applications of life cycle assessment to NatureWorks™ polylactide (PLA) production. *Polym. Degrad. Stab.* 80: 403–419.

Wellenreuther, C., & A. Wolf. 2020. *Innovative Feedstocks in Biodegradable Bio-Based Plastics: A Literature Review*, HWWI Research Paper, No. 194. Hamburg: Hamburg Institute of International Economics (HWWI).

Williams, R. 2025. Advances in microbial fermentation for sustainable production of biobased chemicals. *Ferment. Technol.* 14: 1–2.

Xu, Y.H., C. Li, S. Yang, S. Wang, M. Li, J. Jin, Z. Jiang, & F. Peng. 2025. Metal ion crosslinked cellulose/lignin nanocomposite films: A pathbreaking approach toward high-performance sustainable biomaterials. *ACS Nano* 19: 9801–9813.

Yıldızhan, S., A. Çalık, M. Özcanli, & H. Serin. 2018. Biocomposite materials: A short review of recent trends, mechanical and chemical properties, and applications. *Eur. Mech. Sci.* 2: 83–91.

12 Production of Bioactive Compounds, Pharmaceutical Products, and Antioxidants from Biomass

Lucía Seguí, Cristina Barrera, Claudia Bas-Bellver, and Cristina Calderón-Tapia

12.1 INTRODUCTION

A bioactive compound is a substance that is biologically active, which means that it has an impact on living organisms. Bioactive compounds refer to all compounds, usually without nutritional value, which exert certain bioactive effects on the human body. Although negative effects such as toxicity, allergenicity, and mutagenicity are also a form of bioactivity, biological activity is generally considered positive so that upon consumption at sufficient levels, they provide health benefits as prevention or treatment of several diseases (Câmara et al. 2021).

The increase in dietary-related diseases such as cardiovascular diseases, obesity, diabetes, and cancer has made the development of functional foods a priority in the food industry for non-communicable diseases prevention and has driven the surge of the functional foods market. Functional foods are consumed as part of a normal food pattern and benefit one or more target functions in the body, improving the state of health or reducing health risk. The development of functional ingredients or foods seeks to improve their biological value by incorporating bioactive compounds but also by improving or maintaining their concentration or biological activity. This can be achieved by formulation and blending, cultivation and breeding, or by the use of advanced techniques such as high-pressures homogenization, vacuum impregnation, or encapsulation (Betoret et al. 2022). Natural compounds used in the formulation of new functional food products can be obtained from several sources, but one important source is food discards or waste, which are cheap, abundant, and rich in bioactive compounds. Two basic approaches can be applied to provide added value, either extracting their bioactive constituents using specifically designed procedures, or transforming these byproducts into functional powdered ingredients, stable and concentrated in bioactive constituents such as fiber and antioxidants (Bas-Bellver et al.

DOI: 10.1201/9781003650751-12

2020). In recent years, fermentation has also emerged as a revisited sustainable technology to improve the functional properties of food products, or their byproducts and wastes. Fermentation improves plant-based proteins' digestibility and nutritional value and increases the levels of crucial bioactive compounds such as polyphenols, vitamins, and antioxidants (Abbaspour 2024). Probiotic strains often used in fermentation also contribute to gut health.

Other than the food sector, the pharmaceutical potential of bioactive compounds is widely recognized. Particularly, the plant kingdom is one of the richest sources of bioactive compounds with pharmaceutical potential, and it is considered a well-known source of molecules with demonstrated anticancer therapeutic properties. Nevertheless, bioactive constituents can also be found in other sources such as fungi, microorganisms, animals, and algae. Pharmaceutical applications of bioactive compounds include antioxidant, anticancer, anti-inflammatory, antidiabetic, antimicrobial, and neuroprotective effects, among others (Shrinet et al. 2021). For instance, saponins obtained from *Gynostemma pentaphyllum* may reduce the cholesterol level. Several medicinal plants rich in compounds such as steviosides or polyphenols may exert antidiabetic properties. Some flavones, capsaicin, lycopene, chlorogenic acid, or isothiocyanates (ITCs) have been proposed as having a role in carcinogenesis prevention and treatment. Plant-derived bioactive compounds, such as resveratrol, have also shown potential in restoring central nervous system disorders.

A particular feature of higher plants is their capacity to produce a variety of secondary metabolites, which are generally involved in mechanisms against abiotic stress, such as insects or pathogens. Secondary metabolites have different functions such as competition, defense, attraction, or signaling (Câmara et al. 2021). For instance, flavonoids act in plants as protective agents against free radicals after ultraviolet and light exposure, while terpenoids attract pollinators or inhibit competing plants. Alkaloids have been used to repel herbivores and insects, and ITCs resulting from glucosinolates (GSLs) enzymatic hydrolysis have been proposed for weed control due to their bioherbicide characteristics. These properties unveil the potential of these compounds in the agronomic practice, especially in a context of sustainable development, the promotion of healthy ecosystems, and the European Green Deal, which aims at reducing the dependence on synthetic pesticides and the discovery of new natural compounds and products to be used in crop protection (Bas-Bellver et al. 2025).

12.1.1 Classification of Main Bioactive Compounds

Bioactive compounds belong to different classes (polyphenols, carotenoids, tocopherols, organosulfur compounds, etc.) with different chemical structures (hydro- or lipophilic) and distribution in nature (specific to vegetable or ubiquitous). A classification of major bioactive compounds and their sources is presented in Figure 12.1.

Polyphenols are a family of more than 10,000 different secondary metabolites produced by plants as a response to abiotic factors. They are present in a wide range of plant-based materials, including vegetables, cereals, fruits, legumes, tea, cocoa, herbs, and their derived products. Chemically, polyphenols are characterized by an aromatic ring with at least one hydroxyl group and are divided into two main groups: flavonoids (including flavonols, flavones, isoflavones, flavanols, anthocyanidins, and

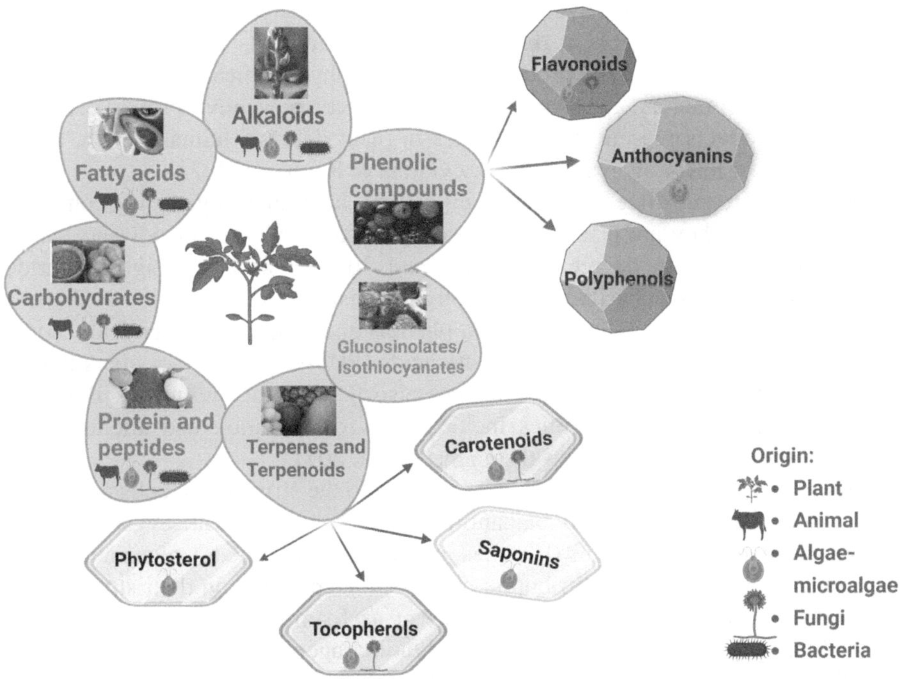

FIGURE 12.1 Classification of bioactive compounds and their sources. All biocompounds originate from plants. Some exclusively from plants (saponins, phytosterol, and anthocyanins), others from bacteria, fungi, and animals (alkaloids, fatty acids, carbohydrates, proteins, peptides, and polyphenols), or from fungi (carotenoids, tocopherols, and flavonoids)

flavanones) and non-flavonoids (including phenolic acids, stilbenes, lignans, and tannins). Beneficial health effects attributed to polyphenols, such as antidiabetic, anti-inflammatory, anticancer, antihypertensive, antibacterial, cardio-protective, and antiseptic properties, vary depending on their chemical structure (Câmara et al. 2021).

Terpenes are organic molecules built from 5 to 10 isoprene units (C_5H_8, 2-methylbuta-1,3-diene), classified into monoterpenes (limonene, linalool, and myrcene), sesquiterpenes (beta-caryophyllene, humulene, and farnesene), diterpenes (gossypol and taxol), triterpenes (squalene, digitoxin, and lupeol), and tetraterpenes (carotenoids). Terpenoids (carvacrol, citronellal, geraniol, linalool, menthol, thymol, etc.) are terpenes containing oxygen molecules by removal or addition of methyl groups. Found mainly in plants such as citrus, cannabis, and mint, they provide aroma, flavor, and medicinal properties (Câmara et al. 2021). Due to their antiplasmodial, antiviral, antidiabetic, neuroprotective, and antimicrobial activities, terpenes and terpenoids are widely applied in pharmaceuticals and as natural alternatives to chemical preservatives in the food industry.

GSLs are secondary metabolites mainly from Brassicaceae. Structurally, GLSs are anions composed of a thiohydroximate-O-sulfonate group linked to glucose, with an alkyl, aralkyl, or indolyl side chain (Câmara et al. 2021). While stable within intact plant cells, GLSs are hydrolyzed upon tissue damage by a β-thioglucosidase (myrosinase), producing ITCs, thiocyanate anions, nitriles, sulfates, and goitrins. ITCs are the major hydrolysis products of GSLs at physiological pH, exhibiting biocidal, anti-inflammatory, neuroprotective, and cancer-preventive properties. These hydrolysis products present different roles in plant defense, human health, and as biopesticides or flavor agents (Rodríguez García & Raghavan 2022).

Prebiotics are substrates selectively utilized by host microorganisms, conferring health benefits. They are primarily plant-derived carbohydrates or polysaccharides with a polymerization degree between 4 and 30 such as fructooligosaccharides (FOS), galactooligosaccharides, mannanoligosaccharides, and xylooligosaccharides. However, human milk oligosaccharides, phenols and other phytochemicals, conjugated linolenic acid, polyunsaturated fatty acids (PUFA), and dietary fibers that are resistant to digestion by host gut enzymes also fit with this definition. Beneficial physiological effects of prebiotics include improved laxation and attenuation of blood cholesterol and glucose levels (Câmara et al. 2021).

Proteins are essential nutrients that contribute to the structure of cells and tissues in the body. Beyond conventional animal-based sources (meat, eggs, milk, and dairy) and plant-based sources (legumes, grains, nuts, and seeds), algae and insects have emerged as valuable contributors of protein. Depending on their structure and physicochemical characteristics, they can act as foaming or antifoaming agents, emulsifiers, gelling agents, water- and fat-binding agents, or texturizers, thus contributing to desirable sensory and functional attributes in a wide range of foods. Bioactive peptides are short sequences of 2–20 amino acid residues released from parent proteins through proteolysis that occurs naturally during digestion, via microbial fermentation, or by enzymatic hydrolysis under laboratory conditions. Whey proteins and cereal grains represent the primary sources of bioactive peptides. Bioactive peptides act as endogenous signals or hormones in organisms by interacting with certain receptors and biomolecules to modulate their biological functions. Biological activities include angiotensin-converting enzyme inhibitory, antidiabetic, cholesterol lowering, antibacterial, immunomodulatory, and antioxidative effects (Sengupta et al. 2020).

Alkaloids are naturally occurring organic compounds that contain nitrogen atoms, usually within heterocyclic rings (Câmara et al. 2021). Major dietary sources of alkaloids include Solanaceae species such as tomato, eggplant, pepper, and potato, as well as tea, coffee, and various spices. Pharmacologically relevant alkaloids include morphine, a potent analgesic for pain relief; quinine, historically used to treat malaria and other febrile conditions; and nicotine, which has been investigated for therapeutic applications in ulcerative colitis. Many alkaloids also exhibit psychotropic effects (e.g. psilocin) and stimulant activity (e.g. cocaine, caffeine, nicotine, and theobromine), and they can be toxic to humans at high doses.

Fatty acids are considered essential bioactive compounds since the human body cannot synthesize them. Plant oils are good sources of fatty acids, and they come mostly from olive, palm fruit, and oilseeds, although other plant materials such as eggplant or avocado are also good sources (Rodríguez García & Raghavan 2022). Fatty acids are aliphatic monocarboxylic acids containing 4–28 carbon atoms. They serve as the building blocks of many lipids. Fatty acids are commonly classified according to chain length and number of double bonds (Djuricic & Calder 2021). PUFAs are long-chain fatty acids containing various double bonds, including the essential omega-3 and omega-6 families. Linoleic acid, the parent compound of the omega-6 family, is mainly found in corn and sunflower oils. It is a precursor of arachidonic acid, which gives rise to bioactive mediators involved in inflammation and lipid metabolism. Alpha-linolenic acid (ALA), the parent compound of the omega-3 family, is mainly found in soybean, rapeseed, and flaxseed oils. ALA is converted into eicosapentaenoic acid and docosahexaenoic acid, which are associated with lowering plasma triglycerides, reducing cardiovascular risk, and exerting anti-inflammatory effects, among other health benefits. Other long-chain fatty acids associated with beneficial effects include conjugated linoleic acid found in ruminant meat and dairy products, and oleic acid, abundant in olive oil and nuts. Butyric acid is a short-chain fatty acid present in milk and one of the main byproducts of fiber fermentation in the colon, with anti-colon cancer properties. Examples of medium-chain fatty acids involved in reducing blood triglyceride levels, plasma insulin level, and fat accumulation include caprylic and capric acids.

12.1.2 Bioaccessibility, Bioavailability, and Bioactivity

Bioactive compounds need to be ingested and released to exert their bioactivity potential. In this context, bioaccessibility is defined as the fraction of the active compound that is released from the food matrix in the gastrointestinal tract and is available for intestinal absorption. Bioavailability is defined as the proportion of a nutrient or a bioactive compound used for normal physiological functions. Bioactivity refers to the specific effect in the body upon exposure to a substance. The bioaccessibility and bioavailability of each bioactive compound differ greatly, and the most abundant compounds in ingested foods are not necessarily those leading to the highest concentrations of active metabolites in target tissues. Besides, several factors interfere with bioaccessibility such as the food source or the chemical interactions among bioactive compounds and other biomolecules present (Galanakis & Drago 2022).

Bioactive compounds are widely incorporated into food products to enhance their sensory properties or contribute to their nutritional and health properties. However, due to several factors such as low solubility in aqueous phases, instability during processing, as well as their low bioaccessibility and bioavailability, the use of these active substances is sometimes limited. To solve this, encapsulation techniques have been proposed to protect and stabilize bioactives, even control their release, and increase their bioavailability. Encapsulation of bioactive compounds applies to both the food and the pharmaceutical industries as well as in other applications, in which release control is important such as in the agrochemical industry. With encapsulation, bioactive compounds are protected from environmental factors such as light, heat, and oxygen, thus enhancing their bioavailability. Encapsulation techniques such as spray-drying (SD), freeze-drying (FD), coacervation, and emulsification stand out due to their effectiveness and versatility (Rezagholizade-Shirvan et al. 2024). Other approaches to enhance stability, controlled release, and targeted delivery of bioactive compounds involve their inclusion within nanometer-sized carriers such as liposomes, hydrogels, or inorganic elements like silica, metal nanoparticles, or clay.

12.2 PRODUCTION OF BIOACTIVE COMPOUNDS FROM BIOMASS SOURCES

Bioactive compounds have been traditionally obtained from plant materials through extraction using simple procedures, such as solvent extraction, or more complex or advanced ones like supercritical fluid extraction. Extraction processes face several difficulties such as low yield, fluctuation of the compounds of interest due to seasonal and climatic conditions, as well as overexploitation of natural resources to satisfy the demand for certain bioactive compounds due to the huge amount of sample required. This has led to an increased interest in the production of bioactive compounds from other organisms such as bacteria, fungi, or microalgae that function as small factories, reducing costs as well as land requirements and other resources, and simplifying processes (Pattnaik et al. 2025). This bioproduction needs to be followed by different separation or downstream processes, which in most cases involve extraction to separate and concentrate the compound of interest. Therefore, bioactive compounds accumulated in either plants, animals, microbial, fungi, or microalgae are recovered from the source by tailored processes, specifically designed for obtaining a product suitable for specific applications, which require certain purity or quality of the separated compound (Figure 12.2). In this way, the obtained products may have applications in the food, pharmaceutical, chemical, or agricultural sectors, among others.

A different approach for bioactive compounds recovery is to integrally valorize the biomass by transforming the raw materials into products rich in a variety of compounds, which make them interesting for the industry. This integral valorization usually consists of the stabilization of fruit and vegetables by dehydration and further powdering to obtain a valuable functional ingredient, which can be used to fortify foods such as bread or pasta. In powder form, fruit and vegetables have been proposed as functional food supplements or ingredients (Kurćubić et al. 2025). Besides, this approach has gained interest for the valorization of surpluses, byproducts, and discards (Ramírez-Pulido et al. 2021; Benvenutti et al. 2025).

In the following sections, production of bioactive compounds will be addressed considering this classification: (a) extraction techniques, (b) bioproduction (fermentation), and (c) integral valorization approaches, as illustrated in Figure 12.2.

12.2.1 Extraction Techniques for the Recovery of Bioactive Compounds

Bioactive compounds can be extracted by different methods, classified into conventional or classical methods and non-conventional or green extraction techniques (Rodríguez García & Raghavan 2022; Agrawal & Nirmal 2024). Table 12.1 shows different examples of bioactive compounds extraction from biomass using conventional and non-conventional extraction techniques.

12.2.1.1 Conventional Techniques

Conventional extraction techniques are classical methods based on solvent extraction or its combination with heat. Main conventional techniques include Soxhlet

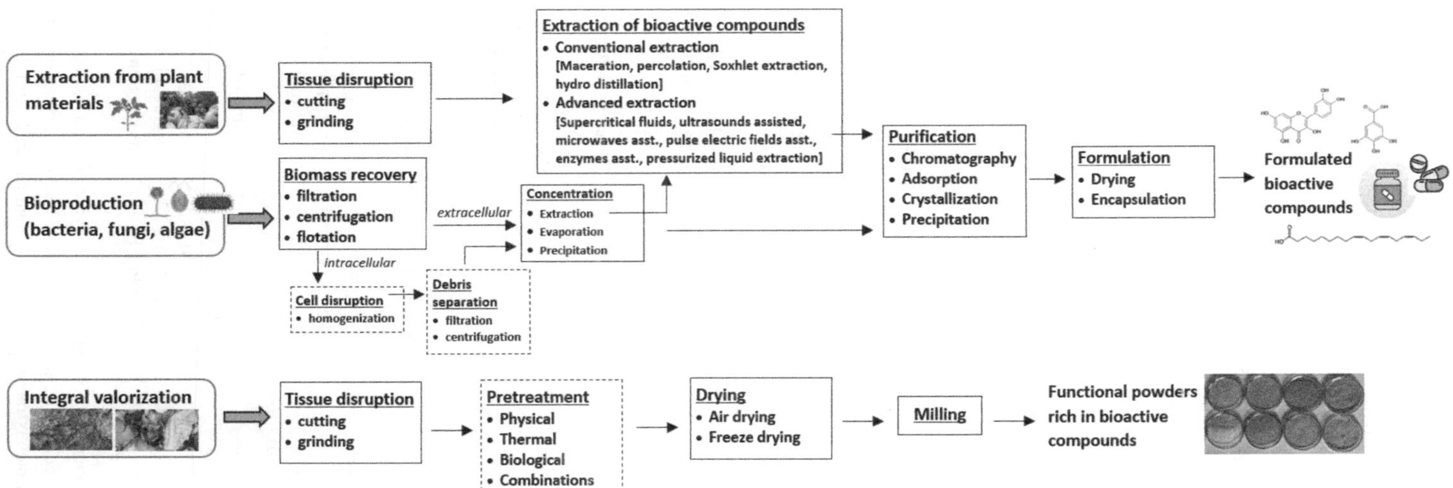

FIGURE 12.2 Production of bioactive compounds from various biomass sources: extraction from plant materials, bioproduction and recovery (micro-organisms, fungi, algae), and integral biomass valorization processes

TABLE 12.1
Extraction Techniques Applied to Obtain Bioactive Compounds from Biomass: Compounds, Source, Extraction Techniques and Main Applications

Bioactive Compounds	Source	Extraction Technique	Main Applications	References
Phenolic compounds	Eggplant peel (anthocyanins)	SE (ethanol at 40°C, 60–80 min)	Food industry (colorants)	Sengupta et al. (2020)
	Blackberry pomace	SFE (50°C, 25 MPa) + UAE (400 W, water/ethanol)	Food industry	Rodríguez García and Raghavan (2022)
	Defatted almond skin	SWE (160°C, 7 bar and 150 rpm for 30 min and 180°C, 15 bar and 150 rpm for 30 min)	Food industry	Freitas et al. (2023)
	Mango "criollo" peel	DESs (β-alanine, DL-malic acid and water (1:1:3 molar ratio))	Food and agro-food industry	Ojeda et al. (2023)
Terpenes and terpenoids	Peppermint (essential oils)	Hydrodistillation	Food industry (ice creams formulation)	Berktas and Cam (2021)
	Carrot peel (β-carotene)	MAE (850 W, acetone in water)	Food industry (colorants)	Rodríguez García and Raghavan (2022)
	Tomato juice	PEFE (at 5 kV/cm)	Beverage industry	Andreou et al. (2020)
	Mango peel and pulp	SFE (using CO_2)	Food, pharmaceutical and cosmetic industries	Villacís-Chiriboga et al. (2021)
Carbohydrate (fructooligo-saccharides)	Litchi seeds	UAE (222 W for 45 min with water)	Pharmaceutical industry	Sengupta et al. (2020)
Fatty acids	Sea buckthorn seed (*Hippophae salicipholia*)	UAE (552 W and 50°C for 8.28 min with n-hexane)	Cosmetical and pharmaceutical industries	Sanwal et al. (2022)
Proteins and peptides	Brewers' spent grain	EAE (10–20 μL alcalase/g_{dm} at pH 8 for 4 h) and combined with UAE or PEFE	Food industry, animal nutrition, bio-degradable film production	Jaeger et al. (2021)

DESs, extraction with deep eutectic solvents; EAE, enzyme-assisted extraction; MAE, microwave-assisted extraction; PEFE, pulsed-electric field extraction; SE, solvent extraction; SFE, supercritical fluid extraction; SWE: subcritical water extraction; UAE, ultrasound-assisted extraction.

extraction, maceration, and hydrodistillation (Rodríguez García & Raghavan 2022; Agrawal & Nirmal 2024; Mungwari et al. 2025).

Extraction efficiency depends mainly on the chosen solvent, which has an influence in mass transfer, molecular affinity, cost, and environmental sustainability (Agrawal & Nirmal 2024). For example, alcohols (ethanol and methanol) are the most used solvents for the extraction of bioactive compounds such as anthocyanins, polyphenols, tannins, terpenoids, or alkaloids (Rodríguez García & Raghavan 2022). Water is used as a solvent to extract tannins, terpenoids, or anthocyanins. Other solvents like chlorinated solvents (chloroform, carbon tetrachloride, and chlorobenzene) and non-chlorinated solvents (acetone and acetonitrile) are employed for the extraction of terpenoids and flavonoids, respectively (Mungwari et al. 2025). Other variables that affect the efficiency of extraction include temperature, particle size, solid to solvent ratio, extraction time, and number of extractions (Rodríguez García & Raghavan 2022; Mungwari et al. 2025).

Soxhlet extraction is one of the most popular extraction methods, often used as a standard method for new techniques. It is widely used for the extraction of phytochemicals, fatty acids, and essential oils from plant materials using a solvent (Agrawal & Nirmal 2024). Soxhlet is a continuous extraction method, in which the solvent repeatedly passes through the dry sample. The solvent is heated until evaporation and finally condensed, allowing the desired compounds to dissolve from the matrix. The process runs continuously until the extraction is completed (Mungwari et al. 2025). Despite the low processing cost and simplicity, this technique has some drawbacks compared to novel techniques. These encompass long extraction times, high solvent and energy consumption, less safety, and poor extraction efficiency (Sengupta et al. 2020; Agrawal & Nirmal 2024). To enhance extraction efficiencies, this technique can be combined with microwave-assisted extraction and ultrasonic extraction (Agrawal & Nirmal 2024).

Maceration or solvent extraction (SE) is a low-cost technique proposed to get essential oils and fundamental bioactive ingredients (Agrawal & Nirmal 2024). In this technique, milled plant material is soaked with the proper solvent, allowing the mixture to remain at room temperature for extended periods (2–3 d), with frequent stirring to facilitate extraction (Mungwari et al. 2025). Despite the low investment cost, low-temperature requirements, and easy-to-use equipment, maceration takes longer time and gives smaller yields (Agrawal & Nirmal 2024). As an example, it has been reported that anthocyanins can be extracted from eggplant peel using acidified ethanol as solvent at 40°C for 60–80 min (Sengupta et al. 2020).

Hydrodistillation is used to extract essential oils and various bioactive compounds from plant matrices. It can be carried out with water or steam before dehydrating the plant material (Agrawal & Nirmal 2024). For that, the solid sample is mixed with water and heated until evaporation. The condensed mixture moves to a separator, where the bioactive compounds and lipids split from the water. This technique permits to automatically separate bioactive compounds from water. However, extraction time can be long, and some volatile components may be lost with the high extraction temperatures (Agrawal & Nirmal 2024). Thereby, this technique is not applicable for heat-labile compounds. Hydrodistillation has been used to extract essential oils (menthol (35%–60%), menthone (2%–44%), and menthyl acetate (0.7%–23%) from peppermint (*Mentha piperita* L.) using distilled water for a period of 2 h (Berktas & Cam 2021).

12.2.1.2 Non-conventional or Green Extraction Techniques

Non-conventional or green extraction techniques have been developed to enhance conventional techniques and deal with their disadvantages. These novel techniques are considered as "green" because they are environmentally sustainable technologies that aim to reduce energy and solvent consumption, hazardous substances, and extraction times as well as to enhance efficiency (Rodríguez García & Raghavan 2022; Agrawal & Nirmal 2024). The main non-conventional techniques are described in this section.

Microwave-assisted extraction (MAE) applies an electromagnetic field ranging from 300 MHz to 300 GHz and wavelengths between 1 mm and 1 m. The principle of microwave heating depends on the direct interactions of microwaves with polar compounds. The heat is generated by dipole rotation and ionic conduction mechanisms (Agrawal & Nirmal 2024; Mungwari et al. 2025). In MAE, the solute molecules are first separated from the matrix by increasing temperature and pressure. The solvent (usually ethanol in water) then diffuses into the sample. Lastly, the liquid containing the solubilized components is separated from the residual solids (Rodríguez García & Raghavan 2022; Agrawal & Nirmal 2024). It must be noted that MAE can be performed with or without the use of a solvent. Factors that should be considered to optimize MAE include frequency, solid to solvent ratio, extraction temperature and pressure, operation time, and microwave power (Agrawal & Nirmal 2024). Phenolic compounds, flavonoids, and carotenoids have been extracted by this method (Rodríguez García & Raghavan 2022). MAE's benefits include increased extraction yield, reduced equipment size, and shorter extraction time (Agrawal & Nirmal 2024). MAE of carrot peel using acetone in water (51% v/v) at 850 W has been used to extract β-carotene for use as a coloring agent in foods.

Ultrasound-assisted extraction (UAE) involves the use of mechanical waves (20 kHz–100 MHz frequency) that pass through a medium, inducing compression and expansion cycles, which produce cavitation phenomena. Cavitation involves the production, growth, and collapse of bubbles, which generate inter-particles collisions and higher diffusion of the extractable compounds into the solvent (Agrawal & Nirmal 2024). Ultrasounds can be applied directly to the medium, or indirectly, with an ultrasonic water bath (Rodríguez García & Raghavan 2022). The main variables of UAE include sonication time, ultrasound power, solvent, liquid to solid ratio, plant material, and frequency (Agrawal & Nirmal 2024). UAE presents advantages such as rapid extraction time, reduced energy and solvent consumption, and instrument simplicity (Rodríguez García & Raghavan 2022; Mungwari et al. 2025). It has been reported as an efficient technique to extract phenolic compounds from agri-food wastes, carotenoids, and oils, among others. UAE at 222 W for 45 min, using water as the solvent, was used to extract FOS from litchi seeds (Sengupta et al. 2020). Sanwal et al. (2022) demonstrated that the application of UAE at 552 W for 8.28 min, at 50°C while using n-hexane as solvent, can enhance the extraction of fatty acids from sea buckthorn seed (*Hippophae salicipholia*). The combination of UAE with other extraction techniques has been proven to increase the extraction yield. This is the case of phenolic extraction from blackberry pomace by supercritical fluid extraction (SFE) (50°C and 25 MPa) and UAE (400 W using water and ethanol as co-solvents), where the extraction was of 9.87%, twice the extraction if only SFE was used (Rodríguez García & Raghavan 2022).

Pulsed-electric field extraction (PEFE) is a non-thermal process, in which a pulse-electric field is applied for a brief period (nanoseconds/milliseconds) through a solvent-solid mixture. This induces cell electroporation, which permeabilizes plant matrix membranes, favoring the mass transfer and the solvent access (Agrawal & Nirmal 2024). Molecules are separated according to their charge (Sengupta et al. 2020; Rodríguez García & Raghavan 2022). The extraction depends on temperature, field strength, number of pulses, energy input, and the material used. Polyphenols, anthocyanins, carotenoids, or flavonoids can be extracted with PEFE (Rodríguez García & Raghavan 2022). PEFE offers some advantages as it results in high extract yield, less extraction time, and low energy consumption (Agrawal & Nirmal 2024). As an example, Andreou et al. (2020) applied PEFE to enhance juice yields and its valorization in industrial tomato processing. They obtained that PEFE processing reduces tomato peel detachment work in 72.3%, and the application of PEFE to chopped tomatoes increased the juice yield up to 20.5%. PEFE also increased carotenoid extraction from residues (seeds and peels) up to 56.4%.

Enzyme-assisted extraction (EAE) can be used as a pretreatment or extraction method. This technique is employed to release compounds bound to cell walls or intercellular ones by the action of specific enzymes (cellulase, hemicellulase, pectinase, α-amylase, and protease) that hydrolyze cell wall components (Rodríguez García & Raghavan 2022). Factors influencing EAE are particle size and chemical composition of the plant material, type and concentration of enzyme, water proportion, and treatment time. This technique is useful for the extraction of oils and phenolic compounds, including flavonoids and anthocyanidins. It can be used in combination with MAE, UAE, and SFE (Agrawal & Nirmal 2024). However, it is an expensive technique for industrial-scale operation (Rodríguez García & Raghavan 2022; Mungwari et al. 2025). EAE has been used for the extraction of proteins like hordein and glutelin from brewers' spent grain using 10–20 mL alcalase/gdm at pH 8 for 4 h and in combination with other techniques like UAE or PEFE (Jaeger et al. 2021).

SFE uses fluids under supercritical conditions. Above the critical point, the fluid acquires gas and liquid-like properties of diffusion, surface tension, viscosity, and density, favoring penetration into the matrix and mass transfer in a short time (Rodríguez García & Raghavan 2022; Mungwari et al. 2025). It depends on many factors such as solvent (fluid), pressure, density, temperature, particle size, moisture content, extraction time, flow rate, and solvent-sample-ratio (Sengupta et al. 2020). Carbon dioxide (CO_2) is the most widely used solvent for SFE. Having low critical conditions (31°C and 7380 kPa) due to its low toxicity, non-flammability, chemical stability, and cost-effectiveness, it is adequate to extract thermolabile compounds using CO_2 (Mungwari et al. 2025). However, CO_2 is used mostly for fats, lipids, and non-polar compounds extraction due to its low polarity, but its combination with a cosolvent (like ethanol) can modify CO_2 polarity, allowing the extraction of polar compounds (Mungwari et al. 2025). SFE using CO_2 as fluid has been used for the extraction of β-carotene from mango peel and pulp with possible uses in cosmetic, pharmaceutical, and food industries (Villacís-Chiriboga et al. 2021).

Extraction with deep eutectic solvents (DESs) refers to the replacement of organic solvents currently used in solid-liquid extractions with non-hazardous and eco-friendly ones. DESs are obtained by combining a hydrogen bond acceptor (e.g. quaternary ammoniums) and a hydrogen bond donor (e.g. urea, carboxylic acids, or amines) and exhibit high chemical and thermal stability, no flammability, low toxicity, high electrical conductivity, and a good solubilizing capacity for several organic compounds (Ojeda et al. 2023). When DESs are made of natural eutectic compounds from plant metabolites or their derivatives (e.g. choline chloride, citric acid, malic acid, maleic acid, acetic acid, glucose, fructose, sucrose, trehalose, terpenoids, or water), they are referred to as NADESs. NADESs are readily biodegradable and have zero toxicity. Most studies using DESs as extraction solvents focus on recovering bioactive phytochemicals from plants, vegetables, fruits, and byproducts from the food and agri-food industry. It has been reported that the extraction of total phenols, flavonoids, and overall antioxidant compounds as well as specific polyphenols (e.g. gallic acid, catequin, and mangiferin) from mango "criollo" peel using a mixture of β-alanine, DL-malic acid, and water (1:1:3 molar ratio) is significantly higher than that achieved with conventional solvents such as methanol or ethanol (Ojeda et al. 2023).

Subcritical water extraction (SWE) employs liquid water as a solvent at temperatures between 100°C and 374°C, under sufficient pressure to prevent evaporation. Increasing the temperature reduces the dielectric constant of water and changes water polarity from strong polar to much less polar, thereby achieving extraction yields comparable to those of conventional organic solvents. As an example, extracts obtained from blanched defatted almond skin using subcritical water at 160°C, 7 bar, 150 rpm, for 30 min and 180°C, 15 bar, 150 rpm, for 30 min contained higher levels of total phenolic compounds than those produced with organic solvent-based methods and showed DPPH radical scavenging capacities comparable to well-known antioxidants such as ascorbic acid and α-tocopherol (Freitas et al. 2023). In this case, the increased antioxidant capacity could be attributed not only to the higher extraction efficiency of antioxidant compounds but also to the formation of new antioxidants via Maillard, caramelization, and thermo-oxidation reactions.

12.2.2 Bioproduction: Microbial and Algae Production Methods

Industrial interest in obtaining biochemical compounds through natural and ecological approaches is growing. It has been observed, particularly in the food and cosmetics industries, that the scientific community is increasingly interested in the origin of products, leaning toward sustainable and healthy options derived from natural sources. This shift moves away from synthetic chemicals, which are not environmentally friendly, can have undesirable effects depending on their application and often involve high production costs and the use of toxic catalyst (Ali et al. 2025). Following this trend, the current focus is on microbial production of bioactive compounds. Bacteria, fungi, algae, and microalgae are being used as small production factories, more economical production plants with simplified processes (Pattnaik et al. 2025). Fermentation is a biotechnological strategy used for the biosynthesis of various compounds such as enzymes, biofuels, fermented foods, antibiotics, and proteins. It also produces secondary metabolites, flavonoids, and therapeutic proteins. Pharmaceutical and food industries have an interest in implementing fermentation processes at an industrial scale using specialized biorefinery techniques, to obtain bioactive compounds, including antioxidants, pigments, enzymes, antibiotics, or nutraceuticals, among others.

The production of bioactive compounds mainly depends on the composition and source of the substrate, the type of microorganism used, and specific operating conditions such as water activity, pressure, and temperature. Two main types of fermentation, that is, solid-state fermentation (SSF) and submerged fermentation (SMF), can be used for the synthesis of bioactive compounds. SMF takes place in liquid medium, in continuous stirred bioreactors, which allow direct control over oxygen transfer through mechanical means, and homogeneously distributed nutrients. In contrast, SSF involves a matrix with low water content and low water activity. In this process, the biomass is used as the solid support and as a source of nutrients, thus reducing the need for chemicals, growth factors, or additives, which results in a more ecofriendly and cheaper process than SMF (Nausad et al. 2024; Frumuzachi et al. 2025). Using SSF, it is important to consider the adequate selection of the microorganism and substrate, optimize particle size which ensures aeration, and adjust the water activity required for microorganism growth. Enough ventilation and airflow are essential to guarantee that oxygen reaches every area of the solid substrate, and the relative humidity required for maintaining the appropriate water activity. Inadequate airflow can result in areas with insufficient oxygen or water accumulation, potentially leading to anaerobic conditions that hinder microbial growth.

A wide range of substrates can be used for SSF, including agricultural and agro-industrial waste, byproducts, lignocellulosic materials, animal waste, and even inert substrates. In contrast, SMF uses more complex media, which combines a variety of carbon and nitrogen sources, in addition to micronutrients. In recent years, the utilization of agro-industrial residues in fermentation has emerged as a sustainable option. This approach aligns with the principles of a circular economy and promotes the recovery of waste, offering environmental and economic benefits. SSF requires low humidity, typically in the range of 12%–70%, which represents a lower risk of contamination. In this process, microorganisms secrete enzymes to break down complex nutrients in the solid substrate, transforming them into absorbable forms. SMF requires higher water content (70%–90%) due to the nature of the culture medium, which implies higher risk of contamination. Complex nutrients are absorbed or diffused through the cell membranes of microorganisms (Gómez-Ramos et al. 2025). Typically, SSF can yield higher production, with a lower raw material cost and simpler equipment. However, its scale-up is a more complex process. It generates less wastewater and has lower energy requirements. In contrast, SMF yields lower production and has a higher raw material cost. Its scale-up is simpler due to better control over online monitoring and automation, as well as more efficient mass transfer for proper microbial growth and metabolism.

TABLE 12.2
Microorganisms and Algae for Production of Bioactive Compounds: Production Methods, Extraction Techniques, and Main Applications

Source	Production Method	Bioactive Compound	Extraction Technique	Main Application	References
Bacteria					
Lactiplantibacillus plantarum	SMF	Gallic, chlorogenic, and caffeic acids	UAE	Food, nutraceutical, pharmaceutical industries	Kalinowska et al. (2023)
Lactobacillus acidophilus, Lactiplantibacillus plantarum	SMF	Ferulic acid, vanillin, and vanillic acid	SE	Food, cosmetic, pharmaceutical industries	Pattnaik et al. (2025)
Lactobacillus fermentum	SMF	Linolenic, behenic, paullinic, and palmitoleic acids		Food industry	Nausad et al. (2024)
Bacillus subtillis	SSF	Aminobutiric acid GABA	SE	Food, pharmaceutical Industries	Gómez-Ramos et al. (2025)
Fungi					
Aspergillus niger	SSF	Phenolic Acids (caffeic, 3- and 5-caffeoylquinic, and 4-feruloylquinic acids), and free aglycones	UAE	Nutraceutical industry	Frumuzachi et al. (2025)
Aspergillus awamori, Aspergillus oryzae	SSF	Glucoamylase, protease, xylanase, cellulase	SE	Food, nutraceutical, pharmaceutical industries	Gómez-Ramos et al. (2025)
Microalgae/algae					
Tetraselmis tetrathele	Salinity-light stress culture	Pigments, antioxidant compounds	SE	Food, aquaculture, pharmaceutical industries	Nezafatian et al. (2023)
Chlorella vulgaris	Culture with CNMs	Carotenoids, lipids	SFE	Pharmaceutical industries	Agarwal et al. (2023)
Padina tetrastromatica, Laminaria ochroleuca Pyropia yezoensis (macroalgae)	Biomass macroalgae from fucoidan extraction	Protein	SWE	Food, pharmaceutical industries	Hans et al. (2024)

CNMs, carbon nanomaterials; MAE, microwave-assisted extraction; SE, solvent extraction; SFE, supercritical fluid extraction; SMF, submerged fermentation; SSF, solid-state fermentation; SWE, subcritical water extraction; UAE, ultrasound-assisted extraction.

The use of microorganisms capable of metabolizing biomass to produce bioactive compounds is currently of great interest. Table 12.2 details some studies on microorganisms and algae, including their extraction methods, production of bioactive compounds, and key industrial applications. It has been documented that some bacteria possess reductase, esterase, and decarboxylase enzymes that enable them to release bioactive compounds such as phenolic acids from complex plant matrices. For instance, *Lactiplantibacillus plantarum* can be used to produce gallic acid, chlorogenic acid, and caffeic acid from apple pomace (Kalinowska et al. 2023). *Lactobacillus acidophilus* and *L. plantarum* have been used to produce ferulic acid, vanillin, and vanillic acid from sugarcane bagasse (Pattnaik et al. 2025).

Fungi and yeasts are also used in the production of bioactive compounds. These microorganisms acquire the capacity to obtain bioactivity via deglycosylation and hydrolysis of glycosidic bonds, and fermentation-induced modifications. Among them, *Aspergillus niger* has been studied to produce caffeic acid, 3- and 5-caffeoylquinic acids, and 4-feruloylquinic acid from a carbohydrate-rich substrate (Frumuzachi et al. 2025).

To align with the objective of supplying natural, environmentally friendly, and functional products for human use, the utilization of algae and microalgae holds much promise for the synthesis of bioactive compounds. Microalgae exhibit rapid growth and adapt well to various substrates like wastewater, seawater, and the innovative use of carbon nanosheets. Exposing them to abiotic stress is known to increase the production of bioactive compounds on MAE, UAE, and EAE. These methods have led to higher production yields in a shorter time with reduced use of solvents, making them more environmentally friendly. Among these techniques, SFE is particularly noteworthy as it can produce an extract without the need for a solvent. This ensures a product that is free from risk for human, animal, and plant consumption (Agarwal et al. 2023).

At an industrial level, the use of microalgae from the classes Euglenophyta, Chrysophyta, Pyrrophyta, Chlorophyta, Rhodophyta, Phaeophyta, and Xanthophyta is particularly noteworthy. For example, *Chlorella vulgaris* is used to produce lipids, carotenoids, and lutein. *Dunaliella salina* is employed to produce fatty acids and indole derivatives. Additionally, when using food wastewater as a medium to produce pigments and increase antioxidant capacity, the microalga *Tetraselmis tetrathele* is utilized (Nezafatian et al. 2023). On the other hand, macroalgae, also known as seaweeds, are classified into three groups: red algae (Rhodophyta), green algae (Chlorophyta), and brown algae (Ochrophyta and Phaeophyceae). These are primarily used for producing polysaccharides, pigments, minerals, and proteins, which are widely utilized in the food, pharmaceutical, and biotechnology industries. Due to their size, they can be grown in large, controlled environments or naturally harvested from coasts and seabeds (Ali et al. 2025).

The extraction of bioactive compounds from algae can be achieved by both conventional and non-conventional techniques, similar to those used for microalgae. However, in recent years, there has been a greater focus on green extraction techniques such as solvent extraction (or maceration) and SWE. These have shown promising results in improving the efficiency of algae such as *Saccharina japonica* for the synthesis of polysaccharides like fucoidan and carrageenan. The red algae

Kappaphycus alvarezii and *Chondrus crispus* are used for the synthesis of carrageenans using solvent extraction. Additionally, *Padina tetrastromatica*, *Laminaria ochroleuca*, and *Pyropia yezoensis* are widely utilized to produce proteins and free amino acids via SWE technique (Hans et al. 2024).

12.2.3 Integral Valorization of Biomass to Produce Functional Powdered Ingredients

Biomass can be transformed into products rich in a variety of bioactive compounds interesting for industrial applications. This integral valorization usually consists of the stabilization of fruit and vegetables through dehydration and further powdering. In powder form, fruit and vegetables have been proposed as functional food ingredients (Kurćubić et al. 2025), among other applications. In recent years, this approach has gained interest in the valorization of byproducts and discards. In addition to being rich in dietary fiber, minerals, vitamins, and antioxidants, the production of flours or powders has the advantage of fully up-cycling these wastes without the generation of residues such as solvent or other side streams (Ramírez-Pulido et al. 2021; Benvenutti et al. 2025).

Probably the most remarkable application of these powders is their use as functional ingredients for new food formulation. Powdered fruits and vegetables have been proposed as functional food supplements that can be used to fortify foods (Kurćubić et al. 2025) (Table 12.3).

One major application extensively reported in the literature consists of partially replacing flours with powders obtained from fruits and vegetables, or their byproducts, in bakery products (Benvenutti et al. 2025). Being gluten-free, fruit and vegetable flours are suitable for gluten-sensitive individuals, people with celiac disease, those who have dietary restrictions, or consumers who are reducing gluten consumption due to a "health halo" effect. Partial replacement of flour by fruit and vegetable powders has been used to enhance the nutritional value of formulated products, thanks to their content in bioactive compounds such as fiber, polyphenols, carotenoids, or GSLs, for instance (Bas-Bellver et al. 2024a).

According to Bas-Bellver et al. (2024a), powders obtained from discards of IV-range carrots production were incorporated into rice-based muffins formulations (0%–30% of replacement). Results confirmed that carrot powder addition enhanced the antioxidant properties of the formulated products, an increase that was proportional to the percentage of replacement. The incorporation of carrot powder modified the pasting properties of the dough, reducing the viscometric parameters (peak, trough, setback, and final viscosities) as compared to rice flour. The modification of dough behavior was partially explained by the water competition with the fiber present in the vegetable powder. The other compounds interfering with starch gelatinization were phenolic constituents, which have also been reported to interact with starch and modify pasting behavior and leavening. In the products, it was evidenced that reduced viscoelastic properties, due to fiber competition for water and interaction with phenolic compounds, led to reduced leavening ability, fewer trapped air bubbles, and increased compactness. Other than increasing the nutritional profile, powders

TABLE 12.3
Formulation of New Food Products by the Addition of Powders or Flours Obtained from Different Biomass Sources

New Food Product	Powder/Flour Source	Percentage of Replacement (%)	Target Functional Properties	References
Gluten-free bakery (muffins)	Carrot byproducts	0–30	Nutritional improvement	Bas-Bellver et al. (2024a)
Bakery (wheat sponge cakes)	Grape pomace	0–10	Nutritional improvement	Nakov et al. (2020)
Bakery (wheat biscuits)	Tropical fruit peels	5–15	Nutritional improvement	Nitisuk et al. (2025)
Fresh pasta	Pomegranate peels, broccoli, and olive oil by-products	3–8	Nutritional improvement Antimicrobial/preservative	Panza et al. (2023)
Fruit salad	Pomegranate peel	2.5–5	Antimicrobial/preservative	Lacivita et al. (2021)
Chocolate-milk products	Avocado pulp	15–60 (in dairy mixes)	Nutritional improvement Antimicrobial/preservative	Kumari et al. (2021)
Cocoa powder	Fermented jackfruit seed	10	Nutritional improvement	Ravindran et al. (2020)
Chicken patties	Lemon, orange, grapefruit, and banana peels	1	Antimicrobial/preservative Shelf-life increase	Abdel-Naeem et al. (2022)
Low-fat pork burgers	Seaweed *Undaria* sp.	3	Techno-functional properties (emulsifying, water binding) Nutritional improvement	Nagai et al. (2022)
Seaweed-yak meat products	Seaweed *Laminaria japonica*	10–70	Techno-functional properties (emulsifying, water binding) Nutritional improvement Antimicrobial/preservative	Wang et al. (2025)

derived from fruit and vegetables and their byproducts can also serve as coloring and flavoring agents (Benvenutti et al. 2025). Besides this application, they have also been incorporated into edible coatings or directly applied to extend the shelf-life of formulated products due to their antimicrobial properties. An example can be found in processed meat products, where fruit peel powders (lemon, orange, grapefruit, and banana) have proved to lengthen their shelf-life (Abdel-Naeem et al. 2022).

Nakov et al. (2020) incorporated grape pomace powders into wheat-based bakery products. Sponge cakes enriched with grape pomace powder exhibited higher content in lipids, proteins, anthocyanins, free polyphenols, dietary fiber, and antioxidant capacity than the control, thus enhancing their nutritional attributes. Specifically, grape pomace significantly improved the content of free phenolics, which are scarce in bread wheat. The main phenolics provided by grape pomace were catechin, gallic acid, quercitin, protocatechuic acid, kaempferol, and apigenin. The authors concluded that the addition of small quantities of grape pomace powder did not worsen the technological characteristics of the cakes and improved sensory qualities.

In a different study by Nitisuk et al. (2025), tropical fruit peels powders (mango, banana, dragon fruit, pineapple, and papaya) were used to formulate wheat-based biscuits. The work showed that tropical fruit peels are potent natural sources of antioxidants, particularly mango peel powder. The addition of 5% of fruit peel powder into biscuits enhanced the fiber content, antioxidant potential, and phytochemical concentration, while maintaining their sensory acceptability. Simulated in vitro digestion revealed that essential bioactive compounds present in mango peel powder remained partially bioaccessible and functionally active post-digestion.

Fruit and vegetable powders have also been proposed to enrich fresh pasta products (Panza et al. 2023). In this work, the authors focused on improving the nutritional properties of the product but also on lengthening their shelf-life due to the antimicrobial properties of the flours. Pomegranate peels were the most effective in extending the shelf life of fresh pasta, as compared to broccoli and olive oil residues. The potential of pomegranate peel powder as a natural preservative was also studied in Lacivita et al. (2021), where its effects were tested on fruit salad quality during refrigerated storage; 2.5% and 5% of pomegranate peel were added to nectarine and pineapple in fructose syrup. The results indicated reduced growth of mesophilic bacteria, total psychrotrophic microorganisms, yeasts and lactic acid bacteria (LAB), without negatively affecting the sensory attributes of fresh-cut fruit.

Fruit powders have also been proposed to improve cocoa products, such as cocoa powder or chocolate-milk products. Avocado pulp powder was included as a part of butter fruit milkshake powder (mixture of avocado pulp and dairy components) to formulate chocolate, with enhanced nutraceutical properties of chocolate (Kumari et al. 2021). The addition of this avocado-milk powder significantly improved the nutritional value of chocolate, in the form of dietary fibers, proteins, and phenolic, flavonoid, and beta-carotene contents. A positive impact on shelf-life was also reported. In a different study, fermented jackfruit seed was used to replace cocoa powder for producing chocolates and diet-chocolates (Ravindran et al. 2020). Jackfruit is an underutilized seasonal fruit, for which this approach contributes to valorize excess fruit and discards. The addition of fermented jackfruit powder up to 10% did not significantly alter the characteristics of chocolate, which showed lower glycemic indexes.

Micro- and macroalgae have also been transformed into powder and incorporated into different food formulations owing to their protein, mineral, and vitamin contents, and fatty acid profiles. The feasibility of using whole algae as an ingredient in various human food products (meat-based, cereal-based, dairy products, and beverages) and their functional and nutritional impacts have been extensively revised (Gullón et al. 2020). For instance, the addition of wakame powder increased the polyphenol and mineral content of meat products such as patties or colloidal meat systems, improved the water and fat-binding properties, created harder and chewier structures, and solved the problems related to low salt content. As a good source of polysaccharides, phenolic compounds, pigments, fatty acids, proteins, peptides, and amino acids, seaweed can be used to design novel tailored meat products and increase their shelf-life, improve their nutritional properties, or promote functional properties (Table 12.3). For instance, *Undaria* sp. (wakame) and *Laminaria japonica* flours have been included in meat formulations such as low-fat burgers or plant-based meat products to improve techno-functional properties such as water binding and emulsifying capacity. Additionally, meat products also showed better nutritional profiles (fatty acids and antioxidants) and improved shelf-life due to the antimicrobial properties of the powders incorporated (Nagai et al. 2022; Wang et al. 2025). However, although macroalgae have been proven to modify fat profiles and prevent oxidative deterioration of meat products, there are still some challenges regarding the organoleptic and sensorial properties of the resulting products that have an impact on consumer acceptability (Gullón et al. 2020).

12.2.3.1 Processes for the Transformation of Biomass into Functional Powders

The general process for powder (or flour) obtention from biomass consists of three basic stages: tissue disruption, dehydration, and milling (Bas-Bellver et al. 2020; Ramírez-Pulido et al. 2021) (Figure 12.2).

Tissue disruption (e.g. chopping, grinding, or enzyme hydrolysis) is an essential step for releasing bioactive compounds due to cell membrane breakage. For instance, GSLs are degraded into ITCs and other hydrolysis products by means of the myrosinase enzyme upon tissue disruption (Ramírez-Pulido et al. 2021). The intensity of disruption determines tissue behavior during drying, which in turn impacts the milling efficiency and powder particle size. In Bas-Bellver et al. (2020), the interdependence between these stages was demonstrated for a series of vegetable wastes. Grinding prior to air drying yielded powders with a smaller particle size than powders obtained from chopped matrices.

Dehydration aims to remove water and extend the shelf life of the product by lowering its water activity. Hot air-drying (HAD), vacuum-drying, FD, and SP are dehydration technologies commonly applied to obtain powdered ingredients, singly or in combination with others. HAD and FD are mostly preferred for solid matrices such as fruit, vegetables, algae, and their byproducts, allowing for their integral use (Ramírez-Pulido et al. 2021). HAD is commonly applied in the food industry due to its low processing and investment costs. It usually takes place in convective dryers, where hot air or stream with low relative humidity is introduced in the drying chamber to promote water evaporation from the product. During air drying, water is evaporated from the sample's surface and transferred to the vapor phase (surrounding air). The surface evaporation modifies the product structure, which undergoes significant changes such as shrinkage and glass transitions. This can lead to case-hardening phenomena, thus hindering water mass transfer inside the product and limiting water evaporation and reducing drying rate. The resultant product might be dried in the surface but wet in the core. This situation reduces both drying and milling efficiency, leading to high moisture content and coarse particle sizes (Bas-Bellver et al. 2022).

Case-hardening phenomena are usually favored by fast initial drying rates, which depends on several factors such as the characteristics of the material, that is, which fruit, vegetable, or algae is being dried; and the history of the sample before drying, that is, the pretreatments to which the sample has been subjected. As discussed previously, chopped vegetable samples lead to fine powders. Another example, in this case attributed to differences in the material origin, is reported in Bas-Bellver et al. (2022), where, as compared to cabbage leaves, broccoli stems exhibited faster initial drying rates, which caused case-hardening and a decrease in the drying rates since the beginning of the process. As a result, coarse powders are obtained. In contrast, cabbage residues exhibited an initial constant drying rate, followed by a falling drying rate, resulting in a more homogeneous drying and fine particle size of the powdered products. Technologies such as ultrasound or microwave can be used to assist drying and enhance drying rates and drying efficiency, thus saving time and energy (Ramírez-Pulido et al. 2021)

FD or lyophilization is a different technique suitable for solid matrices. It consists of three stages: freezing, sublimation or primary drying, and desorption or secondary drying. The purpose of freezing is to maximize the amount of frozen water in the form of ice crystals. Sublimation takes place below the triple point of water (276.16 K and 611.73 Pa), ensuring the direct phase change from ice to water vapor. It can reduce water content to ~15%, a value limited by the water bound to the matrix, which cannot be converted into ice crystals during freezing (non-freezable water). Sublimation takes place under vacuum at a temperature around −45°C. A final desorption stage is required to reduce water content to about 2%–3%. Desorption occurs under vacuum conditions at an increased temperature of approximately 25°C–30°C. The aim is to vaporize and remove part of the non-freezable water, which cannot be sublimated. FD usually yields high-quality products and powders, which strongly preserve their functional properties. However, as for antioxidant capacity, HAD can have a more positive impact due to an increased incidence of temperature-dependent biochemical reactions that generate more bioactive forms (Maillard reactions and isomerization), or the inhibition of oxidative enzymes (Bas-Bellver et al. 2020). FD has limited industrial applications due to its high operating costs, which are 4–8 times higher compared to HAD. SD has gained relevance for the development of functional ingredients, preservation of bioactive compounds, or even encapsulation. It consists of generating small droplets from which water is evaporated almost instantaneously and adiabatically, with no negative thermal impact on the sample. However, SD is not an option for the integral valorization of biomass since, to generate droplets, the inlet stream needs to be in the form of a solution, dispersion, or emulsion, which implies previous extraction of bioactive constituents (Ramírez-Pulido et al. 2021).

After drying, the material is exposed to milling to produce the powder. Milling conditions are set to obtain a specific particle size, although, as discussed, particle size depends on drying performance and pretreatment. Hence, the properties of powders are determined by the material being dried and process stages and conditions, which confer them specific physicochemical properties, including antioxidant properties and bioactive compounds content, and techno-functional properties. These properties are important for defining potential uses of the powders in food formulation or other applications.

The process for obtaining powders can incorporate pretreatment, which aims to improve the functional properties of the flours (Figure 12.2). Thermal and physical pretreatment methods such as ultrasound, microwave, freezing, or autoclave have been proposed to enhance the antioxidant properties of the treated materials and corresponding powders. For instance, ultrasound pretreatment significantly improved the antioxidant properties of broccoli biomass and the corresponding flours obtained after HAD and FD, as compared to control and other thermophysical treatments such as autoclave, pasteurization, and microwave (Bas-Bellver et al. 2024b). Biological methods, including fermentation and enzymatic transformations, are also being investigated as pretreatments. Due to the increasing interest in fermentation as a pathway for the integral valorization of biomass, it will be discussed in the following section.

12.2.3.2 Valorization of Fruit and Vegetable Products Through Fermentation for Flour Obtention

In recent years, fermentation has gained attention for producing functional flours. This stage can be applied as a unique pretreatment (Serna-Barrera et al. 2024) or in combination with other pretreatments like thermophysical ones (e.g. microwave and freezing) (Baldassa et al. 2025). Fermentation by LAB improves digestibility, reduces undesirable compounds such as antinutrients, and may introduce probiotics beneficial to human gut health. The most notable genera of LAB include *Lactobacillus*, *Leuconostoc*, *Pediococcus*, and *Streptococcus*. These are renowned for fermenting several substrates, including vegetables and fruits (Marcelli et al. 2025). Commonly used species as starters are *L. mesenteroides*, *L. plantarum*, *P. acidilactici*, *Levilactobacillu brevis*, *L. pentosus*, *Lacticaseibacillus rhamnosus*, *L. casei*, *Lactococcus lactis*, *L. curvatus*, and *L. delbrueckii* subsp. bulgaricus.

L. plantarum, *L. reuteri*, and *L. salivarius* have been investigated to improve the antioxidant properties of broccoli stem products (Serna-Barrera et al. 2024). Fermenting broccoli stems for 24 h with the three microbial strains led to a substantial increase in total phenols and flavonoids. Subsequent FD and grinding had minimal impact on the microbial population and significantly enhanced the extractability of antioxidant compounds. Fermentation can promote the hydrolysis of phenolics and GSLs through microbial enzymatic activity, leading to the formation of health-beneficial metabolites. LAB fermentation has been found to modify sugar-acid ratio and amino acid content of carrot pulp. *L. plantarum* and *P. pentosaceus* fermentation enhanced the total phenolic and flavonoid content of carrot residues, whereas *L. acidophilus* and *L. paracasei* exhibited the most pronounced changes in odor profiles (Liu et al. 2025).

Fungal fermentation has been proposed for functional flours development. One remarkable example is growing of *Pleurotus ostreatus* (oyster mushroom) on a substrate to improve its nutritional value, reduce anti-nutrients, and improve mycoprotein content, enzymes, and bioactive compounds. This process can be achieved through SSF or SMF. *P. ostreatus* has been successfully grown on legumes and cereals, resulting in flour with high biological protein value (aminoacids profile), antioxidant compounds content, decreased antinutrient factors, and increased bioaccessibility. Due to high protein content, these flours can be utilized to formulate functional bakery foods such as cookies or bread (Espinosa Páez et al. 2023).

Ganoderma lucidum is another fungus that has been proposed to produce functional ingredients. For instance, González-Díaz et al. (2024) cultivated *G. lucidum* on brewer's spent grain to develop functional food ingredients with enhanced techno-functional and nutritional properties, particularly focusing on fiber and protein content. Oven and freeze-dried flours were added to traditional bread recipes. The flours exhibited good techno-functional attributes (water absorption, oil retention, and solubility), which involved enhanced texture, freshness, and overall quality in bread prototypes.

12.3 AN OVERVIEW OF INDUSTRIAL APPLICATIONS OF BIOACTIVE COMPOUNDS

In this section, a general overview of the potential industrial applications of bioactive compounds is provided. The potential of bioactive substances to improve human health and well-being has driven the research to incorporate them in foods and pharma applications, as previously discussed. Nevertheless, bioactive compounds offer enormous potential in other sectors such as in cosmetics, agriculture, textiles, and environmental applications (Figure 12.3) (Chauhan et al. 2023).

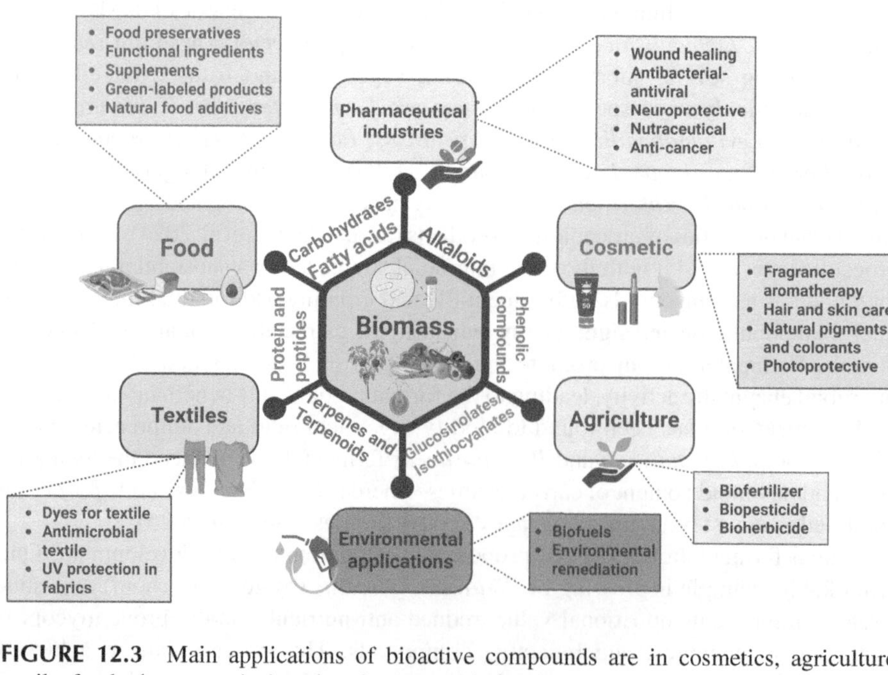

FIGURE 12.3 Main applications of bioactive compounds are in cosmetics, agriculture, textile, food, pharmaceutical, and environmental industries.

Besides nutritional enrichment of formulated products, the incorporation of polyphenols in the food and beverage industries can serve as natural colorants and flavoring agents and increase the shelf-life of the product. In food packaging, their antioxidant and antibacterial qualities can improve food safety and increase the shelf-life of perishable goods. Bioactives incorporation improves packaging's mechanical, barrier, and functional properties (Cazón et al. 2025). Due to their capacity to alter cellular pathways, polyphenols have demonstrated potential in drug discovery in the pharmaceutical industry. Their use in the cosmetic industry can enhance the effectiveness of skincare products by providing defense against sun damage, inflammation, and skin aging (Chauhan et al. 2023). Bioactive compounds can be used in agriculture to control pests and weeds, reducing the need for synthetic products and contributing to sustainable integral pest and weed management practices (Rodríguez García & Raghavan 2022; Chauhan et al. 2023; Bas-Bellver et al. 2025). In the textile industry, bioactive compounds can be used as natural dyes, which satisfies the demand for ecofriendly and non-toxic clothing solutions (Chauhan et al. 2023).

12.4 CONCLUSIONS AND FUTURE PERSPECTIVES

Biomass is an excellent source of bioactive compounds. Their antifungal, antimicrobial, anti-inflammatory, antiviral, anti-carcinogenic, and anti-mutagenic properties make it fit for wide range of industrial applications, including cosmetics, agriculture, food, environmental, textile, and pharmaceuticals.

While traditionally extracted from plant biomass, growing interest is focused on producing them through fermentation. Conventional extraction techniques rely on solvents and heat, but innovative technologies such as microwave, ultrasound, PEFE, and alternative solvents, supercritical fluids, deep eutectic solvents, or subcritical water can improve efficiency and sustainability.

Agro-industrial residues are considered a sustainable, low-cost source for the extraction of bioactive compounds. They can also be used as fermentation substrates for the biosynthesis of bioactive compounds by SSF or SMF. LAB are suitable microorganisms to produce phenolics, whereas fungi can be exploited to produce organic acids. Fermentation with LAB releases phenolics and GSLs and modifies sugar-acid and amino acid profiles, while fungal fermentation reduces anti-nutrients and increases mycoproteins, enzymes, and bioactive compounds in cereals and legumes. Microalgae are proposed to be a primary source of fatty acids, polyphenols, pigments, and polysaccharides. Biomass can also be transformed into functional food powders through disruption, dehydration, and milling, which are applied in new foods development. The application of pretreatment can enhance both drying efficiency and powder functionality.

Despite promising results achieved with innovative extraction techniques and fermentation, further research is still needed to address challenges such as high costs, scalability, and yields comparable to those of conventional methods. As for the integral valorization of biomass, functional food powders have succeeded in modifying target properties of new foods (nutritional, techno-functional, preservative, etc.). However, commercialized products remain limited, probably due to reduced profitability as compared to the nutraceutical or pharmaceutical sectors. Further research is required to improve processing strategies and to maximize both the quality and applicability of these powders.

It is expected that other applications of bioactive compounds in agriculture, textiles, cosmetics, or environmental industries continue to grow due to a rising interest in natural compounds, reduced use of synthetic and toxic substances, and more sustainable processes.

ACKNOWLEDGMENTS

The authors would like to acknowledge the use of BioRender to produce figures

REFERENCES

Abbaspour, N. 2024. Fermentation's pivotal role in shaping the future of plant-based foods: An integrative review of fermentation processes and their impact on sensory and health benefits. *Appl. Food Res.* 4: 100468.

Abdel-Naeem, H.H.S., H.A. Elshebrawy, K. Imre, A. Morar, V. Herman, R. Paşcalău, & K.I. Sallam. 2022. Antioxidant and antibacterial effect of fruit peel powders in chicken patties. *Foods* 11: 301.

Agarwal, A., A. Selvam, M. Majood, O. Agrawal, S. Chakrabarti, & M. Mukherjee. 2023. Carbon nanosheets to unravel the production of bioactive compounds from microalgae: A robust approach in drug discovery. *Drug Discov. Today* 28: 103586.

Agrawal, R.S., & N.P. Nirmal. 2024. Technologies for extraction of bioactive compounds and its applications. In *Bioactive Extraction and Application in Food and Nutraceutical Industries*, ed. T. Sarkar, & S. Pati, 1–20. Humana.

Ali, Md S., M. Haq, S.W. Park, J.M. Han, J.W. Kim, M.S. Choi, S.M. Lee, J.S. Park, M.S. Chun, H.J. Lee, & B.S. Chun. 2025. Recent advances in recovering bioactive compounds from macroalgae and microalgae using subcritical water extraction: Prospective compounds and biological activities. *Food Chem.* 469: 142602.

Andreou, V., G. Dimopoulos, E. Dermesonlouoglou, & P. Taoukis. 2020. Application of pulsed electric fields to improve product yield and waste valorization in industrial tomato processing. *J. Food Eng.* 270: 109778.

Baldassa, S., C. Barrera, M. Muñoz-Ibáñez, & L. Seguí. 2025. Assessing the impact of microwaves and other disruptive pretreatments on *Lactiplantibacillus plantarum* growth and the antioxidant properties of broccoli stalks. *Foods* 14: 1809.

Bas-Bellver, C., C. Barrera, N. Betoret, & L. Seguí. 2020. Turning agri-food cooperative vegetable residues into functional powdered ingredients for the food industry. *Sustainability* 12: 1284.

Bas-Bellver, C., C. Barrera, N. Betoret, & L. Seguí. 2022. Impact of disruption and drying conditions on physicochemical, functional and antioxidant properties of powdered ingredients obtained from brassica vegetable by-products. *Foods* 11: 3663.

Bas-Bellver, C., C. Barrera, N. Betoret, L. Seguí, & J. Harasym. 2024a. IV – Range carrot waste flour enhances nutritional and functional properties of rice-based gluten-free muffins. *Foods* 13: 1312.

Bas-Bellver, C., C. Barrera, & L. Seguí. 2024b. Impact of thermophysical and biological pretreatments on antioxidant properties and phenolic profile of broccoli stem products. *Foods* 13: 3585.

Bas-Bellver, C., N. Melero-Carnero, D. López-González, C. Barrera, M. Verdeguer, & L. Seguí. 2025. Bioproducts obtained from broccoli and cabbage wastes are proposed as functional food ingredients and bioherbicides for sustainable weed management. *Sustainability* 17: 282.

Benvenutti, L., F.M. Moura, G. Zanghelini, C. Barrera, L. Seguí, & A.A.F. Zielinski. 2025. An upcycling approach from fruit processing byproducts: Flour for use in food products. *Foods* 14: 153.

Berktas, S., & M. Cam. 2021. Peppermint leaves hydrodistillation byproducts: Bioactive properties and incorporation into ice cream formulations. *J. Food Sci. Technol.* 58: 4282–4293.

Betoret, E., C. Barrera, L. Seguí, & N. Betoret. 2022. Sustainable strategies in the development of functional foods. In *Innovation Strategies in the Food Industry: Tools for Implementation*, ed. C.M. Galanakis, 145–156. Academic Press.

Câmara, J.S., B.R. Albuquerque, J. Aguiar, R.C.G. Corrêa, J.L. Gonçalves, D. Granato, J.A.M. Pereira, L. Barros, & I.C.F.R. Ferreira. 2021. Food bioactive compounds and emerging techniques for their extraction: Polyphenols as a case study. *Foods* 10: 37.

Cazón, P., A.R. Mateus, & A.S. Silva. 2025. Advances in active packaging using natural biopolymers and fruit by-products for enhanced food preservation. *Food Res. Int.* 213: 116439.

Chauhan, S., N.K. Pandit, A. Mohanty, & S.S. Meena. 2023. Resource recovery of bioactive compounds from food waste and their diverse industrial applications. *Biomass Convers. Bioref.* 15: 27941–27961.

Djuricic, I., & P.C. Calder. 2021. Beneficial outcomes of omega-6 and omega-3 polyunsaturated fatty acids on human health: An update for 2021. *Nutrients* 13: 2421.

Espinosa Páez, E., C.E. Hernández-Luna, S. Longoria-García, C. Torres-Alvarez, C. Velez-Argumedo, & B.E. González-Martínez. 2023. Improving nutritional and functional quality characteristics in bread by using flours obtained from fermentation of kidney beans and oats with *Pleurotus ostreatus*. *CyTA J. Food* 21: 151–158.

Freitas, P.A.V., L. Martín-Pérez, I. Gil-Guillén, C. González-Martínez, & A. Chiralt. 2023. Subcritical water extraction for valorization of almond skin from almond industrial processing. *Foods* 12: 3759.

Frumuzachi, O., A. Nicolescu, G.A. Martău, R. Odocheanu, F. Ranga, A. Mocan, & D.C. Vodnar. 2025. Sustainable valorization of lignocellulosic corn husks via solid-state fermentation: Enhanced recovery of phenolic compounds and organic acids. *Bioresour. Technol.* 436: 132959.

Galanakis, C.M., & S.R. Drago. 2022. Introduction. In *Nutraceutical and Functional Food Components*, ed. C.M. Galanakis, 1–18, Academic Press.

Gómez-Ramos, G.A., L.A. Prado-Barragán, B.A. Rivera, S. Huerta-Ochoa, V.M. Rivera, M. Couder-García, R.d.l.Á. García-Hernández, & C.O. Castillo-Araiza. 2025. Bioreactor engineering for circular economy: Bioactive compound production in solid-state fermentation - Review. *Chem. Eng. Technol.* 48: e202400289.

González-Díaz, R.L., S. Saldarriaga-Hernandez, M.D. Ramírez-Aguirre, S. Guerrero-Higareda, T. García-Cayuela, L.E. Garcia-Amezquita, & D. Carrillo-Nieves. 2024. Valorization of brewer's spent grains through *Ganoderma lucidum* cultivation: Functional food ingredients and bread prototypes. *LWT* 214: 117131.

Gullón, B., M. Gagaoua, F.J. Barba, P. Gullón, W. Zhang, & J.M. Lorenzo. 2020. Seaweeds as promising resource of bioactive compounds: Overview of novel extraction strategies and design of tailored meat products. *Trends Food Sci. Technol.* 100: 1–18.

Hans, N., D. Solanki, T. Nagpal, H. Amir, S. Naik, & A. Malik. 2024. Process optimization and characterization of hydrolysate from underutilized brown macroalgae (*Padina Tetrastromatica*) after fucoidan extraction through subcritical water hydrolysis. *J. Environ. Manage.* 349: 119497.

Jaeger, A., E. Zannini, A.W. Sahin, & E.K. Arendt. 2021. Barley protein properties, extraction and applications, with a focus on brewers' spent grain protein. *Foods* 10: 1389.

Kalinowska, M., E. Gołebiewska, M. Zawadzka, R. Choińska, K. Koronkiewicz, K. Piasecka-Jóźwiak, & M. Bujak. 2023. Sustainable extraction of bioactive compound from apple pomace through lactic acid bacteria (LAB) fermentation. *Sci. Rep.* 13: 1–16.

Kumari, A., S. Pandey, A.K. Chauhan, & M. Singh. 2021. Optimization and characterization of functional chocolate with addition of butter fruit milkshake powder as a source of phenolic, flavonoid and carotenoid. *Indian J. Dairy Sci.* 74: 54–60.

Kurćubić, V., N. Miletić, S. Stajić, & M. Petković. 2025. Fruit and vegetable powder as a functional food supplement. In *Healthier Food: Design and Development of Foods with Improved Functional and Nutritional Properties*, ed. V. Kurćubić, N. Miletić, S. Stajić, & M. Petković, 341–390. Springer.

Lacivita, V., A.L. Incoronato, A. Conte, & M.A.D. Nobile. 2021. Pomegranate peel powder as a food preservative in fruit salad: A sustainable approach. *Foods* 10: 1359.

Liu, Z., X. Huang, Q. Liu, J. Yang, J. Li, M. Xiao, T. Xiong, & M. Xie. 2025. Lactic acid bacteria fermentation improves sensory properties, bioactivity, and metabolic profiles of carrot pulp. *Food Biosci.* 66: 106303.

Marcelli, A., A. Osimani, & L. Aquilanti. 2025. Vegetable byproducts from industrial processing: From waste to functional ingredient through fermentation. *Foods* 14: 2704.

Mungwari, C.P., C.K. King'ondu, P. Sigauke, & B.A. Obadele. 2025. Conventional and modern techniques for bioactive compounds recovery from plants: Review. *Sci. Afr.* 27: e02509.

Nagai, N.F., J.M. Lorenzo, N. Ranalli, J.Á. Pérez-Álvarez, N. Sepulveda, R. Domínguez, E.M. Santos, A. Teixeira, & S.C. Andrés. 2022. Use of seaweed powder (*Undaria* sp.) as a functional ingredient in low-fat pork burgers. *Algal Res.* 67: 102862.

Nakov, G., A. Brandolini, A. Hidalgo, N. Ivanova, V. Stamatovska, & I. Dimov. 2020. Effect of grape pomace powder addition on chemical, nutritional and technological properties of cakes. *LWT* 134: 109950.

Nausad, M., H. Kumar, G. Sharma, K. Dulta, A. Dviwedi, S. Guyot, & S. Sharma. 2024. Characterization of oil extracted from wild apricot seeds kernel using submerged alcoholic fermentation and its quality characteristics. *Process Biochem.* 142: 175–183.

Nezafatian, E., O. Farhadian, A. Yegdaneh, M. Safavi, E. Daneshvar, & A. Bhatnagar. 2023. Enhanced production of bioactive compounds from marine microalgae *Tetraselmis tetrathele* under salinity and light stresses: A two-stage cultivation strategy. *Bioresour. Technol.* 376: 128899.

Nitisuk, P., P. Wanyo, T. Chamsai, & K. Charoenjit. 2025. Sustainable valorization of tropical fruit peels for sustainable production of natural antioxidants and functional food ingredients. *Sustain. Food Technol.* 3: 1189–1202.

Ojeda, G.A., M.M. Vallejos, S.C. Sgroppo, C. Sánchez-Moreno, & B. de Ancos. 2023. Enhanced extraction of phenolic compounds from mango byproducts using deep eutectic solvents. *Heliyon* 9: e16912.

Panza, O., V. Lacivita, F. Tarantino, A. Manzi, A. Conte, & M.A.D. Nobile. 2023. Fruit and vegetable byproducts as source of bioactive compounds to preserve handmade fresh pasta. *LWT* 190: 115584.

Pattnaik, B.P., D. Gupta, D. Deb, M. Selvaraj, M.A. Assiri, S.R. Mohapatra, H.P. Sahoo, S. Tapas, & P.K. Sarangi. 2025. Fermentation of sugarcane bagasse for production of value-added phenolic compounds using potential bacterial strains: A comparative analysis. *Biomass Bioenergy* 202: 108161.

Ramírez-Pulido, B., C. Bas-Bellver, N. Betoret, C. Barrera, & L. Seguí. 2021. Valorization of vegetable fresh-processing residues as functional powdered ingredients. A review on the potential impact of pretreatments and drying methods on bioactive compounds and their bioaccessibility. *Front. Sustain. Food Syst.* 5: 654313.

Ravindran, A., M. Raman, N. Babu, A. Dinakaran, T.V. Sankar, & T.K.S. Gopal. 2020. Diet chocolates and replacement of cocoa powder with jackfruit seed powder. *Food Nutr. Sci.* 11: 220–233.

Rezagholizade-Shirvan, A., M. Soltani, S. Shokri, R. Radfar, M. Arab, & E. Shamloo. 2024. Bioactive compound encapsulation: characteristics, applications in food systems, and implications for human health. *Food Chem.: X* 24: 101953.

Rodríguez García, S.L., & V. Raghavan. 2022. Green extraction techniques from fruit and vegetable waste to obtain bioactive compounds - A review. *Crit. Rev. Food Sci. Nutr.* 62: 6446–6466.

Sanwal, N., S. Mishra, J.K. Sahu, & S.N. Naik. 2022. Effect of ultrasound-assisted extraction on efficiency, antioxidant activity, and physicochemical properties of sea buckthorn (*Hippophae salicipholia*) seed oil. *LWT* 153: 112386.

Sengupta, S., D. Bhattacharya, & M. Mukhopadhyay. 2020. Bioactive compounds production from vegetable biomass: A biorefinery approach. In *Biorefinery Production Technologies for Chemicals and Energy*, ed. A. Kuila, & M. Mukhopadhyay, 241–258. John Wiley & Sons, Ltd.

Serna-Barrera, M.A., C. Bas-Bellver, L. Seguí, N. Betoret, & C. Barrera. 2024. Exploring fermentation with lactic acid bacteria as a pretreatment for enhancing antioxidant potential in broccoli stem powders. *AIMS Microbiol.* 10: 255–272.

Shrinet, K., R.K. Singh, A.K. Chaurasia, A. Tripathi, & A. Kumar. 2021. Bioactive compounds and their future therapeutic applications. In *Natural Bioactive Compounds: Technological Advancements*, ed. R.P. Sinha, & D.-P. Häder, 337–362. Academic Press.

Villacís-Chiriboga, J., S. Voorspoels, M. Uyttebroek, J. Ruales, J.V. Camp, E. Vera, & K. Elst. 2021. Supercritical CO_2 extraction of bioactive compounds from mango (*Mangifera indica* L.) peel and pulp. *Foods* 10: 1–22.

Wang, Y., X. Chang, Y. Wang, J. Xie, G. Han, & H. Qi. 2025. Seaweed, used as a water-retaining agent, improved the water distribution and myofibrillar protein properties of plant-based yak meat burgers before and after freeze-thaw cycles. *Foods* 14: 2541.

13 Unlocking Rural Economic Value Through Biogas Digestate in Limpopo, South Africa

Thilivhali Rasimphi, David Tinarwo, and Sophie T. Mulaudzi

13.1 INTRODUCTION

Smallholder farmers in South Africa face persistent challenges, including rising input costs, soil degradation, and limited access to affordable fertilizers (Mutungwazi et al. 2018; Bruno et al. 2022). In response, digestate, a nutrient-rich byproduct of anaerobic digestion (AD), has gained attention as a sustainable alternative that not only supplies essential nutrients but also improves soil structure and long-term fertility (Wainaina et al. 2020; IEA 2023). While commercial agriculture often relies on synthetic fertilizers, smallholder farmers require solutions that are both cost-effective and environmentally sound. This study explores how biogas digestate can unlock rural economic value in Limpopo by reducing production costs, enhancing soil health, and creating opportunities for sustainable farming and local livelihoods. By assessing farmer practices and perceptions, identifying technological and market pathways, and addressing policy and institutional barriers, the study aims to position digestate as a driver of circular bioeconomy transitions and inclusive rural development in Limpopo.

The growing demand for sustainable energy and effective waste management highlights the need for integrated approaches that simultaneously address environmental, agricultural, and socio-economic challenges (Roopnarain & Adeleke 2017; IEA 2023). Biogas technology, particularly through AD, has emerged as an approach capable of converting organic waste into renewable energy while generating digestate as a potentially valuable byproduct (Muchara et al. 2014; Rasimphi et al. 2024). International research shows that digestate is rich in essential nutrients such as nitrogen (N), phosphorus (P), and potassium (K), making it an effective substitute for chemical fertilizers (Herrera et al. 2022). While international studies confirm

DOI: 10.1201/9781003650751-13

digestate's agronomic benefits and commercial potential, little is known about its role in semi-arid, smallholder farming systems in South Africa, where soil infertility, water scarcity, and economic marginalization are defining challenges.

Most existing research emphasizes nutrient recovery but pays limited attention to digestate use's economic, cultural, and institutional dimensions (Arellano-Yasaca & Chu 2025). Studies in Kenya and Tanzania demonstrate that digestate application improves maize and vegetable yields, reduces reliance on costly synthetic fertilizers, and strengthens climate resilience in semi-arid contexts (Mutungwazi et al. 2018; Wainaina et al. 2020). Similarly, South African studies on organic amendments confirm the potential of biofertilizers to address soil nutrient depletion and support smallholder productivity. Despite this evidence, digestate uptake in South Africa remains limited due to inadequate awareness, poor technical training, and weak institutional and policy support.

Beyond agricultural use, emerging research highlights the potential of digestate in new value chains, such as the production of bioplastics, soil conditioners, and construction materials (Mayer et al. 2020; Singh et al. 2022). These innovations reduce waste and create opportunities for green entrepreneurship and local employment, aligning with South Africa's National Development Plan (NPC 2012) and just energy transition objectives. However, the extent to which digestate can realistically be valorized within rural South African contexts remains underexplored. Furthermore, South African biogas initiatives have largely overlooked digestate valorization in policy and practice, leaving its potential contribution to circular bioeconomy transitions underdeveloped. This gap makes it difficult to design strategies that integrate digestate into rural livelihood systems, create local markets, or address barriers to adoption.

Digestate is one of the resulting products generated from biogas systems and is rich in nutrients. Digestate has the capacity to compete favorably with inorganic fertilizers for better crop productivity, yield, and enhancement of soil health (Kemausuor et al. 2014). The mineral contents and characteristics of digestate depend, most of the time, on the characteristics of the substrate and the modality of digestion (O'Connor et al. 2022). These defined attributes, such as organic matter content, ammonium (NH_4), the C/N ratio, and N content present in the different substrates or feedstocks that form the digestates, will show differences in efficiencies and productivity in plants and soils, therefore making it the main source of this research (Njenga et al. 2021; Bian et al. 2023). However, a good proportion of nutrients such as N, P, and K are retained. One of the benefits of utilizing digestate is the higher nutrient content than its respective feedstock. Although, from the resulting digestate, a considerable amount of N in the NH_4 form is emitted during the AD process, C is removed as methane and carbon dioxide.

Digestate holds multiple functions in its beneficial roles to both the soil and the plants/crops (Wang et al. 2023). In the first instance, digestate is known to have fertilizing attributes that help in the productivity of the plants due to the availability of important nutrients necessary for plant growth (Smith et al. 2020). Secondly, their influence on soil health cannot be overemphasized as they play huge roles that promote soil efficiency through nutrient cycling in the soil, carbon transformation, and soil structure maintenance (Deng et al. 2023). The application of digestate in the field could have fewer shorter-term results due to the slow mineralization rate or action from microbes. Nitrogen is an essential element for plant growth and soil microbial activity and is the nutrient taken up in the largest amount by plants and the most common limiting factor for plant growth.

In Limpopo Province, South Africa, where agriculture is central to rural livelihoods, communities continue to face persistent problems, such as energy poverty, dependence on firewood, soil degradation, and weak waste management infrastructure (Rasimphi et al. 2024; LDP 2025). Although biogas projects have been introduced in the province, their implementation has mainly focused on cooking, heating, and electricity production. Digestate has often been treated as waste rather than a resource, representing a missed opportunity to enhance soil fertility, reduce fertilizer costs, and stimulate rural economies.

The overarching research question guiding this study:

- How can digestate from rural biogas systems be valorized to support sustainable agriculture and stimulate inclusive rural economic development in Limpopo Province, South Africa?

To address this question, the study pursues the following objectives:

1. Assess current practices and perceptions of digestate use among Limpopo Province households, smallholder farmers, and cooperatives.
2. Identify technological and market pathways for processing digestate into value-added products, such as biofertilizers, compost, and other commercial applications.
3. Evaluate the environmental and socio-economic benefits of digestate valorization, including soil fertility improvement, input cost reduction, waste management, and livelihood opportunities.
4. Examine institutional, policy, and awareness-related barriers that limit digestate uptake and commercialization in South Africa.

By achieving these objectives, the study contributes to the broader discourse on circular economy transitions and sustainable rural development while offering insights for policy and practice on integrating digestate valorization into national energy and agricultural strategies.

13.2 UNLOCKING THE POWER OF DIGESTATE

Digestate plays a crucial role in the ecological cycle. Unlike traditional compost, digestate is characterized by its high nutrient content, which can enrich soil and improve agricultural productivity (Haque et al. 2023). When used as an organic soil amendment, it promotes better plant growth and enhances soil vitality. By repurposing and reutilizing organic waste, digestate exemplifies a holistic approach to sustainability (Czekala et al. 2023). One of the most significant advantages of digestate is its nutrient composition. Figure 13.1 is a schematic representation of the biogas production process and its applications. Organic substrates, such as animal manure, agricultural residues, food waste, and wastewater, are fed into an anaerobic digester where they undergo microbial decomposition (Guan et al. 2024). This process generates two outputs: biogas and digestate. Biogas can be utilized as a renewable energy source for heating, cooking, and lighting, thereby reducing reliance on conventional fuels.

Digestate, the nutrient-rich byproduct, has multiple applications, including use as organic fertilizer, biopesticide, for seed priming, aquaculture, livestock management, soil remediation, irrigation, and other bioprocesses. Figure 13.1 illustrates the dual benefits of biogas systems in producing clean energy while contributing to sustainable agriculture and environmental management.

Biogas technology provides a dual advantage by addressing both energy and agricultural needs through the production of biogas and digestate. Biogas serves as a renewable energy source that can be used for cooking, heating, and lighting, reducing reliance on firewood, charcoal, and other fossil-based fuels (Rizzioli et al. 2023). This not only improves household energy security but also contributes to reducing deforestation and greenhouse gas emissions. On the other hand, digestate, the nutrient-rich residue from the AD process, offers multiple agricultural and environmental benefits (Bruno et al. 2022). It can be applied as an organic fertilizer to improve soil fertility, used as a biopesticide to reduce chemical inputs, and employed in seed priming to enhance crop germination. Additionally, digestate supports aquaculture, livestock management, soil remediation, irrigation, and other bioprocesses that

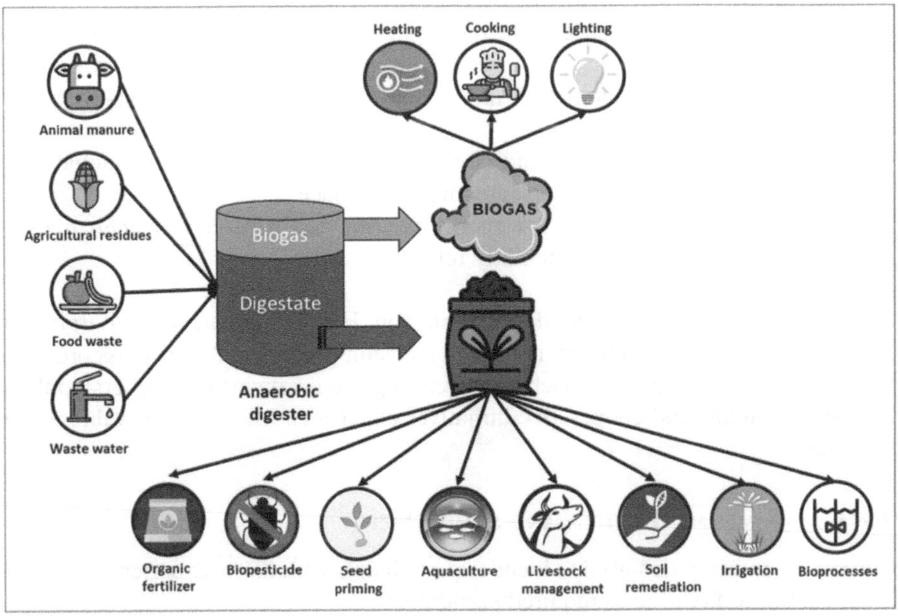

FIGURE 13.1 Unlocking the value of the digestate.

(Adapted from Guan et al. 2024.)

promote circular economy practices. Together, these outputs highlight the significance of biogas systems in enhancing rural livelihoods, fostering sustainable agriculture, and contributing to climate change mitigation.

This study explored the untapped potential of digestate valorization in rural Limpopo Province. The findings show that digestate offers clear agronomic benefits, particularly in improving soil fertility, enhancing water retention, and reducing dependence on costly chemical fertilizers. At the same time, the research highlights economic opportunities through small-scale entrepreneurship and youth-led enterprises as well as environmental gains linked to waste reduction and circular economy practices.

13.3 METHODOLOGY

13.3.1 Research Design

The study employed a qualitative research design grounded in an interpretivist paradigm. This approach was chosen to capture how rural communities in Limpopo perceive, use, and assign value to biogas digestate and related byproducts. A qualitative design enabled rich, contextual insights into practices, beliefs, and challenges often overlooked in quantitative studies. In addition, a case study strategy was adopted to provide an in-depth understanding of local realities across selected communities.

13.3.2 Study Area and Site Selection

Research was conducted in the Vhembe and Sekhukhune Districts of Limpopo Province, focusing on villages and smallholder farming communities where small-scale biogas installations had been introduced. The districts were selected for three reasons:

- It has a high prevalence of mixed crop-livestock farming systems that provide feedstock for biogas.
- Several renewable energy and circular agriculture interventions, including household-level biodigesters, are already in place.
- Communities face persistent energy poverty and soil fertility challenges, making digestate use particularly relevant.

Three villages, that is, Maila, Tiyani, and Emkhondweni, were purposively selected in consultation with agricultural extension officers and project records. The sites represented a mix of households operating biogas digesters, farmers applying digestate to fields, and cooperative initiatives experimenting with renewable energy solutions.

13.3.3 Sampling Strategy and Sample Size

Purposive sampling was used to identify participants with direct experience of biogas systems. Eligibility criteria required participants to:

- Have operated or interacted with a biogas system for at least six months
- Be actively involved in agricultural or entrepreneurial activities linked to digestate or slurry

In total, 15 participants were interviewed, comprising:

- Household biogas users (n = 7)
- Smallholder farmers applying digestate (n = 5)
- Local stakeholders, such as cooperative leaders, extension officers, and technicians (n = 3)

Sampling continued until data saturation was reached, when no new themes were emerging.

13.3.4 Data Collection Methods

Data were collected between August and October 2024 using two methods:

- **Semi-structured interviews:** Conducted in Tshivenda, Isiswathi, or English, depending on participant preference. Topics covered include perceptions of digestate, applications, challenges, and potential income opportunities. With consent, all interviews were audio-recorded and later transcribed.
- **Field observations:** Undertaken during site visits to document practical aspects, such as digestate storage, spreading techniques, and visible crop responses. These observations served to corroborate interview data.

13.3.5 Data Analysis

Interview transcripts and field notes were analyzed thematically, following Braun and Clarke's (2006) six-step process: familiarization, coding, theme development, review, definition, and reporting. NVivo software was used to support systematic coding and data management. Themes were developed inductively but interpreted with reference to existing literature on circular agriculture, rural entrepreneurship, and renewable energy adoption.

13.3.6 Limitations

Certain limitations shape the study's findings. The small sample size (15 participants) restricts generalization beyond the selected sites. The purposive focus on Vhembe District means results may not fully represent other rural contexts in South Africa. In addition, reliance on self-reported data introduces potential bias, although triangulation and member checking helped mitigate this.

13.4 RESULTS AND DISCUSSION

This section integrates empirical findings from participant interviews, field observations, and case studies with insights from the literature to critically examine the opportunities and barriers surrounding digestate use in Limpopo Province. Eight major themes were identified through NVivo-assisted thematic analysis, cutting across agronomic, economic, social, technical, environmental, and institutional dimensions. Each theme is discussed below, emphasizing how the findings confirm, extend, or challenge existing knowledge.

Table 13.1 presents the eight major themes that emerged from the thematic analysis of interview transcripts and field observations. The findings highlight the multiple dimensions of biogas digestate utilization and its role in rural development.

- On the agricultural side, digestate contributes to soil fertility, crop performance, and water retention.
- Economically, it offers opportunities for income generation and enterprise development. However, awareness remains uneven, shaped by misconceptions and cultural attitudes.
- Technical and operational challenges, such as storage and transport, further constrain adoption.

Despite these limitations, digestate provides significant environmental and circular economy benefits through waste reduction, soil restoration, and climate adaptation. Institutional gaps in policy and municipal involvement point to the need for stronger enabling frameworks, while capacity building through training, demonstrations, and youth engagement is essential for scaling up. Finally, gender and social inclusion remain critical as women are underrepresented but present opportunities for targeted participation and empowerment.

The thematic analysis demonstrates that the use of digestate is multidimensional, with implications for agriculture, economics, society, and the environment. From an agricultural perspective, digestate enhances soil fertility, crop productivity, and water retention, reinforcing its role as a valuable organic input for smallholder farming

TABLE 13.1
Major Themes Identified During Thematic Analysis

Theme	Key Insights
Agricultural value of digestate	Improvements in soil fertility, enhanced crop performance, and increased water retention.
Economic and commercial opportunities	Potential for income generation, job creation, and rural enterprise development.
Awareness and perceptions	Misconceptions about technology, cultural attitudes influencing adoption, and uneven levels of awareness.
Technical and operational challenges	Difficulties related to digestate storage, transportation, and system maintenance.
Environmental and circular economy benefits	Contributions to waste reduction, soil restoration, and climate change adaptation.
Policy and institutional support	Gaps in standards, weak municipal involvement, and lack of enabling frameworks.
Capacity building and knowledge transfer	Importance of training, demonstration projects, and engaging youth in biogas initiatives.
Gender and social inclusion	Underrepresentation of women, with opportunities for targeted participation and empowerment.

Source: Authors field data (2024).

systems (Möller & Müller 2012; Smith et al. 2020). Economically, participants recognized opportunities for income generation and rural enterprise development, suggesting that digestate could support local value chains if commercialization pathways are developed (FAO 2011). However, awareness remains uneven across communities, shaped by misconceptions, cultural attitudes, and limited exposure to the technology, which constrains uptake.

Operational challenges were frequently highlighted, particularly regarding digestate storage, transport, and the maintenance of biogas systems, all of which affect reliability and user confidence (Amigun & von Blottnitz 2010). Despite these challenges, strong environmental benefits were reported, including waste reduction, soil restoration, and contributions to climate change adaptation, framing digestate as a tool for advancing circular economy practices in rural settings. This aligns with Holm-Nielsen et al. (2009) and Meyer & Priefer (2012). Policy and institutional gaps emerged as significant barriers, with participants noting limited municipal involvement, inadequate standards, and weak enabling frameworks (Domenech & Bahn-Walkowiak 2019). This indicates that without supportive governance, the broader adoption of digestate will remain constrained. At the same time, respondents emphasized the importance of capacity building and knowledge transfer, particularly through training, and youth engagement, as essential to overcoming technical and awareness barriers. Lwiza et al. (2017) highlighted the importance of training as key, especially for women and youth.

Gender and social inclusion also surfaced as cross-cutting themes, with women underrepresented in current initiatives but identified as key stakeholders who could be empowered through targeted participation strategies (Clancy et al. 2012; Sovacool et al. 2015). Taken together, these themes suggest that the successful integration of digestate into rural livelihoods depends on a combination of technical innovation, institutional support, social inclusivity, and sustained awareness-building.

13.4.1 Shared Findings Across Sites

Across the three villages, participants consistently described digestate as a valuable soil amendment. Farmers noted improved crop growth, healthier soils, and reduced dependence on chemical fertilizers. Field observations confirmed that soils treated with digestate retained moisture longer and supported stronger plant growth than untreated plots. These findings align with international evidence highlighting digestate's role in nutrient recycling and soil fertility improvement (Möller & Müller 2012; Wainaina et al. 2020). These results suggest that digestate is a practical substitute for costly synthetic fertilizers and a context-appropriate strategy for enhancing soil resilience in semi-arid smallholder systems.

Participants also emphasized cost-saving benefits, noting that the high price of synthetic fertilizers made digestate an attractive alternative. This reflects studies in Kenya and Tanzania, where digestate reduced input costs while sustaining yields (Mutungwazi et al. 2018; Singh & Mohapatra 2020). At the same time, participants in Limpopo viewed digestate as a tool for resilience, particularly against drought, echoing Nkoa (2014) on the importance of organic amendments for soil water retention in semi-arid environments. They also highlight the dual value of digestate in strengthening household livelihoods while enhancing the long-term sustainability of smallholder agriculture in Limpopo.

13.4.2 Site-Specific Differences

Despite the aforementioned shared benefits, each village emphasized different aspects of digestate use:

- **Maila:** Farmers were primarily concerned with input costs. They highlighted digestate as a cost-effective substitute for chemical fertilizers, which are often unaffordable or unavailable. This aligns with research that links digestate adoption to economic sustainability in resource-constrained farming systems (Surendra et al. 2014).
- **Tiyani:** Respondents emphasized digestate's role in waste management. Households reported that separating organic waste for biodigesters reduced open dumping and improved local hygiene. This extends findings by Geissdoerfer et al. (2017) who argued that circular practices can emerge from household-level behavior change.
- **Emkhondweni:** Participants focused on the potential of digestate for entrepreneurship. Youth in particular expressed interest in producing composted fertilizers for sale. This reflects global evidence that digestate valorization can stimulate rural bioeconomies (Wainaina et al. 2020) while highlighting untapped entrepreneurial potential in Limpopo.

13.4.3 Implications for Policy and Practice

The case studies demonstrate that while digestate has strong agronomic and environmental benefits, uptake is shaped by local priorities and constraints. A one-size-fits-all approach is unlikely to succeed. Instead, tailored interventions are required:

- In **Maila**, policies should emphasize input cost reduction and farmer training.
- In **Tiyani**, digestate could be integrated school-linked farming into municipal waste management strategies.
- In **Emkhondweni**, youth entrepreneurship and market development support could unlock new value chains.

By situating digestate within diverse local realities, the case study, as shown in Table 13.2, the findings confirm that digestate valorization is not only a technical issue but also a socio-economic and cultural one. This supports calls for systems-based approaches integrating technology, policy, and community practices (Sovacool et al. 2019).

13.4.3.1 Agricultural Value of Digestate

Research in semi-arid contexts underscores the importance of organic amendments in improving soil water-holding capacity, reducing vulnerability to drought, and strengthening climate resilience (Nkoa 2014). In this regard, the Limpopo findings are highly relevant, as participants repeatedly emphasized water retention as

TABLE 13.2
Comparative Insights from the Case Study

Site	Main Benefits Reported	Key Barriers	Policy/Practice Implications
Maila	• Digestate is seen as a cost-saving substitute for chemical fertilizers. • Improved soil fertility and crop yields.	• High cost of biodigester installation. • Limited access to technical training.	• Strengthen farmer training on digestate application. • Introduce subsidies/ finance for digesters.
Tiyani	• Digestate is valued for reducing waste dumping, improving hygiene, and retaining soil moisture.	• Cultural perceptions of slurry as "dirty." • Limited awareness of agronomic benefits.	• Integrate digestate into municipal waste management strategies. • Launch awareness campaigns with local champions.
Emkhondweni	• Youth expressed interest in entrepreneurship (packaging and selling compost/ fertilizer) perceived market opportunities for digestate products.	• Lack of standards/ certification for digestate. • Transport and storage difficulties.	• Develop pilot projects for branded digestate products. • Support youth-led enterprises with training and finance.

a critical benefit of digestate use. Participants consistently reported that digestate improved crop growth, soil quality, and water retention. Farmers highlighted visible benefits, such as healthier plants and soils holding water better which reduced the need for expensive fertilizers. These accounts were corroborated during field visits as shown in Figure 13.2 where soils treated with digestate displayed better structure and moisture retention than untreated plots. Such observations resonate with international studies demonstrating that digestate enhances nutrient recycling and soil fertility by returning N, P, and K to the soil.

Figure 13.2 illustrates field observations of farmers engaged in vegetable production activities. In the left image, farmers are applying liquid (digestate) inputs using buckets, to enhance crop growth and soil fertility. The right image shows a group of farmers collectively working in a vegetable field, primarily engaged in weeding and crop management. These practices highlight the use of labor-intensive methods in rural farming systems, emphasizing both the role of organic soil amendments and the importance of collective participation in sustaining agricultural livelihoods. This expands on existing literature, which often focuses on nutrient recovery as the primary value of digestate (Möller & Müller 2012), by showing that in dryland farming systems, its contribution to soil moisture conservation may be equally or even more significant. Similar outcomes have been observed in East Africa, where digestate

FIGURE 13.2 Farmers utilizing the slurry as a liquid fertilizer for their crops.

(Adapted from Field data.)

application improved maize and vegetable yields while enhancing drought tolerance (Mutungwazi et al. 2018; Wainaina et al. 2020). In South Africa, where fertilizer prices remain volatile and rural farmers often struggle with declining soil fertility, digestate provides a low-cost and locally available alternative.

Studies on biofertilizers and organic amendments in the region have shown their potential to reverse soil nutrient depletion and support smallholder productivity (Muchara et al. 2014; Rasimphi et al. 2024). The findings from Limpopo confirm this, while extending the discussion by situating digestate within a climate adaptation framework. Digestate emerges as a dual-purpose resource supporting household economic security and long-term environmental sustainability by reducing input costs and improving soil resilience.

Overall, this evidence suggests that while international literature highlights nutrient recovery as the main agronomic value of digestate, context-specific findings from Limpopo indicate that water retention and drought resilience are equally crucial benefits. These insights reinforce the importance of tailoring digestate promotion strategies to local ecological conditions, especially in semi-arid regions, where water scarcity is a defining constraint for smallholder agriculture.

13.4.3.2 Economic and Commercial Opportunities

Respondents identified digestate as a potential source of income through composting, biofertilizer sales, and small-scale agro-enterprises. Youth in particular expressed interest in product development and entrepreneurship, reflecting a growing recognition of digestate as more than a household input. These perspectives align with global evidence that digestate-based enterprises can anchor rural bioeconomies by creating localized value chains and stimulating green jobs (Surendra et al. 2014; Wainaina et al. 2020). In Kenya, for example, farmer cooperatives have successfully packaged digestate into branded biofertilizers, opening new markets and strengthening farmer livelihoods (Singh & Mohapatra 2020). Similar initiatives in East Africa and Asia have demonstrated how small-scale processing and commercialization can generate income streams while reinforcing circular economy practices.

However, the literature also highlights that successful market uptake depends on regulatory and institutional frameworks establishing quality standards, certification systems, and consumer trust (Mayer et al. 2020). Without these safeguards, commercialization risks being undermined by inconsistent product quality, limited consumer awareness, and skepticism about waste-derived fertilizers. The experiences of respondents in Limpopo echo these concerns. While enthusiasm for entrepreneurship was evident, many participants noted the absence of local standards or government guidance, which made it difficult to envision scaling digestate beyond informal household use.

These findings suggest that digestate valorization in Limpopo can catalyze rural entrepreneurship, particularly among youth and cooperatives. However, such initiatives may stall without supportive policies and market development mechanisms. Lessons from Kenya and other regions show that pilot projects, branding strategies, and farmer-led enterprises can help demonstrate product reliability and build market confidence. For Limpopo, fostering youth-led enterprises around digestate could simultaneously address unemployment and enhance the sustainability of farming systems, provided that institutional barriers are addressed through certification frameworks and targeted support.

13.4.3.3 Awareness and Perception

Although some participants understood digestate's benefits, many held misconceptions associating it with dirt, unpleasant odors, or health risks. In several cases, traditional norms shaped adoption patterns, with some communities considering the reuse of waste materials culturally inappropriate. Despite observed agronomic benefits, these perceptions directly influenced household willingness to apply digestate to crops. Such findings mirror evidence from Nepal, India, and Ethiopia, where awareness gaps and cultural beliefs have slowed the adoption of biogas byproducts (Ghimire 2013; Johnson et al. 2014; Mengistu et al. 2015). The persistence of these attitudes suggests that informational campaigns alone are insufficient. As Roubik et al. (2018) argued, practical demonstrations, farmer field schools, and peer-to-peer exchanges are far more effective than brochures or top-down awareness drives in shifting mindsets.

This is particularly relevant in rural South Africa, where knowledge transfer is often most effective through community networks and trusted local champions rather than external interventions. Observations from this study showed that farmers who had already experienced positive results with digestate became informal advocates, encouraging neighbors to experiment with application methods. This grassroots form of knowledge diffusion highlights the potential for scaling adoption through community-led learning rather than relying solely on formal extension services.

Overall, these findings underline that adoption barriers are both technical and deeply cultural. Addressing them requires a strategy beyond scientific information, incorporating culturally sensitive communication, locally respected champions, and experiential learning platforms. For Limpopo, embedding digestate promotion within participatory extension approaches is essential to building trust and acceptance.

13.4.4 Environmental and Circular Economy Benefits

Participants in the study highlighted the multiple environmental benefits of digestate, particularly its role in diverting organic waste from landfills, reducing uncontrolled dumping, and contributing to climate change mitigation. Improved waste segregation at the household level was noted, suggesting that biogas systems can foster more sustainable household waste management practices. These perceptions are consistent with global findings where digestate is recognized as a valuable byproduct that closes nutrient cycles by returning organic matter to soils, thereby reducing the need for synthetic fertilizers and enhancing soil fertility (Geissdoerfer et al. 2017). At the household level, biogas systems have been associated with measurable reductions in greenhouse gas emissions.

For example, studies in Vietnam reported reductions of up to 16 tons of CO_2 equivalent per year per household due to the replacement of firewood and charcoal with biogas for cooking and heating (Hoang et al. 2021; Ntaganda et al. 2024). Similarly, in Rwanda, biogas adoption has lessened pressure on local forests where most households traditionally rely on firewood. Reduced demand for firewood mitigates deforestation and decreases indoor air pollution, improving household health outcomes (Johnson & Bryden 2012; Nzeyimana et al. 2020). Beyond direct environmental impacts, digestate contributes to circular economy objectives by transforming waste into a resource. When applied to agricultural land, digestate provides nutrients, such as N, P, and K, that enhance crop productivity while reducing dependence on chemical fertilizers.

This recycling of nutrients exemplifies a closed-loop system where waste products are valorized, supporting both sustainable agriculture and climate resilience (Möller & Müller 2012; Li et al. 2020). Furthermore, integrating digestate use into farming practices can create ancillary economic benefits, including reduced fertilizer costs and potential income generation through the sale of surplus compost, reinforcing the link between environmental sustainability and local livelihoods (Muto et al. 2017; Zhang et al. 2021).

Overall, these findings reinforce that biogas and digestate utilization contribute to reducing greenhouse gas emissions and promote circular economy principles by turning waste into a productive agricultural input, supporting both environmental and socio-economic sustainability in rural communities.

13.4.5 Policy and Institutional Support

Respondents highlighted several institutional and regulatory barriers that limit the effective use of digestate, including the absence of standardized quality guidelines, weak government support, and minimal municipal engagement. These gaps undermine farmers' confidence in applying digestate safely and consistently and constrain opportunities for scaling biogas adoption (De Clercq et al. 2017; Sovacool et al. 2019). The lack of digestate standards is particularly critical because, without clear guidance on nutrient content, pathogen limits, or safe application methods, farmers may be hesitant to use it or apply it incorrectly, reducing both agricultural and environmental benefits.

This situation contrasts sharply with European experiences where digestate management benefits from robust regulatory frameworks, quality assurance standards, and financial incentives. Countries, such as Germany, Denmark, and the Netherlands, provide clear guidance on digestate composition, storage, and field application, often supported by subsidies and extension services that facilitate farmer adoption and confidence (Meyer et al. 2018; Kumar et al. 2023). Such frameworks ensure environmental safety and promote a circular economy by encouraging nutrient recycling and reducing reliance on chemical fertilizers.

In Sub-Saharan Africa, however, the absence of enabling policies reflects broader structural and institutional challenges. Limited technical guidance, low awareness of digestate benefits, and minimal municipal oversight reduce the likelihood of consistent and safe digestate use (De Clercq et al. 2017; Sovacool et al. 2019). This regulatory gap is compounded by weak financial and institutional support for post-installation services, including training, monitoring, and maintenance, which are critical for sustaining biogas systems. Without such support, farmers may abandon digestate utilization, revert to traditional waste disposal practices, or fail to reap biogas technology's potential economic and environmental benefits.

Addressing these challenges requires multi-level interventions. At the policy level, governments can develop digestate quality standards, introduce subsidies or incentive schemes, and integrate biogas and digestate management into broader agricultural and environmental programs. At the municipal level, local authorities can support farmers through extension services, training programs, and monitoring initiatives. Strengthening these institutional and regulatory structures would improve digestate utilization and enhance public confidence in biogas systems, support circular economy practices, and maximize both environmental and socio-economic benefits for rural communities.

13.4.6 Capacity Building and Knowledge Transfer

Training emerged as a critical factor in the successful adoption of digestate, with many youth participants viewing it as a resource for sustainable agriculture and an opportunity to develop skills and generate income. Practical capacity-building initiatives, such as farmer field schools, hands-on training, and demonstration plots, were repeatedly cited as effective strategies for improving both the adoption of biogas technologies and the productive use of digestate (Surendra et al. 2014; Wainaina et al. 2020). Such approaches enable participants to gain technical knowledge on digestate handling, storage, and application while fostering problem-solving skills essential for maintaining digesters and overcoming operational challenges.

In addition to technical training, engaging youth in biogas and digestate-related enterprises has broader socio-economic benefits. Evidence suggests that involving young people in green technologies encourages innovation, supports local entrepreneurship, and enhances the uptake of sustainable energy solutions (Berhe et al. 2017; Rahman et al. 2021). Youth involvement also strengthens intergenerational knowledge transfer, as trained individuals can act as local champions, promoting best practices within their communities. By linking capacity-building with income-generating opportunities, training programs create immediate and long-term incentives for adoption, ensuring that biogas systems and digestate management are technically sustainable and socially and economically viable.

13.4.7 Gender and Social Inclusion

Women were notably underrepresented in digestate initiatives, reflecting persistent gendered barriers that limit access to training, decision-making, and income-generating opportunities in rural renewable energy and agricultural systems (Bonjour et al. 2013; Armah et al. 2019). Cultural norms often assign women primary household and caregiving responsibilities, restricting their ability to participate in technical activities or attend training programs. Limited access to education, information, and social networks further compounds these constraints, leaving women less involved in biogas and digestate management despite their central role in household energy use and agricultural production.

Despite these challenges, some women were engaged in related activities, such as composting, small-scale fertilizer preparation, and household-level application of digestate (Surendra et al. 2014; Wainaina et al. 2020). These examples illustrate that, when provided with appropriate support, women can effectively contribute to sustainable waste management and nutrient recycling (Njenga et al. 2021). Studies have consistently shown that targeted interventions, such as women-focused training, mentorship programs, and support for micro-enterprises, significantly enhance participation, adoption rates, and local enterprise development in biogas and related green technologies.

Furthermore, including women in digestate initiatives can have cascading benefits beyond immediate technical adoption. Women's involvement often leads to more efficient household energy management, improved food security, and broader community engagement. Evidence suggests that women's active participation strengthens knowledge transfer within households and communities, fosters innovation in small-scale enterprises, and contributes to sustainable agricultural practices by integrating digestate into soil fertility management (Oparaocha & Dutta 2011). By addressing structural, cultural, and educational barriers, programs that intentionally empower women can enhance social equity and biogas systems' long-term sustainability.

Overall, promoting women's engagement in digestate and biogas initiatives is a matter of equity and a strategic approach to maximize the technical, environmental, and socio-economic impacts of renewable energy interventions in rural areas. Integrating gender-responsive strategies into training, extension, and enterprise development is essential for creating inclusive, resilient, and productive biogas value chains.

13.5 FUTURE OUTLOOK AND RESEARCH DIRECTIONS

The future of the biogas industry lies in an integrated, multi-faceted approach that situates biogas not only as an energy solution but also as a driver of the circular bioeconomy. Achieving this vision requires technological innovation, economic viability, strong policy support, and societal engagement, each reinforcing the other. Technological innovation remains central. Improvements in AD efficiency, codigestion strategies, and feedstock pretreatment have significantly enhanced methane yields (Appels et al. 2011; Khan et al. 2017). Breakthroughs in upgrading and purification, particularly membrane separation, adsorption, and biological scrubbing, are critical for producing high-quality biomethane that can be injected into gas grids or used as transport fuel (Budzianowski & Budzianowska 2015; Sun et al. 2015). At the same time, research should explore decentralized, low-cost upgrading methods to make adoption more feasible in rural and low-income contexts where advanced purification technologies remain prohibitively expensive.

Figure 13.3 illustrates a framework that connects the challenges of limited digestate adoption to the outcomes of increased utilization through targeted interventions. On the left, limited adoption is linked to perception gaps and technical challenges, which align with the themes of Awareness and Perceptions and Technical and Operational Challenges. These barriers are consistent with findings in other studies where lack of awareness, misconceptions, and limited technical capacity constrained the use of biogas byproducts (Parawira 2009).

The framework proposes interventions, such as shifting mindsets through demonstrations and improving digestate handling, storage, and application. These approaches directly address the need for Capacity Building and Knowledge Transfer, which has been identified as critical for building farmer confidence and scaling biofertilizer adoption (FAO 2018; Smith et al. 2020). The development of quality standards and incentives reflects the importance of Policy and Institutional Support, aligning with calls for stronger regulatory frameworks and market mechanisms to promote organic soil amendments (UNEP 2019). Furthermore, the empowerment of

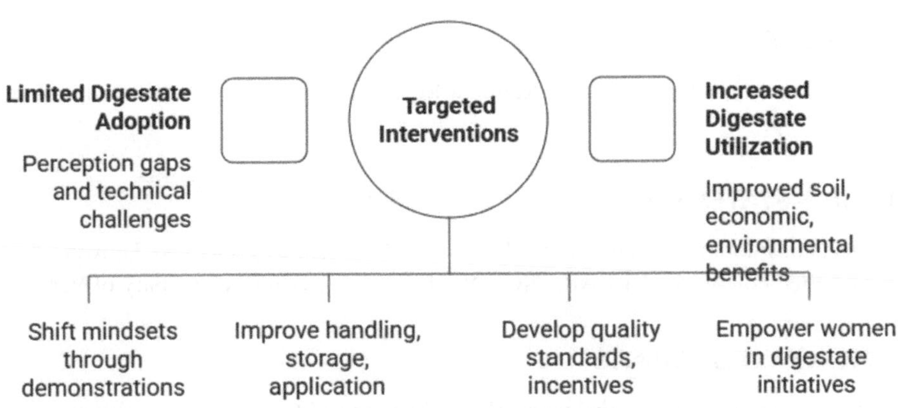

FIGURE 13.3 Enhancing digestate adoption framework.

women in digestate initiatives responds to the theme of Gender and Social Inclusion, as gender-sensitive interventions have been shown to increase technology uptake and enhance livelihood outcomes in rural energy systems (Clancy et al. 2012). On the right, the anticipated outcomes of these interventions include improved soil fertility, economic opportunities, and environmental sustainability.

These correspond to the themes of Agricultural Value of Digestate, Economic and Commercial Opportunities, and Environmental and Circular Economy Benefits. Similar results have been reported in sub-Saharan Africa and Asia where digestate use contributed to soil restoration, increased yields, and reduced reliance on synthetic fertilizers while supporting circular economy practices (Holm-Nielsen et al. 2009; Arthur et al. 2011). Taken together, the framework demonstrates how integrated actions can translate into practical benefits, linking technical, social, economic, and policy dimensions to create a pathway for sustainable digestate utilization.

13.6 CONCLUSION

However, the study was exploratory and based on a limited sample, and it did not test the long-term scalability or commercial viability of digestate valorization. Further research is required to validate these insights through larger studies, pilot projects, and longitudinal assessments of soil and livelihood impacts. Policy frameworks also need to establish quality standards, certification mechanisms, and financial incentives to enable digestate commercialization. In conclusion, digestate valorization is not a standalone solution, but it is an important component of integrated rural development strategies. Digestate can contribute to South Africa's just energy transition and National Development Plan objectives by linking renewable energy, agriculture, and local enterprise, provided that future efforts build on these findings with broader validation and stronger institutional support.

CONFLICT OF INTEREST

There are no conflicts of interest.

FUNDING

This research was funded by the National Research Foundation (NRF) Postgraduate Scholarship.

ETHICS STATEMENT

Ethical considerations were followed, and verbal consent was obtained. Univen Ethics Committee ETHICAL CLEARANCE No: FSEA/23/IRD/06, University of Venda.

ACKNOWLEDGMENTS

During the preparation of this work, the author(s) used Grammarly to improve readability and flow. After using this tool/service, the author(s) reviewed and edited the content as needed and take full responsibility for the content of the publication.

REFERENCES

Amigun, B., & H. von Blottnitz. 2010. Capacity-cost and location-cost analyses for biogas plants in Africa. *Resour. Conserv. Recycl.* 55: 63–73.

Appels, L., J. Lauwers, J. Degrève, L. Helsen, B. Lievens, K. Willems, J.V. Impe, & R. Dewil. 2011. Anaerobic digestion in global bioenergy production: Potential and research challenges. *Renew. Sustain. Energy Rev.* 15: 4295–4301.

Arellano-Yasaca, D.V., & C.-Y. Chu. 2025. Insights into nutrients recovery from food waste liquid digestate: A critical review and systematic analysis. *Waste Manage.* 200: 114743.

Armah, F.A., D.O. Yawson, G.T. Yengoh, J.O. Odoi, & E.K.A. Afrifa. 2019. Gender and biogas adoption in sub-Saharan Africa: A review. *Renew. Sustain. Energy Rev.* 103: 93–103.

Arthur, R., M.F. Baidoo, & E. Antwi. 2011. Biogas as a potential renewable energy source: A Ghanaian case study. *Renew. Energy* 36: 1510–1516.

Berhe, T.G., D. Hoag, G. Tesfay, & T. Tadesse. 2017. The role of renewable energy in youth employment creation: The case of Ethiopia. *Renew. Sustain. Energy Rev.* 74: 423–430.

Bian, X., K. Wang, & H. Gong. 2023. Biochar-enhanced agricultural application of liquid digestate from food waste anaerobic digestion for celery cultivation. *Sci. Total Environ.* 869: 161562.

Bonjour, S., H. Adair-Rohani, J. Wolf, N.G. Bruce, S. Mehta, A. Prüss-Ustün, M. Lahiff, E.A. Rehfuess, V. Mishra, & K.R. Smith. 2013. Solid fuel use for household cooking: Country and regional estimates for 1980–2010. *Environ. Health Perspect.* 121: 784–790.

Braun, V., & V. Clarke. 2006. Using thematic analysis in psychology. *Qual. Res. Psychol.* 3: 77–101.

Bruno, M., M. Marini, E. Angouria-Tsorochidou, F.M. Pulselli, & M. Thomsen. 2022. Ex ante life cycle assessment and environmental cost-benefit analysis of an anaerobic digester in Italy. *Clean. Waste Syst.* 3: 100021.

Budzianowski, W.M., & D.A. Budzianowska. 2015. Economic analysis of biomethane and bioelectricity generation from biogas using different support schemes and plant configurations. *Energy* 88: 658–666.

Clancy, J., M. Matinga, S. Oparaocha, & T. Winther. 2012. *Social Influences on Gender Equity in Access to and Benefits from Energy*. World Bank Report Background paper. http://hdl.handle.net/10986/9207 License: CC BY 3.0 IGO.

Czekała, W., M. Nowak, & G. Piechota. 2023. Sustainable management and recycling of anaerobic digestate solid fraction by composting: A review. *Bioresour. Technol.* 375: 128813.

De Clercq, D., Z. Wen, & F. Fan. 2017. Performance evaluation of state-led and industry-led circular economy programs in China. *Waste Manage.* 68: 46–56.

Deng, X., Y. Huang, W. Yuan, W. Zhang, P. Ciais, W. Dong, P. Smith, & Z. Qin. 2023. Building soil to reduce climate change impacts on global crop yield. *Sci. Total Environ.* 903: 166711.

Domenech, T., & B. Bahn-Walkowiak. 2019. Transition towards a resource efficient circular economy in Europe: Policy lessons from the EU and the Member States. *Ecol. Econ.* 155: 7–19.

FAO. 2011. *Energy-Smart Food for People and Climate: Issue Paper*. Rome: FAO.

FAO. 2018. *Upscaling Climate Smart Agriculture: Lessons for Extension and Advisory Services*. Rome: FAO.

Geissdoerfer, M., P. Savaget, N.M. Bocken, & E.J. Hultink. 2017. The circular economy – A new sustainability paradigm? *J. Clean. Prod.* 143: 757–768.

Ghimire, P.C. 2013. SNV supported domestic biogas programs in Asia and Africa. *Renew. Energy* 49: 90–94.

Guan, D., J. Zhao, Y. Wang, Z. Fu, D. Zhang, H. Zhang, J. Xie, Y. Sun, J. Zhu, & D. Wang. 2024. A critical review on sustainable management and resource utilization of digestate. *Process Saf. Environ. Prot.* 183: 339–354.

Haque, F., C. Fan, & Y.-Y. Lee. 2023. From waste to value: Addressing the relevance of waste recovery to agricultural sector in line with circular economy. *J. Clean. Prod.* 415: 137873.

Herrera, A., G. D'Imporzano, M. Zilio, A. Pigoli, B. Rizzi, E. Meers, O. Schouman, M. Schepis, F. Barone, A. Giordano, & F. Adani. 2022. Environmental performance in the production and use of recovered fertilizers from organic wastes treated by anaerobic digestion vs synthetic mineral fertilizers. *ACS Sustainable Chem. Eng.* 10: 986–997.

Hoang, L.A., T.H. Do, & L. Pham. 2021. Environmental and economic impacts of household biogas systems in Vietnam. *Renew. Energy* 178: 212–222.

Holm-Nielsen, J.B., T. Al Seadi, & P. Oleskowicz-Popiel. 2009. The future of anaerobic digestion and biogas utilization. *Bioresour. Technol.* 100: 5478–5484.

IEA. 2023. *Renewables 2023: Analysis and Forecast to 2028*. Paris: IEA Bioenergy. https://www.iea.org/reports/renewables-2023

Johnson, N.G., & K.M. Bryden. 2012. Energy supply and use in a rural Kenyan village. *Energy* 37: 601–612.

Johnson, N.G., K.M. Bryden, & A.L. Hammett. 2014. Modeling and evaluation of stove intervention impacts. *Energy Sustain. Dev.* 21: 73–81.

Kemausuor, F., A. Kamp, S.T. Thomsen, E.C. Bensah, & H. Østergård. 2014. Assessment of biomass residue availability and bioenergy yields in Ghana. *Resour. Conserv. Recycl.* 86: 28–37.

Khan, I.U., M.H.D. Othman, H. Hashim, T. Matsuura, A.F. Ismail, M. Rezaei-DashtArzhandi, & I. Wan Azelee. 2017. Biogas as a renewable energy fuel – A review of biogas upgrading, utilisation and storage. *Energy Convers. Manag.* 150: 277–294.

Kumar, A., J. Singh, & P.S. Panesar. 2023. Advances in digestate management: A step toward circular bioeconomy. *Bioresour. Technol. Rep.* 21: 101314.

LDP. 2025. *Limpopo Development Plan 2025–2030*. Polokwane: Office of the Premier.

Li, Y., S.Y. Park, & J. Zhu. 2020. Solid-state anaerobic digestion for methane production from organic waste. *Waste Manage.* 68: 455–465.

Lwiza, F., J. Mugisha, P.N. Walekhwa, J. Smith, & B. Balana. 2017. Dis-adoption of household biogas technologies in Central Uganda. *Energy Sustain. Dev.* 37: 124–132.

Mayer, F., R. Bhandari, S.A. Gäth, H. Himanshu, & N. Stobernack. 2020. Economic and environmental life cycle assessment of organic waste treatment by means of incineration and biogasification. Is source segregation of biowaste justified in Germany? *Sci. Total Environ.* 721: 137731.

Mengistu, M.G., B. Simane, G. Eshete, & T.S. Workneh. 2015. A review on biogas technology and its contributions to sustainable rural livelihood in Ethiopia. *Renew. Sustain. Energy Rev.* 48: 306–316.

Meyer, A.K.P., E.A. Ehimen, & J.B. Holm-Nielsen. 2018. Future European biogas: Animal manure, straw and grass potentials for a sustainable European biogas production. *Biomass Bioenergy* 111: 154–164.

Meyer, R., & Priefer, C. 2012. *Organic Farming and Bioenergy Production – Conflicting Goals and Approaches to a Solution*, Working Report No. 151. Karlsruhe, Germany: Institut für Technikfolgenabschätzung und Systemanalyse (ITAS).

Möller, K., & T. Müller. 2012. Effects of anaerobic digestion on digestate nutrient availability and crop growth: A review. *Eng. Life Sci.* 12: 242–257.

Muchara, B., G.F. Ortmann, E. Wale, & M.A. Darroch. 2014. Socio-economic determinants of smallholder farmers' adoption of conservation agriculture in KwaZulu-Natal, South Africa. *Agrekon* 53: 1–25.

Muto, M., I.A. Omondi, & K. Takahashi. 2017. Adoption of on-farm waste composting and its impact on household income and food security in rural Kenya. *Agric. Econ.* 48: 91–103.

Mutungwazi, A., P. Mukumba, & G. Makaka. 2018. The impact of biogas digestate on crop production: A case study in Zimbabwe. *Energy Sustain. Dev.* 43: 1–7.

Njenga, M., N. Karanja, M. Iiyama, & R. Jamnadass. 2021. Gendered barriers and opportunities for sustainable biogas adoption in sub-Saharan Africa. *Renew. Sustain. Energy Rev.* 151: 111520.

Nkoa, R. 2014. Agricultural benefits and environmental risks of soil fertilization with anaerobic digestates: A review. *Agron. Sustain. Dev.* 34: 473–492.

NPC. 2012. *National Development Plan 2030: Our Future – Make It Work.* Pretoria: NPC.

Ntaganda, A., M. Nkundimana, & A. Mutabazi. 2024. Household-level biogas adoption and its impact on climate mitigation in Rwanda. *Energy Policy* 186: 113478.

Nzeyimana, I., A.E. Hartemink, & C.J. Ritsema. 2020. Benefits of biogas in Rwanda: A review. *Renew. Sustain. Energy Rev.* 117: 109480.

O'Connor, J., B.S. Mickan, J. Rinklebe, H. Song, K.H.M. Siddique, H. Wang, M.B. Kirkham, & N.S. Bolan. 2022. Environmental implications, potential value, and future of food-waste anaerobic digestate management: A review. *J. Environ. Manage.* 318: 115519.

Oparaocha, S., & S. Dutta. 2011. Gender and energy for sustainable development. *Curr. Opin. Environ. Sustain.* 3: 265–271.

Parawira, W. 2009. Biogas technology in sub-Saharan Africa: Status, prospects and constraints. *Rev. Environ. Sci. Biotechnol.* 8: 187–200.

Rahman, S., M. Miah, & M. Hoque. 2021. Youth engagement in renewable energy: Opportunities and challenges. *Renew. Energy.* 179: 1420–1430.

Rasimphi, T.E., D. Tinarwo, B. Kilonzo, & P. Nyamukondiwa. 2024. Biogas technology implementation in rural areas: A case study of Capricorn District in Limpopo Province, South Africa. https://ssrn.com/abstract=4822374 or http://dx.doi.org/10.2139/ssrn.4822374

Rizzioli, F., D. Bertasini, D. Bolzonella, N. Frison, & F. Battista. 2023. A critical review on the techno-economic feasibility of nutrients recovery from anaerobic digestate in the agricultural sector. *Sep. Purif. Technol.* 306: 122690.

Roopnarain, A., & R. Adeleke. 2017. Current status, hurdles and future prospects of biogas digestion technology in Africa. *Renew. Sustain. Energy Rev.* 67: 1162–1179.

Roubik, H., J. Mazancová, & P.L. Dinh. 2018. Awareness and perceptions of biogas technology among rural households in Central Vietnam. *Energy Sustain. Dev.* 43: 68–75.

Singh, P., & S. Mohapatra. 2020. Composting and commercialization of biogas digestate in smallholder settings. *Int. J. Recycl. Org. Waste Agric.* 9: 225–234.

Singh, R., H. Pathak, & N. Jain. 2022. Valorization of biogas digestate for sustainable resource management. *J. Environ. Manag.* 315: 115183.

Smith, P., K. Calvin, J. Nkem, D. Campbell, F. Cherubini, G. Grassi, V. Korotkov, A.L. Hoang, S. Lwasa, P. McElwee, E. Nkonya, N. Saigusa, J.F. Soussana, M.A. Taboada, F.C. Manning, D. Nampanzira, C. Arias-Navarro, M. Vizzarri, J. House, S. Roe, A. Cowie, M. Rounsevell, & A. Arneth. 2020. Which practices codeliver food security, climate change mitigation and adaptation, and combat land degradation and desertification? *Glob. Change Biol.* 26: 1532–1575.

Sovacool, B.K., M.M. Lipson, & R. Chard. 2019. Temporality, vulnerability, and energy justice in household low carbon innovations. *Energy Policy* 128: 495–504.

Sovacool, B.K., S.E. Ryan, P.C. Stern, K. Janda, G. Rochlin, D. Spreng, M.J. Pasqualetti, H. Wilhite, & L. Lutzenhiser. 2015. Integrating social science in energy research. *Energy Res. Soc. Sci.* 6: 95–99.

Sun, Q., H. Li, J. Yan, L. Liu, Z. Yu, & X. Yu. 2015. Selection of appropriate biogas upgrading technology – A review of biogas cleaning, upgrading and utilization. *Renew. Sustain. Energy Rev.* 51: 521–532.

Surendra, K.C., D. Takara, A.G. Hashimoto, & S.K. Khanal. 2014. Biogas as a sustainable energy source for developing countries: Opportunities and challenges. *Renew. Sustain. Energy Rev.* 31: 846–859.

UNEP. 2019. The huge potential of agriculture to slow climate change. https://www.unep.org/news-and-stories/story/huge-potential-agriculture-slow-climate-change

Wainaina, S., M.K. Awasthi, S. Sarsaiya, H. Chen, E. Singh, A. Kumar, B. Ravindran, S.K. Awasthi, T. Liu, Y. Duan, S. Kumar, Z. Zhang, & M.J. Taherzadeh. 2020. Resource recovery and circular economy from organic solid waste using aerobic and anaerobic digestion technologies. *Bioresour. Technol.* 301: 122778.

Wang, N., X. Bai, D. Huang, Q. Chen, M. Shao, & Q. Xu. 2023. Impacts of digestate-based compost on soil property and nutrient availability. *Environ. Res.* 234: 116551.

Zhang, C., M. Shao, H. Wu, N. Wang, Q. Chen, & Q. Xu. 2021. Management and valorization of digestate from food waste via hydrothermal. *Resour. Conserv. Recycl.* 171: 105639.

14 Regulatory Bottlenecks in Global Biorefinery Expansion

Challenges and Pathways for Policy Reform

Thilivhali Rasimphi, David Tinarwo, Beata Kilonzo, and Sophie T. Mulaudzi

14.1 INTRODUCTION

Over the past few decades, the urgency to shift away from carbon-intensive economies toward sustainable systems has grown, driven by escalating climate change, biodiversity loss, and resource depletion (Liu 2021; Newell 2021). Rising global populations and higher living standards continue to intensify pressure on ecosystems, with projections showing resource extraction could rise by more than 90% by 2060, if current practices persist (Maja & Ayano 2021). In this context, biorefineries have emerged as a cornerstone of the low-carbon bioeconomy, offering pathways to reduce reliance on fossil fuels, enhance resource efficiency, and create value-added products from renewable biomass. However, the expansion of biorefineries worldwide faces significant regulatory bottlenecks.

Fragmented policies, inconsistent sustainability standards, lengthy approval processes, and unclear land-use regulations hinder investment and delay project implementation. These challenges undermine the role of biorefineries in advancing the goals of the Paris Agreement and the UN Sustainable Development Goals (SDGs), which envision a global transition toward decarbonization and regenerative economies (Gomez-Echeverri et al. 2018). Addressing these barriers requires policy reforms that create a more enabling environment. Streamlined regulatory frameworks, harmonized sustainability standards, transparent approval processes, and incentives for innovation are essential (Moussa 2022). Therefore, strengthening international cooperation and aligning national policies with global sustainability commitments can ensure that biorefineries contribute meaningfully to climate mitigation, biodiversity protection, and sustainable development.

The expansion of biorefineries has been widely recognized as a cornerstone for advancing the global bioeconomy and meeting sustainability targets. However, the

DOI: 10.1201/9781003650751-14

literature consistently reveals that their large-scale deployment faces multidimensional barriers (ARE 2021). Scholars argue that while technological progress has been considerable, systemic challenges continue to restrict the commercial viability of these facilities (Guan et al. 2025). One major obstacle concerns technological complexity. Biorefineries depend on diverse feedstocks, such as agricultural residues, forestry byproducts, and organic waste. Yet their heterogeneous composition complicates processing, conversion efficiency, and product standardization (Cherubini 2010; Adegoke et al. 2025). Although progress has been made in biomass valorization, scaling pilot plants to industrial levels remains limited due to high operational risks and uncertainty in process optimization (Allain et al. 2022). This demonstrates that the technological frontier, while advancing, is not yet sufficient to drive widespread commercialization.

Biorefineries face steep capital investment costs and operate in markets dominated by fossil-based industries that benefit from entrenched infrastructure and subsidies (Adelakun et al. 2024; Sravan et al. 2026). As Makepa and Chihobo (2024) point out, the lack of competitive pricing and unstable biomass supply chains create financial risks that discourage private investors. Consequently, many projects stall at demonstration phases (Rasimphi et al. 2024), reinforcing the argument that without stronger market incentives and de-risking mechanisms, economic viability will remain elusive. The most frequently cited barrier, however, lies in regulatory bottlenecks (IFC 2021). Fragmented and inconsistent policy frameworks slow down project approvals, complicate compliance, and reduce investor confidence (Gomez-Echeverri et al. 2018; IEA 2025). In particular, weak coordination around sustainability certification, land-use governance, and bio-based product standards undermines the ability of biorefineries to operate across borders. Scholars argue that such regulatory uncertainty not only delays expansion but also discourages the integration of biorefineries into global trade systems.

Finally, social and political acceptance issues shape the discourse, especially on the concerns over land competition with food crops, deforestation risks, and the potential for greenwashing (Liu 2021). Without inclusive governance and transparent communication, projects risk community resistance, which further delays their scaling. Taken together, the literature suggests that while technological and economic factors are significant, the most pressing challenges are regulatory and socio-political in nature (Newell 2021; IEA 2024). Addressing these requires harmonized policies, clearer sustainability standards, and frameworks that balance economic growth with ecological and social safeguards. This reinforces the argument that the future of biorefineries depends less on technical feasibility, which is steadily advancing, and more on creating enabling institutional and policy environments aligned with international climate and sustainability commitments.

The regulatory and institutional complexity surrounding these projects is a significant barrier to widespread adoption. Regulatory bottlenecks, including policy instability, fragmented permitting processes, unclear sustainability standards, and a lack of cross-sectoral alignment, often introduce delays, increase project costs, and reduce investor confidence (Philp et al. 2013; Scarlat et al. 2015). In many countries, bioeconomy policies are not yet fully integrated across agriculture, energy, waste management, and industry sectors. This lack of policy coherence can result in conflicting mandates, overlapping responsibilities, and limited support mechanisms for biorefinery development.

Furthermore, inconsistent or absent sustainability criteria for biomass sourcing and end-product certification create uncertainties around compliance and market access (Camia et al. 2018). These challenges are especially pronounced in emerging economies, where institutional capacity and inter-agency coordination are still evolving (Mountraki et al. 2016). Addressing these regulatory hurdles is essential for unlocking the full potential of biorefineries. Forward-looking strategies include the development of coherent national bioeconomy frameworks, streamlined permitting processes, harmonized sustainability standards, and strengthening multi-stakeholder governance structures. Enhanced coordination among public authorities, private investors, communities, and research institutions will be vital in building the regulatory environment necessary to support innovation, de-risk investments, and accelerate the global deployment of biorefinery technologies.

This chapter highlights that while biorefineries hold significant promise as enablers of the global bioeconomy, their expansion is constrained less by technical feasibility and more by regulatory and institutional bottlenecks. The fragmented policies, inconsistent sustainability standards, and complex approval processes identified across the literature not only slow down deployment but reduce investor confidence and weaken international competitiveness. These challenges undermine the capacity of biorefineries to deliver on global sustainability commitments, such as the Paris Agreement and the UN SDGs. The importance of this chapter lies in drawing attention to the regulatory dimension of the debate, which is often overshadowed by discussions of technology and finance.

By systematically analyzing how regulatory bottlenecks shape the pace and direction of biorefinery development, this work underscores the need for coherent policy reforms and stronger institutional coordination. Such reforms are essential for unlocking private sector investment, ensuring social legitimacy, and aligning biorefinery expansion with broader environmental and development goals. Ultimately, addressing these barriers is not only about facilitating the scale-up of biorefineries but also about creating an enabling environment, where technological innovation, economic viability, and sustainability can converge. In doing so, biorefineries can move from being promising concepts to transformative pillars of the global transition toward low-carbon, circular, and inclusive economies.

14.2 REGULATORY BOTTLENECKS IN BIOREFINERY EXPANSION

Regulatory bottlenecks represent one of the most significant obstacles to the global expansion of biorefineries, as fragmented, unstable, and poorly coordinated policies often undermine investor confidence and delay project implementation. The causes are discussed in this section.

14.2.1 Policy Instability and Uncertainty

To foster investment in the biorefineries, there is a need for long-term policy, typically capital-intensive ventures with extended development, construction, and payback periods (Newell 2021). Due to their technological complexity and reliance on consistent biomass supply chains, biorefineries demand high upfront investment and long-term operational planning (Batchu et al. 2022). Investors and project developers seek predictable, transparent, and supportive policy environments that can ensure the viability of projects over time (Philp et al. 2013; Kurki & Ahola-Launonen 2021). However, policy inconsistency and unpredictability remain major restrictions to investment in many regions. Shifts in government priorities, sudden changes in subsidies, feed-in tariffs, tax credits, or renewable energy mandates can rapidly alter the economics of biorefinery projects.

For instance, removing or reducing production incentives or biofuel blending mandates can erode profit margins and undermine previously sound business cases (Lamers et al. 2011; Marangwanda et al. 2020). This uncertainty introduces a level of risk that makes investors hesitant, especially in emerging markets, where institutional support for biobased industries is still developing (Stegmann et al. 2020). Moreover, the lack of long-term national bioeconomy strategies and sector-specific investment roadmaps often leaves biorefinery developers without clear signals regarding future regulatory direction (Scarlat et al. 2015; Carus et al. 2016; Makepa & Chihobo 2024). This makes it difficult to measure return on investment, secure financing, or form long-term partnerships. In contrast, countries with stable and integrated policy frameworks, such as Finland, the Netherlands, and Brazil, have seen stronger growth in their biobased sectors due to clear investment signals and consistent government support.

Additionally, policy unpredictability undermines innovation (Bahadorestani et al. 2020). Investors and companies are less willing to invest in research and development, scale-up technologies, or build infrastructure, when there is uncertainty about future regulations or support mechanisms (Makepa & Chihobo 2024). This slows the deployment of advanced biorefinery technologies and hampers the transition to a sustainable, low-carbon economy. To encourage long-term investments, governments must establish stable, coherent, and forward-looking policy frameworks that provide clear and consistent incentives, de-risk investment, and align with broader sustainability and climate goals. Regulatory certainty is a key enabler for mobilizing private capital and accelerating the development of biobased industries.

14.2.2 Outdated Policies and Regulatory Frameworks

Many regulatory frameworks governing energy production and distribution were developed decades (Akpuokwe et al. 2024; Oduro et al. 2024). As a result, these policies may not adequately account for the unique characteristics and benefits of renewable energy sources (Ochuba et al. 2024). Outdated regulations can create barriers to market entry, hinder investment, and stifle innovation in the renewable energy sector (Oduro et al. 2024). Obtaining permits for renewable energy projects can be a time-consuming and expensive process, often involving multiple layers of regulatory approval at the local, state, and federal levels. Complex permitting requirements, bureaucratic delays, and inconsistent application of regulations can deter developers and investors, delaying project deployment and increasing costs.

14.2.3 Conflicting Regulations Across Jurisdictions

Renewable energy projects often span multiple jurisdictions (Akpuokwe et al. 2024; Eneh et al. 2024). Inconsistencies and conflicts between regulations in different jurisdictions can create uncertainty and ambiguity for project developers, complicating compliance efforts and increasing legal risks. Harmonizing regulations and streamlining permitting processes across jurisdictions can help facilitate renewable energy expansion and promote investment certainty.

Renewable energy projects typically require access to land for siting solar panels, wind turbines, or other infrastructure. However, navigating property rights, land leases, and zoning regulations can be complex and contentious, especially in densely populated or environmentally sensitive areas (Adama et al. 2024; Ochuba et al. 2024; Oduro et al. 2024). Legal uncertainties surrounding land use and property rights can delay project development, trigger legal disputes, and undermine community support for renewable energy initiatives (Akagha et al. 2023; Ochuba et al. 2024; Oduro et al. 2024). Regulatory barriers pose significant challenges to the expansion of renewable energy, hindering the transition to a sustainable and low-carbon energy future. Understanding these obstacles is crucial for policymakers, industry stakeholders, and advocates seeking to overcome them and accelerate the adoption of renewable energy technologies.

14.2.4 Fragmented Permitting and Compliance Processes

Biorefineries often require numerous permits from various regulatory agencies. These permits typically span various domains, including environmental impact assessments (EIAs), air emissions, wastewater discharge, hazardous material handling, land use zoning, and waste management approvals (Scarlat et al. 2015). Each of these regulatory processes may be governed by different departments or authorities, often with overlapping, inconsistent, or unclear mandates. This fragmented permitting landscape presents a major hurdle for project developers. Many countries have no centralized or harmonized permitting framework tailored to biorefineries' specific needs and risks, which are often mistakenly categorized under generic industrial or energy regulations.

The global shift toward renewable energy sources is crucial for combating climate change and ensuring a sustainable future. However, the expansion of renewable energy faces various challenges, including regulatory barriers (Adelakun et al. 2024; Ochuba et al. 2024; Oduro et al. 2024). One of the most successful examples of promoting renewable energy expansion through legal frameworks is Germany Renewable Energy Act - EEG. Enacted in 2000, the EEG guaranteed fixed, above-market rates for renewable energy producers, providing long-term investment security. This strategy led to a significant increase in renewable energy capacity, with renewables accounting for over 40% of Germany's electricity generation by 2020. California's Renewables Portfolio Standards (RPS) mandates that a certain percentage of electricity sold by utilities must come from renewable sources. Since its implementation in 2002, the RPS has been instrumental in driving renewable energy development in the state (Eneh et al. 2024; Zhou et al. 2024; Adegoke et al. 2025).

As a result, developers must navigate a complex web of administrative requirements, leading to longer project timelines, increased compliance costs, and, in some cases, project abandonment (Philp et al. 2013; Camia et al. 2018). The absence of streamlined permitting processes also introduces a high level of uncertainty. Regulatory agencies may lack technical expertise related to biorefinery operations, leading to delays in decision-making or overly cautious approaches to risk management.

For instance, differing interpretations of biomass classification or byproduct handling can trigger disputes over applicable regulations or require multiple rounds of review and revision. These bottlenecks discourage investment, especially from private-sector actors seeking clarity and predictability in project development. Some countries have attempted to address this issue by establishing one-stop shops or integrated licensing systems to reduce bureaucratic friction. For example, Finland and the Netherlands have implemented more coordinated permitting mechanisms for biobased industries, significantly improving efficiency and stakeholder confidence. However, in many regions, particularly in the Global South, such mechanisms are still lacking, and institutional fragmentation remains a persistent barrier. Creating biorefinery-specific regulatory guidelines, improving inter-agency coordination, and building regulatory capacity are critical to accelerating deployment. Streamlined, transparent, and time-bound permitting frameworks would lower entry barriers and signal strong governmental support for innovation in the bioeconomy sector.

14.2.5 Cross-Sectoral Regulatory Misalignments

Biorefineries operate at the nexus of multiple sectors, for example, energy, agriculture, waste management, and environmental protection. Policies, regulations, and institutional frameworks typically govern these sectors (Makepa et al. 2023). While this cross-sectoral nature is a strength regarding resource integration and sustainability potential, it also creates substantial governance and coordination challenges. Unaligned or contradictory objectives across departments in many jurisdictions can significantly complicate biorefinery development (Philp et al. 2013; Scarlat et al. 2015). For example, a biorefinery that uses agricultural residues as feedstock may be affected by energy regulations (e.g. biofuel blending mandates or renewable energy targets), agricultural policies (e.g. land use and crop residue management), and environmental rules (e.g. emissions limits and waste disposal requirements). These policies are often developed in silos, without a coherent strategy that considers the overlapping needs and impacts of biorefinery operations. As a result, developers may face conflicting mandates, such as encouragement to use biomass for energy under one policy and restrictions on biomass harvesting under another.

This lack of institutional coordination can lead to overlapping regulations, duplicated permitting processes, and unclear compliance requirements. It also introduces ambiguity around responsibilities, such as which ministry or agency should lead the promotion, regulation, or support of biorefineries (Marangwanda et al. 2020). This confusion has sometimes led to regulatory delays, investment hesitation, and even project cancellations, especially in countries without a unified bioeconomy strategy (Kleinschmit et al. 2025). Moreover, misalignment between national, regional, and local policies can create additional hurdles. Local governments may not have the authority, capacity, or technical knowledge to implement national bioeconomy plans, while national policies may overlook regional environmental or land-use conditions (Carus et al. 2016). This disconnect hampers effective implementation and adds another layer of complexity for project developers navigating the regulatory landscape (Camia et al. 2018; OECD 2019; FAO 2021). To address these challenges, many scholars and international agencies advocate for integrated bioeconomy governance frameworks that bring together stakeholders from energy, agriculture, and environmental sectors. Cross-sectoral coordination platforms, inter-ministerial committees, and harmonized regulatory instruments are key tools to improve policy coherence and foster a more enabling environment for biorefinery investment and development.

14.2.6 Regional Disparities in Regulatory Frameworks

In regions like Africa, the regulatory landscape for biofuels and biorefineries is often marked by complexity, inconsistency, and fragmentation across national boundaries (Kunatsa & Xia 2022). While several African countries have identified bioenergy as a key component of their national development and energy strategies, the absence of harmonized regulatory frameworks has significantly hindered the growth and integration of biofuel and biorefinery industries across the continent (Karekezi & Kithyoma 2012). One of the primary challenges is the lack of clear and coordinated policy guidelines. In many cases, biofuels fall into a regulatory grey area, neither fully classified under energy nor agricultural regulation, leading to uncertainty in project approvals, licensing, and compliance procedures. Furthermore, overlapping mandates between different ministries (e.g. energy, agriculture, environment, and industry) often result in bureaucratic delays and inefficient decision-making.

For example, developers in countries like Kenya, Uganda, and Ghana report significant delays in securing permits due to unclear processes and insufficient institutional capacity (EC 2018; Moller et al. 2020). Burdensome compliance requirements, especially in countries with underdeveloped infrastructure or limited technical expertise, further complicate the situation. Small- and medium-scale producers struggle to meet the requirements for EIAs, fuel quality testing, or sustainability certifications (Ochuba et al. 2024). Inconsistent enforcement of these standards across regions creates a non-level playing field, discouraging investment and stalling innovation. Additionally, variations in regulatory frameworks and enforcement mechanisms between countries prevent the establishment of regional biofuel markets. While some nations, such as South Africa and Mozambique, have made progress in developing biofuel policies, others lack even basic legal instruments or implementation plans. This patchwork of regulations restricts cross-border collaboration, limits economies of scale, and discourages foreign direct investment in the sector.

14.3 CIRCULAR ECONOMY AND BIOREFINERIES

Biorefineries are integral components of the circular economy paradigm, embodying principles of resource efficiency, waste valorization, and sustainable material cycles. Within this framework, biorefineries contribute significantly to minimizing waste and maximizing resource use by converting diverse biomass feedstocks, such as agricultural residues, forestry waste, and organic byproducts, into valuable biofuels, biochemicals, and bioproducts (Allain et al. 2022). This process not only transforms waste materials into high-value outputs but also reduces reliance on virgin resources, promoting a circular approach to resource utilization. Biorefineries optimize resource efficiency through the utilization of various conversion technologies, including thermochemical and biochemical processes, which extract multiple products from a single biomass source (Makepa et al. 2023). This approach maximizes the use of biomass components, minimizes waste generation, and enhances overall resource productivity. Biorefineries operate within closed-loop systems as shown in Figure 14.1 by integrating waste recycling and byproduct utilization, ensuring that residues and byproducts generated during biomass processing are recycled back into the system as inputs for secondary processes or converted into additional value-added products.

This closed-loop approach not only minimizes environmental impact but fosters a circular material flow within the biorefinery ecosystem. Moreover, biobased products produced in biorefineries contribute significantly to carbon sequestration and climate

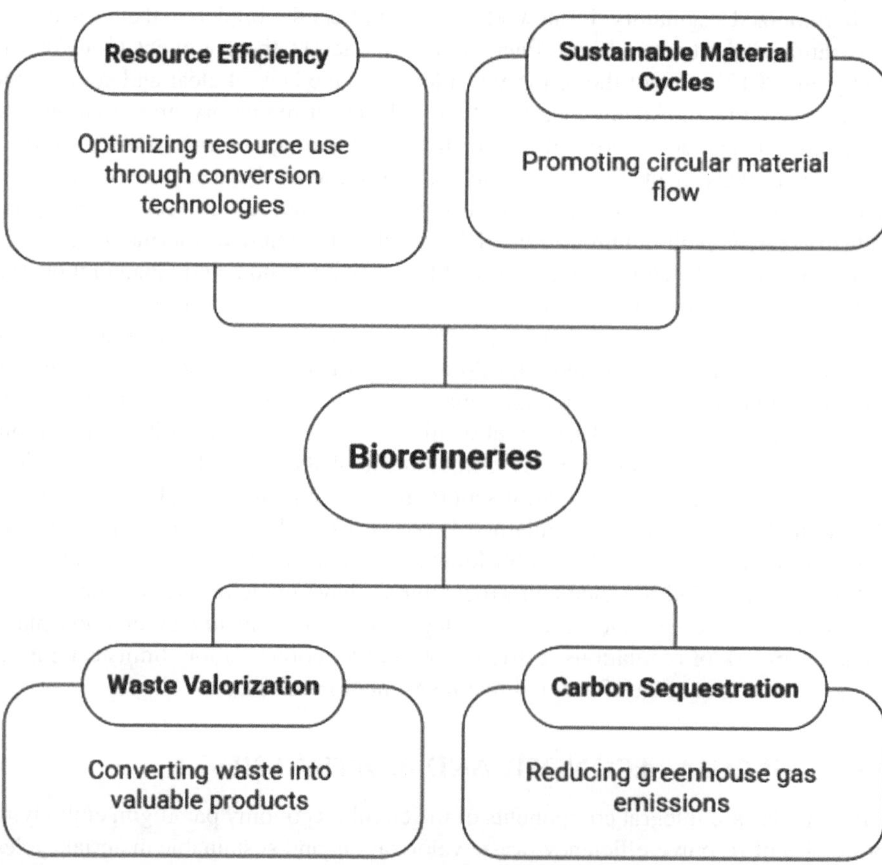

FIGURE 14.1 Biorefinery circular economy.

change mitigation by replacing fossil-based alternatives. Biofuels, bioplastics, and biomaterials derived from sustainable biomass sources help reduce greenhouse gas emissions and support a more sustainable carbon cycle. In essence, the symbiotic relationship between biorefinery operations and circular economy principles underscores the pivotal role of biorefineries as key enablers of sustainable and regenerative industrial practices, driving the transition toward a circular economy model.

14.4 CHALLENGES AND GAPS IN BIOREFINERIES DEPLOYMENT

There are several other regulatory barriers that hinder the expansion of renewable energy. Traditional electricity markets were designed around centralized, fossil fuel-based generation, which may not adequately value the attributes of renewable energy, such as its low marginal cost and environmental benefits (Akpuokwe et al. 2024; Igbinenikaro et al. 2024; Oyewole et al. 2024). This results in market structures and pricing mechanisms that disadvantage renewable energy providers, making it harder for them to compete with conventional energy sources. Lack of long-term policy certainty: The renewable energy sector relies heavily on stable and predictable policy frameworks to attract investment.

However, changes in government policies, subsidies, and incentives can create uncertainty for investors and project developers, leading to delays or cancellations of renewable energy projects (Akpuokwe et al. 2024; Zhou et al. 2024). Permitting and land access for transmission infrastructure: Building the necessary transmission infrastructure to connect renewable energy projects to the grid often requires access to land and permits from multiple regulatory bodies. Delays in obtaining permits or securing land rights can significantly increase project costs and timelines.

Interconnection standards and grid access:

- Integrating renewable energy into the grid requires compliance with technical standards and regulations for grid interconnection (Akpuokwe et al. 2024; Popoola et al. 2024).
- Delays or inconsistencies in these standards can hinder the connection of renewable energy projects to the grid, limiting their ability to deliver electricity to consumers.
- Technological maturity represents a significant obstacle to the successful implementation and commercialization of biorefineries.

While biorefinery technologies hold great promise for converting biomass into valuable products, many of these technologies are still in the early stages of development and lack the necessary maturity for large-scale implementation (Popoola et al. 2024). One of the primary challenges is the need for further technological advancements and optimization. Biorefineries employ complex processes involving biomass pretreatment, enzymatic hydrolysis, fermentation, and product separation. Each step requires continuous improvement and fine-tuning to enhance efficiency, yield, and overall process economics (Zhou et al. 2024). Scaling up these processes from the laboratory to commercial scale can introduce new technical challenges and operational complexities that need to be addressed. Another aspect of technological immaturity is the limited integration of multiple conversion pathways. Biorefineries aim to produce a diverse range of products, such as biofuels, biochemicals, and biomaterials.

However, the integration of various conversion pathways within a single biorefinery is a complex task that requires sophisticated process design and optimization. The lack of mature and integrated technologies for the simultaneous production of multiple products poses challenges in terms of process efficiency, product quality, and economic viability (Mutsengerere et al. 2019). Furthermore, the lack of standardized protocols and benchmarking criteria for biorefinery technologies hampers the assessment of their readiness for commercial deployment. The absence of consistent evaluation metrics and performance indicators makes it difficult to compare and select the most suitable technologies for specific applications (Kunatsa & Xia 2022). This can lead to uncertainty and reluctance among investors and potential stakeholders, hindering the commercialization of biorefineries. To overcome these technological challenges, increased research and development efforts are required to advance the maturity of biorefinery technologies. Collaborative initiatives between industry, academia, and government institutions are necessary to foster innovation and accelerate the development of robust, scalable, and economically viable biorefinery platforms (Xu et al. 2022). Additionally, greater emphasis on pilot-scale and demonstration projects can help bridge the gap between laboratory-scale proof-of-concept and commercial implementation, enabling the validation and optimization of emerging technologies.

Key challenges and gaps include:

- **Policy instability:** Frequent changes in subsidies, renewable energy mandates, and biofuel blending requirements were observed in several jurisdictions, particularly in emerging markets (Philp et al. 2013; Scarlat et al. 2015). These shifts create uncertainty for investors and hinder long-term planning.
- **Fragmented permitting:** Biorefineries encounter complex approval processes spanning environmental, agricultural, and industrial regulations (Camia et al. 2018). Many countries lack a coordinated framework tailored to biorefineries, resulting in extended project timelines and higher compliance costs.
- **Inconsistent sustainability standards:** Criteria for biomass sourcing, life cycle emissions, and product certification vary widely across countries (Jumbe et al. 2009; Scarlat & Dallemand 2011), restricting market access and discouraging international investment.
- **Cross-sectoral misalignment:** Policies across energy, agriculture, and environmental sectors often conflict, with overlapping responsibilities and unclear regulatory authority (FAO 2021; Kleinschmit et al. 2025). This is particularly evident in emerging economies, where institutional coordination is limited.

The establishment and operation of biorefineries require substantial investments in infrastructure, equipment, and technology, making them highly capital-intensive ventures. Capital intensity represents a significant obstacle to the successful implementation and commercialization of biorefineries (Nurdiawati & Urban 2022). One of the primary challenges is the high upfront capital costs associated with biorefinery construction. The scale and complexity of biorefinery facilities, which typically involve multiple processing units and integrated systems, contribute to significant capital requirements. The need for specialized equipment, such as biomass pretreatment units and fermentation tanks, further adds to the capital costs. These high initial investments pose financial risks and can deter potential investors and project developers.

Moreover, the long payback period of biorefinery projects presents a challenge in attracting financing and securing adequate returns on investments (Liu et al. 2021). Biorefineries often require several years to reach full production capacity and generate revenues, primarily due to the time-consuming process of technology development, scale-up, and optimization (Makepa et al. 2023). The extended payback period increases the financial risks associated with biorefinery investments, as the return on investment may be uncertain or delayed. In addition, the uncertainty surrounding the market demand and pricing for biorefinery products can further exacerbate the capital intensity challenge. The market dynamics for biofuels, biochemicals, and biomaterials are influenced by various factors, including policy support, competition, and consumer preferences.

The volatility and unpredictability of these markets can impact the profitability and long-term financial sustainability of biorefinery projects, making it challenging to attract the necessary capital investments. To address the capital intensity challenge, various strategies can be employed. Collaborative efforts between industry, government, and financial institutions can help establish supportive financing mechanisms, such as subsidies, tax incentives, and loan guarantees, to alleviate the financial burden on biorefinery projects. Additionally, innovative financing models, such as public-private partnerships and strategic alliances, can facilitate access to capital and mitigate the risks associated with biorefinery investments. Furthermore, continuous efforts to improve the cost-effectiveness and efficiency of biorefinery technologies are essential. Technological advancements, process optimization, and economies of scale can help reduce the capital intensity of biorefineries by enhancing production yields, lowering operational costs, and improving overall process economics. Research and development initiatives focused on cost reduction and process simplification are crucial for making biorefineries more financially viable.

14.4.1 Comparative Case Studies

Analysis of four case study countries: Germany, Brazil, the United States, and South Africa, revealed differences in regulatory maturity and their impact on biorefinery development:

- **Germany:** High regulatory clarity and stable incentives for bioenergy projects have facilitated commercial-scale biorefineries (Carus et al. 2016). Coordinated permitting and supportive sustainability criteria reduce uncertainty for investors.
- **Brazil:** Advanced bioethanol and biodiesel policies provide clear guidance for biomass sourcing and biofuel blending mandates (Scarlat et al. 2015). However, regional disparities in enforcement and licensing still create bottlenecks for smaller-scale developers.
- **United States:** State-level variation in energy and environmental regulations introduces opportunities and complexity (Philp et al. 2013). Some states provide strong incentives, while others impose stringent environmental requirements, affecting project viability.

- **South Africa:** Emerging policies and fragmented regulatory frameworks hinder investor confidence (Kamusoko & Mukumba 2025). Permitting delays, unclear sustainability standards, and lack of coordination among ministries are significant barriers.

14.4.2 Integrated Findings

Triangulating the data sources revealed a consistent pattern: regulatory bottlenecks, policy instability, fragmented permitting, inconsistent sustainability criteria, and cross-sectoral misalignments. Countries with stable, coordinated, and transparent regulatory frameworks (e.g. Germany and Brazil) demonstrate higher adoption rates and investor confidence, while emerging markets face significant delays and reduced private-sector engagement. Addressing these bottlenecks through adaptive, forward-looking reforms can enable broader biorefinery deployment and support the transition toward a circular, low-carbon bioeconomy (Table 14.1).

TABLE 14.1
Summary of Regulatory Challenges from Document Analysis

Regulatory Challenge	Description	References	Impact on Biorefinery Expansion
Policy instability	Frequent changes in subsidies, renewable energy mandates, and biofuel blending requirements	Philp et al. (2013); Scarlat et al. (2015)	Increased investment risk, project delays
Fragmented permitting	Multiple approvals across environmental, agricultural, and industrial authorities	Camia et al. (2018)	Higher compliance costs, prolonged timelines
Inconsistent sustainability standards	Variation in biomass sourcing criteria, life-cycle emissions, and product certification	Jumbe et al. (2009); Scarlat and Dallemand (2011)	Reduced market access, investor hesitation
Cross-sector misalignment	Conflicting policies across the energy, agriculture, and environment sectors	Kleinschmit et al. (2025); FAO (2021)	Regulatory confusion, project cancellations

14.5 PATHWAYS FOR REGULATORY REFORM

14.5.1 Establishing Stable and Predictable Policies

Biorefineries require policy environments that are stable, predictable, and long-term in orientation. Their capital-intensive nature and extended payback periods make investors highly sensitive to policy volatility (Philp et al. 2013). Stable support measures, such as renewable energy targets, feed-in tariffs, tax incentives, and subsidies reduce risk perceptions and provide the market certainty necessary for scaling bio-based industries (Cherubini 2010; U.S. DOE 2023). Evidence from Germany and Brazil illustrates how blending mandates and feed-in tariffs created reliable markets and revenue streams, directly enabling sector growth (Lamers et al. 2011; Scarlat et al. 2015). In the United States, instruments such as the Renewable Fuel Standard and the Biomass Crop Assistance Program demonstrate how sustained policy commitments can mobilize private capital for second-generation technologies.

By contrast, abrupt withdrawal of subsidies or inconsistent mandate renewals has led to project cancellations and delayed investments across multiple regions (IRENA 2022). These episodes reveal a persistent tension between short-term political cycles and the long-term horizons of biorefinery investments (Moller et al. 2020; FAO 2021). Current frameworks often fail to integrate energy, agriculture, and environmental objectives, resulting in fragmented policy landscapes. When embedded within broader development agenda, national bioeconomy strategies and roadmaps can mitigate this gap by aligning multiple sectoral priorities, while assuring policy continuity.

South Africa's experience illustrates both opportunities and challenges in this regard. While the country has introduced renewable energy procurement programs, its bioeconomy strategy has been criticized for limited implementation and policy inconsistency (AUC-ECA 2013). Shifts in government priorities and the absence of long-term, enforceable biofuel mandates have stalled investment in large-scale projects, despite clear potential for biomass utilization in rural areas (IRENA 2022). The lack of alignment between national energy policy and agricultural development frameworks has further limited momentum. Regional commitments under the AU and SADC could help stabilize investor expectations, but such initiatives remain underdeveloped.

14.5.2 Streamlining Permitting and Compliance Processes

Regulatory ambiguity remains a critical barrier to biorefinery development. Because biorefineries generate multiple outputs, fuels, materials, and chemicals, they are often subjected to overlapping industrial, agricultural, and energy regulations not designed for integrated systems (Philp et al. 2013; IEA 2022). This regulatory fragmentation increases administrative costs, delays project approvals, and places disproportionate burdens on small and medium enterprises. European evidence demonstrates that targeted reforms can significantly reduce these barriers (Camia et al. 2018). Finland and the Netherlands have introduced centralized "one-stop-shop" regulatory platforms, which consolidate oversight functions and provide developers a single point of contact (Carus et al. 2016; Kleinschmit et al. 2025). This approach has reduced duplication, lowered compliance costs, and improved regulatory predictability. In contrast, many jurisdictions rely on generic industrial regulations, leading to uncertainty over feedstock classification, waste handling, and emission thresholds.

The Organization for Economic Co-operation and Development (OECD) emphasizes that governments should complement regulatory streamlining with technical support mechanisms, including digital permitting tools, advisory services, and clear guidance documents (AUC-ECA 2013). These resources improve transparency and expand participation in the bioeconomy by reducing barriers for new entrants. A key gap remains the absence of biorefinery-specific regulatory frameworks, which would provide clarity, while recognizing the sector's integrated and multi-output character. The EIA processes in South Africa are often lengthy and duplicative, with responsibilities spread across different departments. This has delayed renewable energy projects and discouraged small-scale innovators. Without a streamlined biorefinery-specific permitting system, projects face high transaction costs that limit both domestic and foreign investment.

14.5.3 Enhancing Cross-Sectoral Coordination

The multi-sectoral character of biorefineries linking energy, agriculture, environment, and industry requires integrated governance structures rather than siloed. However, in many countries, fragmented governance persists, producing overlapping mandates, conflicting regulations, and duplicative procedures (Philp et al. 2013; Scarlat et al. 2015). This lack of coherence weakens investment signals and complicates implementation (Carus et al. 2016; Kleinschmit et al. 2025). Examples from Finland and the Netherlands show how cross-sectoral coordination strengthens institutional support for bio-based industries. Finland's bioeconomy strategy was developed through multi-ministerial collaboration, aligning renewable energy and environmental objectives with industrial development (FAO 2021). Similarly, Dutch interagency platforms have coordinated research, funding, and regulation, creating clearer pathways for investors.

Integrated governance also facilitates resolution of policy conflicts, such as food versus fuel trade-offs or renewable energy expansion versus land conservation (Camia et al. 2018; IRENA 2022). Harmonized sustainability criteria for biomass sourcing and integrated environmental impact protocols provide common frameworks that reduce administrative burdens and improve regulatory transparency (AUC-ECA 2013). In South Africa, bioenergy initiatives are often caught between the Departments of Energy, Agriculture, and Environmental Affairs mandates. Limited inter-ministerial collaboration has slowed implementation of the Biofuels Industrial Strategy and impeded coordination on biomass resource mobilization. Without institutional mechanisms to align objectives across ministries, biorefineries remain constrained by fragmented governance structures.

14.5.4 Harmonizing Regional Regulatory Frameworks

Regional regulatory fragmentation compounds the challenges faced by biorefineries. Divergent sustainability standards, product definitions, and certification schemes across national borders increase transaction costs and deter cross-border investment

(Moller et al. 2020; IRENA 2022). Harmonization through shared standards, common product definitions, and unified sustainability schemes can reduce these frictions and create regional value chains.

The European Union's Renewable Energy Directive (RED II) illustrates how regional alignment can improve investment climates by providing consistent certification frameworks, harmonized sustainability criteria, and predictable permitting processes (IRENA 2022). While African regional bodies, such as the AU, SADC, and ECOWAS, have initiated bioenergy strategies, implementation has remained fragmented and often lacks enforceability. For South Africa, closer alignment with SADC frameworks could provide an opportunity to build regional biobased value chains, particularly in biomass trade and shared certification standards. However, limited institutional capacity and competing national priorities continue to slow progress.

14.6 BIOREFINERIES IN RURAL SOUTH AFRICA

South Africa is committed to the SDGs as various efforts are undertaken by the different stakeholders (industries and government agencies) to achieve the country's carbon-neutral goals (Stats SA 2023). The South African landscape shows the key factors promoting the development of biorefineries and how these initiatives could help fast-track the advancement of biorefinery-based technologies (Stats SA 2022; Sekoai et al. 2023). The specific biowaste valorization methods that could be adopted by South Africa are also included. Scientific studies that discuss the development of biorefinery-based technologies in South Africa, especially those that are presented, have not yet been published in the literature.

Rural South Africa presents both the need and the opportunity for advancing biorefinery systems. Many communities face unreliable energy supply, limited waste management infrastructure, and underutilization of agricultural residues, such as crop stalks, manure, and agro-industrial byproducts (Burger 2022). If channeled into small-scale biorefineries, these biomass resources could provide clean energy, organic fertilizers, and other value-added products, while reducing reliance on firewood and fossil fuels. Such initiatives align with national goals for expanding renewable energy, reducing greenhouse gas emissions, and promoting rural economic development (DEFF 2020). However, progress has been constrained by barriers, including insufficient technical skills, weak investment incentives, and limited integration of local communities in project design and ownership. Strengthening rural-focused biorefinery development could therefore contribute simultaneously to sustainable energy access, circular waste management, and rural livelihoods.

On the global scale, biorefineries are increasingly recognized as central to the transition toward a low-carbon bioeconomy. Countries, such as Brazil, the United States, and members of the EU, have advanced integrated systems that convert diverse biomass feedstocks into biofuels, biochemicals, heat, and power (Cherubini 2010; IEA 2021). These models highlight the potential of biorefineries to promote resource efficiency through cascading use of biomass and to stimulate industrial symbiosis. However, worldwide deployment continues to face challenges related to high capital costs, technological complexity, and the sustainability of feedstock supply chains. Addressing these issues is vital, if biorefineries are to deliver environmental and socio-economic benefits at scale. For South Africa, integrating lessons from global practice with local realities may open pathways for more inclusive and resilient biobased systems.

14.7 FUTURE RESEARCH DIRECTIONS

The literature highlights important gaps that warrant further investigation.

- Firstly, more research is needed on policy resilience, that is, how regulatory frameworks can be designed to withstand political and economic fluctuations, while maintaining investor confidence.
- Secondly, comparative studies could explore institutional capacity-building models across different regions, identifying best cross-sectoral and cross-border governance practices.
- Thirdly, limited empirical evidence exists on the impacts of regulatory streamlining (e.g. one-stop permitting platforms) in developing contexts, particularly in sub-Saharan Africa.
- Finally, greater attention should be paid to the social dimensions of regulatory reform, including how inclusive governance and participatory policy design can ensure that biorefinery development contributes to rural development, job creation, and social equity.

14.8 CONCLUSION

Regulatory bottlenecks remain a critical constraint on the expansion of biorefineries. Four areas emerge as priorities: establishing stable policies, streamlining permitting systems, enhancing cross-sectoral governance, and harmonizing regional frameworks. While international experience demonstrates the benefits of coherent and predictable regulation, South Africa and other developing regions continue to face persistent gaps in implementation, institutional coordination, and long-term policy stability. Addressing these challenges, alongside future research on policy resilience and inclusivity, is essential if biorefineries are to realize their potential as engines of sustainable development and drivers of the global bioeconomy.

FUNDING

This research was funded by the National Research Foundation (NRF) Postgraduate Scholarship, Energy and Water SETA, South African National Energy Institute (SANEDI), and United Nations Industrial Organisation (UNIDO).

DECLARATION

During the preparation of this work, the authors used Grammarly to improve readability and flow. After using this tool/service, the authors reviewed and edited the content as needed and take full responsibility for the publication's content.

REFERENCES

Adama, H.E., O.A. Popoola, C.D. Okeke, & A.E. Akinoso. 2024. Economic theory and practical impacts of digital transformation in supply chain optimization. *Int. J. Adv. Econ.* 6: 95–107.

Adegoke, A.S., R.B. Abidoye, & R.Y. Sunindijo. 2025. Stakeholders' awareness of the benefits of passive retrofit in Nigeria's residential building sector. *Sustainability* 17: 6582.

Adelakun, B.O., J.K. Nembe, B.B. Oguejiofor, C.U. Akpuokwe, & S.S. Bakare. 2024. Legal frameworks and tax compliance in the digital economy: A finance perspective. *Eng. Sci. Technol. J.* 5: 844–853.

Akagha, O.V., J.O. Coker, N.S. Uzougbo, & S.S. Bakare. 2023. Company secretarial and administrative services in modern Irish corporations: A review of the strategies and best practices adopted in company secretarial and administrative services. *Int. J. Manag. Entrep. Res.* 5: 793–813.

Akpuokwe, C.U., A.O. Adeniyi, S.S. Bakare, & N.E. Eneh. 2024. Legislative responses to climate change: A global review of policies and their effectiveness. *Int. J. Appl. Res. Soc. Sci.* 6: 225–239.

Allain, A., J.-F. Ruault, M. Moraine, & S. Madelrieux. 2022. The 'bioeconomics vs bioeconomy' debate: Beyond criticism, advancing research fronts. *Environ. Innov. Soc. Trans.* 42: 58–73.

ARE. 2021. *Annual Report 2021.* Brussels: ARE. https://www.ruralelec.org/wp-content/uploads/2023/11/Annual-Report-2021.pdf

AUC-ECA. 2013. *Africa Bioenergy Policy Framework and Guidelines: Towards Harmonizing Sustainable Bioenergy Development in Africa.* Addis Ababa: AUC-ECA.

Bahadorestani, A., N. Naderpajouh, & R. Sadiq. 2020. Planning for sustainable stakeholder engagement based on the assessment of conflicting interests in projects. *J. Clean. Prod.* 242: 118402.

Batchu, S.P., B. Hernandez, A. Malhotra, H. Fang, M. Ierapetritou, & D.G. Vlachos. 2022. Accelerating manufacturing for biomass conversion via integrated process and bench digitalization: A perspective. *React. Chem. Eng.* 7: 813–832.

Burger, S. 2022. *Continuing Skills Development Critical to Shaping South Africa's Waste Sector.* https://www.engineeringnews.co.za/article/continuing-skills-development-critical-to-shaping-south-africas-waste-sector-2022-11-08

Camia, A., J. Giuntoli, & R. Jonsson. 2018. *Biomass Production, Supply, Uses, and Flows in the European Union - First Results from an Integrated Assessment.* Publications Office of the European Union. https://doi.org/10.2760/181536

Carus, M., A. Raschka, K. Iffland, L. Dammer, R. Essel, & S. Piotrowski. 2016. How to shape the next level of the European biobased economy? http://www.renewablematter.eu/art/170/How_to_Shape_The_Next_Level_of_The_European_BioBased_Economy (accessed on 24 May 2017).

Cherubini, F. 2010. The biorefinery concept: Using biomass instead of oil to produce energy and chemicals. *Energy Convers. Manage.* 51: 1412–1421.

DEFF. 2020. *A Circular Economy Guideline for the Waste Sector - A Driving Force towards Sustainable Consumption and Production.* DEFF. https://www.dffe.gov.za/sites/default/files/docs/circulareconomy_guideline.pdf

EC. 2018. *Scaling-Up Energy Investments in Africa for Inclusive and Sustainable Growth: Report of the Africa-Europe High-Level Platform for Sustainable Energy Investments in Africa.* EC. https://international-partnerships.ec.europa.eu/system/files/2020-05/report-africa-europe-high-level-platform-sei_en.pdf

Eneh, N.E., S.S. Bakare, C.U. Akpuokwe, & A.O. Adeniyi 2024. Cross-jurisdictional disaster preparedness: A Nigeria-USA data-analytical approach. *World J. Adv. Res. Rev.* 21: 1822–1829.

FAO. 2021. *Towards Sustainable Bioeconomy Guidelines.* Rome: FAO. https://openknowledge.fao.org/server/api/core/bitstreams/793c0cea-d69a-4b92-b8f5-affc475108f3/content

Gomez-Echeverri, L., D. Mitchell, M.R. Allen, J.W. Hall, B. Mueller, L. Rajamani, & C. Le Quéré. 2018. Climate and development: Enhancing impact through stronger linkages in the implementation of the Paris Agreement and the Sustainable Development Goals (SDGs). *Philos. Trans A: Math. Phys. Eng. Sci.* 376: 20160444.

Guan, J., A. Khan., Y. Zhang., Y. Zhou., M.M.-J. Li., R.D. Patria., & S.-Y. Leu. 2025. Strategic design principles for greener biorefineries: A substrate-process matrix emphasizing complete lignocellulose utilization from various biomass feedstocks. *Green Chem.* 27: 11581–11606.

IEA. 2021. *IEA Bioenergy Annual Report 2021.* IEA. https://www.ieabioenergy.com/wp-content/uploads/2022/04/IEA-Bioenergy-Annual-Report-2021.pdf

IEA. 2022. *Global Biorefinery Status Report 2022.* IEA. https://www.ieabioenergy.com/wp-content/uploads/2022/09/IEA-Bioenergy-Task-42-Global-biorefinery-status-report-2022-220712.pdf

IEA. 2024. *Integration of Biorefineries and Green Hydrogen. Techno-Economic Feasibility and Case Studies.* IEA. https://www.ieabioenergy.com/wp-content/uploads/2025/04/20250326_report-IEA-2022-2024.pdf

IEA. 2025. *Renewables 2025. Analysis and Forecasts to 2030.* IEA. https://iea.blob.core.windows.net/assets/76ad6eac-2aa6-4c55-9a55-b8dc0dba9f9e/Renewables2025.pdf

IFC. 2021. *IFC Annual Report 2021: Meeting the Moment.* IFC. http://hdl.handle.net/10986/36334 License: CC BY-NC-ND 3.0 IGO

Igbinenikaro, O.P., O.O. Adekoya, & E.A. Etukudoh. 2024. Conceptualizing sustainable off-shore operations: Integration of renewable energy systems. *Int. J. Front. Sci. Technol. Res.* 6: 31–43.

IRENA. 2022. *Global Bioenergy Supply and Demand Projections.* IRENA. https://www.irena.org/publications

Jumbe, C.B.L., F.B. Msiska, & M. Madjera. 2009. Biofuels development in sub-Saharan Africa: Are the policies conducive? *Energy Policy* 37: 4980–4986.

Kamusoko, R., & P. Mukumba. 2025. Pineapple waste biorefinery: An integrated system for production of biogas and marketable products in South Africa. *Biomass* 5: 17.

Karekezi, S., & W. Kithyoma. 2012. Renewable energy in Africa: Prospects and limits. Prepared for the *Workshop for African Energy Experts on Operationalizing the NEPAD*, 2–4 June, 2003, Dakar, Senegal.

Kleinschmit, D., A. Giurca, R. Lehmann, F. Rodríguez, H. Sinaga. 2025. Bioeconomy governance in the global South: State of the art and the way forward. *For. Policy Econ.* 171: 103403.

Kunatsa, T., & X. Xia. 2022. A review on anaerobic digestion with focus on the role of biomass co-digestion, modelling and optimization on biogas production and enhancement. *Bioresour. Technol.* 344: 126311.

Kurki, S., & J. Ahola-Launonen. 2021. Bioeconomy in maturation: A pathway towards a "good" bioeconomy or distorting silence on crucial matters?. In *Bio#Futures*, ed. E. Koukios, & A. Sacio-Szymańska, 165–199. Cham: Springer.

Lamers, P., C. Hamelinck, M. Junginger, & A. Faaij. 2011. International bioenergy trade-A review of past developments in the liquid biofuel market. *Renew. Sustain. Energy Rev.* 15: 2655–2676.

Liu, Y., Y. Tang, H. Gao, W. Zhang, Y. Jiang, F. Xin, & M. Jiang. 2021. Challenges and future perspectives of promising biotechnologies for lignocellulosic biorefinery. *Molecules* 26: 5411.

Liu, Z.H. 2021. Challenges and perspectives of biorefineries. In *Emerging Technologies for Biorefineries, Biofuels, and Value-Added Commodities*, ed. Z.H. Liu, & A. Ragauskas, 1–21. Cham: Springer.

Maja, M.M., & S.F. Ayano. 2021. The Impact of population growth on natural resources and farmers' capacity to adapt to climate change in low-income countries. *Earth Syst. Environ.* 5: 271–283.

Makepa, D.C., & C.H. Chihobo. 2024. Barriers to commercial deployment of biorefineries: A multi-faceted review of obstacles across the innovation chain. *Heliyon* 10: e32649.

Makepa, D.C., C.H. Chihobo, W.R. Ruziwa, & D. Musademba. 2023. A systematic review of the techno-economic assessment and biomass supply chain uncertainties of biofuels production from fast pyrolysis of lignocellulosic biomass. *Fuel Commun.* 14: 100086.

Marangwanda, G.T., D.M. Madyira, & T.O. Babarinde. 2020. Combustion models for biomass: A review. *Energy Rep.* 6: 664–672.

Moller, L., L. Kitzing, & S. Steinhilber. 2020. Regulatory risk and investment in renewable energy: Evidence from Africa. *Energy Policy* 138: 111234.

Mountraki, A., M. Tsakalova, A. Panteli, A.I. Papoutsi, & A.C. Kokossis. 2016. Integrated waste management in multiproduct biorefineries: Systems optimization and analysis of a real-life industrial plant. *Ind. Eng. Chem. Res.* 55: 3478–3492.

Moussa, A. 2022. Scale-up to pilot and demonstration plants (TRL 6 and TRL 7). In *Systematic Process Development: From Idea to Value via Technology Readiness Levels*, ed. A. Moussa, 169–181. Münchwilen: American Chemical Society.

Mutsengerere, S., C.H. Chihobo, D. Musademba, & I. Nhapi. 2019. A review of operating parameters affecting bio-oil yield in microwave pyrolysis of lignocellulosic biomass. *Renew. Sustain. Energy Rev.* 104: 328–336.

Newell, P. 2021. *Power Shift: The Global Political Economy of Energy Transitions*. Cambridge: Cambridge University Press.

Nurdiawati, A., & F. Urban. 2022. Decarbonizing the refinery sector: A socio-technical analysis of advanced biofuels, green hydrogen and carbon capture and storage developments in Sweden. *Energy Res. Soc. Sci.* 84: 102358.

Ochuba, N.A., D.O. Olutimehin, O.G. Odunaiya, & O.T. Soyombo. 2024. The evolution of quality assurance and service improvement in satellite telecommunications through analytics: A review of initiatives and their impacts. *Eng. Sci. Technol. J.* 5: 1060–1071.

Oduro, P., P. Simpa, & D.E. Ekechukwu. 2024. Renewable energy expansion: Legal strategies for overcoming regulatory barriers and promoting innovation. *World J. Adv. Eng. Technol. Sci.* 12: 168–186.

OECD. 2019. *Policy Coherence for Sustainable Development 2019: Empowering People and Ensuring Inclusiveness and Equality*. Paris: OECD.

Oyewole, O.L., N.I. Nwulu, & E.J. Okampo. 2024. Optimal design of hydrogen-based storage with a hybrid renewable energy system considering economic and environmental uncertainties. *Energy Convers. Manage.* 300: 117991.

Philp, J., R.J. Ritchie, & K. Guy. 2013. Biobased chemicals: The convergence of green chemistry and industrial biotechnology. *Trends Biotechnol.* 31: 219–222.

Popoola, O.A., H.E. Adama, C.D. Okeke, & A.E. Akinoso. 2024. Cross-industry frameworks for business process reengineering: Conceptual models and practical executions. *World J. Adv. Res. Rev.* 22: 1198–1208.

Rasimphi, T., B. Kilonzo, D. Tinarwo, & P. Nyamukondiwa. 2024. Challenges, opportunities, and possible interventions in the biogas sector in rural areas of Limpopo, South Africa. *Energy Strat. Rev.* 56: 101562.

Scarlat, N., & J.F. Dallemand. 2011. Recent developments of biofuels/bioenergy sustainability certification: A global overview. *Energy Policy* 39: 1630–1646.

Scarlat, N., J.F. Dallemand, F. Monforti-Ferrario, & M. Banja. 2015. Renewable energy policy framework and bioenergy contribution in the European Union – An overview from national renewable energy action plans and progress reports. *Renew. Sustain. Energy Rev.* 51: 969–985.

Sekoai, P.T., V. Chunilall, K. Msele, L. Buthelezi, J. Johakimu, J. Andrew, M. Zungu, K. Moloantoa, N. Maningi, O. Habimana, & A. Swartbooi. 2023. Biowaste biorefineries in South Africa: Current status, opportunities, and research and development needs. *Renew. Sustain. Energy Rev.* 188: 113870.

Sravan, J.S., S. Sahota, O. Sarkar, M.V. Reddy, S.V. Mohan, & Y.-C. Chang. 2026. Technology advancements in future waste biorefineries: Focus on low carbon fuels and renewable chemicals. *Fuel* 404: 136184.

Stats SA. 2022. *Non-Financial Census of Municipalities for the Year Ended 30 June 2020*. Stats SA. https://www.statssa.gov.za/publications/P9115/P91152020.pdf (Accessed 8 June 2022)

Stats SA. 2023. *Sustainable Development Goals: Country Report 2023*. Stats SA. https://www.dws.gov.za/Projects/sdg/docs/SA%20SDG%20Country%20Report%202023.pdf

Stegmann, P., M. Londo, & M. Junginger. 2020. The circular bioeconomy: Its elements and role in European bioeconomy clusters. *Resour. Conserv. Recyc. X* 6: 100029.

U.S. DOE. 2023. *Bioenergy Technologies Office: Multi-Year Program Plan*. U.S. DOE. https://www.energy.gov/sites/default/files/2023-03/beto-mypp-fy23.pdf

Xu, M., M. Yang, H. Sun, M. Gao, Q. Wang, & C. Wu. 2022. Bioconversion of biowaste into renewable energy and resources: A sustainable strategy. *Environ. Res.* 214: 113929.

Zhou, S., B.D. Solomon, & M.A. Brown. 2024. The spillover effect of mandatory renewable portfolio standards. *Proc. Natl. Acad. Sci. USA* 121: e2313193121.

Index

Pages in *italics* refer to figures and pages in **bold** refer to tables

D

E

F

G

H

I

K

L

M

U

V

W

X

Y

Z

For Product Safety Concerns and Information please contact our EU representative GPSR@taylorandfrancis.com Taylor & Francis Verlag GmbH, Kaufingerstraße 24, 80331 München, Germany